W0254294

Macromolecular Symposia

Symposium Editors: G. J. Marosi, A. Michel

202

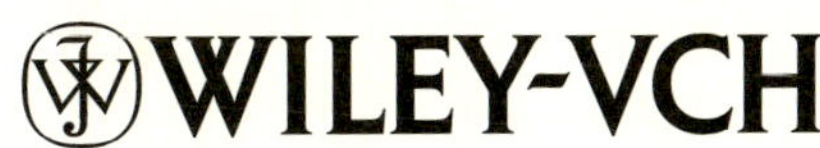

pp. 1–337

September 2003

Macromolecular Symposia publishes lectures given at international symposia and is issued irregularly, with normally 14 volumes published per year. For each symposium volume, an Editor is appointed. The articles are peer-reviewed. The journal is produced by photo-offset lithography directly from the authors' typescripts.
Further information for authors can be obtained from:
Editorial office "Macromolecular Symposia"
Wiley-VCH Verlag GmbH & Co. KGaA,
Boschstrasse 12, 69469 Weinheim,
Germany
Tel. +49 (0) 62 01/6 06-2 38 or -5 81; Fax +49 (0) 62 01/6 06-3 09 or 5 10;
E-mail: macro-symp@wiley-vch.de
http://www.ms-journal.de
Suggestions or proposals for conferences or symposia to be covered in this series should also be sent to the Editorial office at the address above.

Macromolecular Symposia:
Annual subscription rates 2003 (print only or online only)*
Germany, Austria € 1318; Switzerland SFr 2168; other Europe € 1318; outside Europe US $ 1568.
Macromolecular Package, including Macromolecular Chemistry & Physics (18 issues), Macromolecular Bioscience (12 issues), Macromolecular Rapid Communications (18 issues), Macromolecular Theory & Simulations (9 issues) is also available. Details on request.
* For a 5 % premium in addition to **Print Only** or **Online Only**, Institutions can also choose both print and online access.
Packages including Macromolecular Symposia and Macromolecular Materials & Engineering are also available. Details on request.
Single issues and back copies are available. Please inquire for prices.

Orders may be placed through your bookseller or directly at the publishers:
WILEY-VCH Verlag GmbH & Co. KGaA, P.O. Box 10 11 61, 69451 Weinheim, Germany, Tel. +49 (0) 62 01/6 06-400, Fax +49 (0) 62 01/60 61 84. E-mail: service@wiley-vch.de

Macromolecular Symposia (ISSN 1022-1360) is published with 14 volumes per year by WILEY-VCH Verlag GmbH & Co. KGaA, P.O. Box 10 11 61, 69451 Weinheim, Germany. Air freight and mailing in the USA by Publications Expediting Inc., 200 Meacham Ave., Elmont, NY 11003, USA. Application to mail at Periodicals Postage rate is paid at Jamaica, NY 11431, USA. US POSTMASTER please send address changes to: Macromolecular Symposia, c/o Wiley-VCH, III River Street, Hoboken, NJ 07030, USA.

Printing: Strauss Offsetdruck, Mörlenbach. Binding: J. Schäffer, Grünstadt

Selected lectures from the

2nd International Conference on *Mo*dification, *De*gradation and *St*abilization of Polymers (MoDeSt 2002)

Budapest, Hungary

Symposium Editors

György J. Marosi

Department of Organic Chemical Technology
Budapest University of Technology and Economics
H-1111 Budapest, Megyetem rkp. 3
Hungary

Alain Michel

Universite Claude Bernard Lyon I – UMR CNRS 5627
LMPB – Bat. ISTIL – 15 Bd Latarjet – 69622 Villeurbanne
France

ISBN 3-527-30706-0

The **tables of contents** of the published issues are displayed on the WWW.

This service as well as further information on our journals can be found at the following WWW address:

http://www.ms-journal.de

Contents of Macromolecular Symposia 202

2nd International Conference on *Mo*dification, *De*gradation and *St*abilization of Polymers (MoDeSt 2002)
Budapest (Hungary), 2002

* Asterisks indicate the name(s) of the author(s) to whom inquiries should be addressed.

Author Index

Preface

This issue of *Macromolecular Symposia* contains selected papers, focusing on functionalization, (reactive) processing, stabilization of polymers, blends and reinforced systems, presented at the second International *MoDeSt* Conference on *Mo*dification, *De*gradation and *St*abilisation of Polymers held in Budapest.

The MoDeSt society (http://modest.unipa.it) established the regular forum MoDeSt conferences realizing the overlapping interest of scientists and technologists dealing with degradation, stabilization, modification and recycling of polymers and biopolymers. The objectives of the Society are to encourage communication and collaboration between young and senior scientists, researchers from west and east, academia and industry in all aspects of modification affecting the life cycle of polymers.

The sessions of the meeting in Budapest included the thermal degradation, aging, stabilization, combustion and fire retardance, blends, functionalization and reactive processing, biodegradation, biomaterials and drug release, recycling, material and energy recovery. Due to the large agenda of the meeting, more than 350 people from 43 countries attended. There were seven plenary lectures, 185 oral presentations, more or less evenly split among the six sessions of the congress, and 162 posters.

The general lectures provided an opportunity for the participants to hear the most important recent advances in all the areas within the scope of the conference. The plenary lecture was presented by Walter Kaminsky (University of Hamburg) on "Feedstock recycling of polymer by pyrolysis in a fluidized bed." He described a new economic process for producing monomers from polymer wastes like polystyrene and poly(methyl methacrylate). Bill Starnes (College of William & Mary) presented recent results on "New non-metallic additives for the thermal stabilization of PVC." These are sulfur-containing materials that are very useful and can be easily produced. Their mode of action was discussed within the context of the PVC degradation scheme. Frank Karasz (University of Massachusetts) spoke on the "Electro-optical properties of polymer blends." The charge carrier characteristics of the developed blends are supported by the morphologies that are achieved in these polymer blends. Francesco La Mantia (Universita di Palermo & the current chair of MoDeSt) spoke on "Can recycling improve the properties of post-consumer plastic materials?" He spoke about the possibilities of upcycling using appropriate stabilizers for preventing, or at least reducing the degradation that usually occurs during recycling. Hartwig Höcker (University of Technology, Aachen) spoke on "Biomaterials – from drug release systems to surface modified implants with enhanced biocompatibility." The biocompatible and biodegradable nature of the polymers used in drug-delivery systems control the kinetics of drug release and the same characteristics are important in the field of tissue engineering, which presents a particular challenge because of the complex requirement for a porous degradable polymer containing cell adhesion mediators, growth factors, and so on. Gianni Camino (Polytechnio di Torino) spoke on "Thermal degradation and combustion behaviour of polymer-layered silicate nanocomposites." This talk was a comprehensive introduction to the topic of polymer-clay nanocomposites, with special emphasis on fire retardancy of polymers. Miklós Zrínyi (Budapest University of Technology and Economics) talked about "Smart polymer materials and systems." Among the systems that are responsive to external stimuli, such as pH, temperature, and so on, the presentation focused on the magnetic/electric field controlled behavior of soft materials. The applications included robotic, biomedical and optical fields. The final general lecture was given by Norman Billingham (University of Sussex) on "Surface modification of polymers by migrating additives." Some of the additives are required to migrate to the surface, for which a model was developed and tested considering the solubility and diffusion rates of additives in polymers.

The sessions were worthwhile not only for those interested in these areas; others with different interests attended them as well. The large number of presentations did not allow publication of all of them together, therefore the sessions were dedicated to a different journal. This issue focuses mainly on the papers presented in the field of modification of multicomponent polymer systems. The results of this issue confirm that the reactive processing and functionalization of polymer interfaces are of great benefit not only in polymer blends but also in other multicomponent systems such as pigmented, flame retarded and reinforced plastics.

The next MoDeSt international conference will be held in Lyon, France, in 2004.

Gy. J. Marosi, A. Michel

Grafting of Polyolefins with Maleic Anhydride: Alchemy or Technology?

Martin van Duin

DSM Research, P.O. Box 18, 6160 MD, Geleen, The Netherlands

Summary: Nowadays, the process of maleic anhydride (MA) grafting and the application of MA-grafted polyolefins are viewed as mature technologies. The chemistry and technology of modifying apolar polyolefins with the polar and reactive MA either in solution or in the melt were already explored as far back as the 1950s. Commercial applications exploit the improved adhesion of polyolefins to polar materials, both at the macroscopic scale and on the microscopic scale. However, it is hardly recognised that, from a scientific point of view, grafting has still a strong resemblance to alchemy. Both process and application technologies have been developed in a trial and error fashion. Only in the last decade the structure of MA-grafted polyolefins has been elucidated and attempts to "look" inside the extruder during grafting were only recently successful. The first steps towards the development of sound chemical models are currently made. An overview will be given of the progress made in the various areas mentioned.

Keywords: extruder; grafting; maleic anhydride; NMR; polyolefin

Introduction

Grafting of maleic anhydride (MA) onto polyolefins is probably the classical example of free-radical grafting and of reactive extrusion.[1,2] In the 1950s both the first scientific papers and the first patents on MA-grafted polyolefins have been published and filed, respectively. The numbers of papers and patents amount roughly to 750 and 2000, respectively. A large number of companies (amongst others Atofina, Crompton, DSM, DuPont, Eastman, Equistar, Exxon and Mitsui) are grafting MA onto a variety of polyolefins, such as PE, PP, EP(D)M, EBM, EOM and SEBS, on a commercial scale. The total world production of these modified polyolefins is estimated at 150 kton/year. The price of these high-performance polyolefins is determined by the costs of the starting polyolefin, the MA graft content and the production technology.

MA-grafted polyolefins are usually obtained via a free-radical process using preferably as initiators, but also cationic or thermal („ene") reactions are possible. Preferably, MA grafting is

 DOI: 10.1002/masy.200351201

performed in the melt, but also routes via solution, the solid state or the vapour phase (surface modification) are known and/or applied. The applications of MA-modified polyolefins are usually based on the enhancement of adhesion between apolar polyolefins and polar substrates (polyamides, polyesters, coatings, glass, inorganics, metals, and paper) either on the macroscopic level (co-extrusion, over-moulding and adhesives) or on the microscopic level (compatibiliser and dispersion and coupling agent).

It is frequently expressed that grafting of MA onto polyolefins is a mature technology and that all aspects are fully understood. However, it is the author's opinion that grafting of MA bears still a close resemblance to alchemy. The recent progress with respect to the (supra)molecular structure of MA-grafted polyolefins is hardly recognised. Reactive extrusion is a very complicated process, involving melting, dispersion, reaction, mass and heat transport etc., and is still performed as a black box process, i.e. the extrudate characteristics are directly interpreted in terms of the grafting recipe and the conditions. Application development of MA-grafted polyolefins is carried out via trial and error. The goal of this paper is to review the recent developments with respect to the production, the structure and the application of MA-grafted polyolefins.

Production

Recently, devices have been developed that allow sampling along the screw axis of a co-rotating twin-screw extruder in operation. About 2 gram of representative polymer melt sample can be taken within about 5 seconds.[3] A large series of MA grafting experiments has been performed, varying the polyolefin type (PE, EPMs and PP), the grafting recipe (amount of MA and type and amount of peroxide) and the grafting conditions (screw configuration and speed and temperature profile).[4,5] The MA graft content of the products was determined using FT-IR spectroscopy; rheometry and sol/gel extractions were applied to study cross-linking or degradation phenomena of the polyolefin backbone. Figure 1 shows that for a variety of grafting recipes the MA graft content follows a convex profile along the screw axis, independently of the polyolefin type used. The final MA graft content is determined by the MA dosage, the type and the amount of peroxide and the polyolefin composition. For PE and EPM copolymers it was shown that at 200 °C the MA graft content is not affected by the propene content of the polyolefin up to 60 wt.%, but then rapidly decreases to low values for PP. Thermal grafting, i.e. in the absence of peroxide, of these saturated polyolefins occurs hardly.

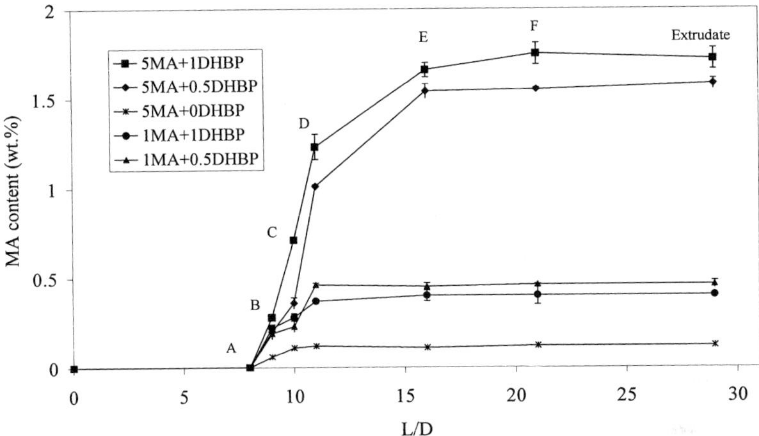

Fig. 1. MA graft content along screw axis as a function of grafting recipe [co-rotating twin-screw extruder: HDPE; DHBP: 2,5-bis(tert-butylperoxy)-2,5-dimethylhexane; 5 kg/hr, 200 °C, 75 rpm].[4]

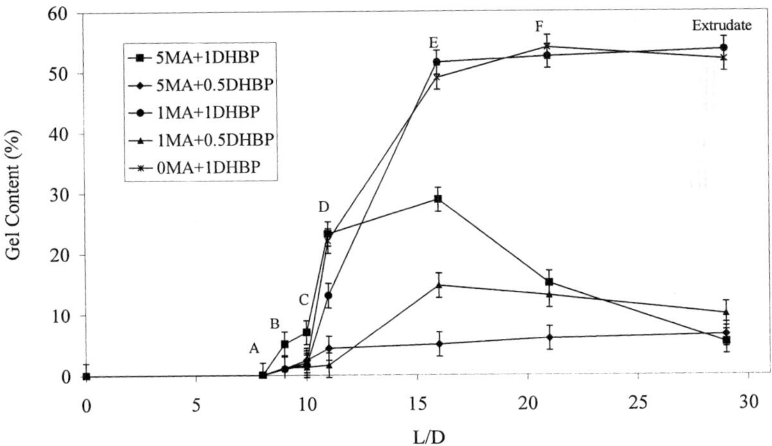

Fig. 2. Gel content along screw axis as a function of grafting recipe [co-rotating twin-screw extruder: HDPE, DHBP: 2,5-bis(tert-butylperoxy)-2,5-dimethylhexane 5 kg/hr, 200 °C, 75 rpm;].[4]

In the case of PE, grafting of MA is accompanied by branching or cross-linking. As can be seen in Figure 2, the PE gel content is determined by the grafting recipe. Actually, the gel content may go through a maximum, which stresses the use of sampling devices for studying the grafting process or for its optimisation. For PP, grafting is accompanied by chain scission. For EPMs with intermediate compositions there is hardly a viscosity change, which is due to compensating effects of branching versus degradation.

In a semi-quantitative approach the experimental MA graft content along the screw axis can be correlated with the calculated decomposition of the peroxide. The latter is calculated using the Arrhenius' equation combining the melt temperature and residence time – both have been measured using the same series of sampling devices – along the screw axis. As for the MA graft content, convex profiles are calculated for the peroxide decomposition. Figure 3 shows that there is a fair correlation between the experimental MA graft content and the calculated peroxide decomposition.

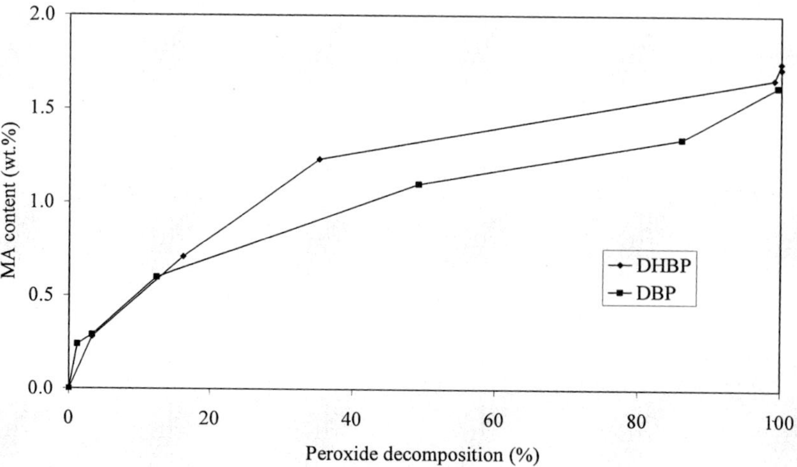

Fig. 3. Experimental MA graft content versus calculated peroxide decomposition [DHBP: 2,5-bis(tert-butylperoxy)-2,5-dimethylhexane; DBP: di-tert-butyl peroxide] along the screw axis (co-rotating twin-screw extruder: HDPE, 5 kg/hr, 200 °C, 75 rpm).[4]

Clearly, this kind of studies results in true insight in the physico-chemical phenomena occurring within the extruder during MA grafting. Eventually, it should result in a quantitative process description and allow proper process optimisation.

Structure

With respect to the molecular structure of MA-grafted polyolefins, many suggestions have been made in the scientific literature, but few of them have experimentally been proven. A recent liquid-state ^{13}C NMR study on polyolefins grafted with ^{13}C-labeled MA in the melt and in solution has identified a series of graft structures as shown in Figure 4.[6,7] A series of low-molecular-weight alkylsuccinic anhydrides had to be synthesised to interpret the NMR spectra.[8] Only monomeric MA grafts have been demonstrated with the exception of HDPE grafted in the melt: mixtures of monomeric, dimeric and trimeric grafts are then formed. Grafting occurs only on secondary C-atoms in PE and ethene-rich EPMs and on tertiary C-atoms in PP and propene-rich EPMs. For EPMs with intermediate compositions grafting occurs both on secondary and tertiary C-atoms. Only saturated MA grafts have been demonstrated with the exception of PP grafted with MA: terminal MA grafts with an exo-cyclic unsaturation are formed as the result of PP chain degradation at a tertiary MA graft site: so first grafting, then β-scission. In summary, it can be stated that only a few of the structures suggested in the literature could be proven.

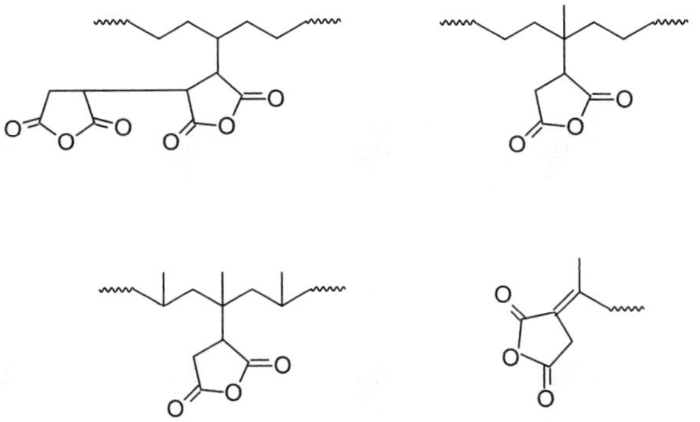

Fig. 4. Chemical structures of MA-grafted polyolefins (PE, EPM and PP).[6,7]

MS on MA-grafted alkanes (C_8 upto C_{30}), used as representative low-molecular-weight models for polyolefins, showed the presence of relatively large amounts of products with 2, 3 and even 4 MA grafts per alkane molecule.[9,10] Such multiple graft structures suggest the occurrence of intramolecular H-transfer (Figure 5). The formation of multiple MA graft structures may be enhanced when MA has not (yet become) dissolved in the polyolefin melt.

Finally, it has been shown using a variety of techniques that MA-grafted polyolefins have a tendency to form rigid, polar clusters (Figure 6), similar to those occurring in ionomers. The difference in polarity between the polar MA grafts and the apolar polyolefin backbones is the driving force for the formation of these MA-graft-rich clusters. It is enhanced by H-bonding, if hydrolysis of the MA grafts into maleic acid grafts has occurred. Dynamic and steady-state light scattering and fluorescence spectroscopy on solutions of EPM-g-MA in apolar hexane and polar THF have shown strong cluster formation in the former apolar solvent.[11,12] It was shown that for a particular EPM-g-MA sample, approximately 70% of the MA grafts is present as multiple grafts structures (cf. Figure 5).[13] SAXS of EPM-g-MA in the solid state has shown the presence of strongly scattering domains of 40 to 45 nm diameter. Upon increasing the MA graft content of EPM-g-MA the number of domains remains more or less constant, but the number of MA graft units per polar domain increases.[14] Finally, these MA-rich clusters (30 to 90 nm diameter) have actually been „visualised“ via AFM on very thin EPM-g-MA films cast on mica.[15]

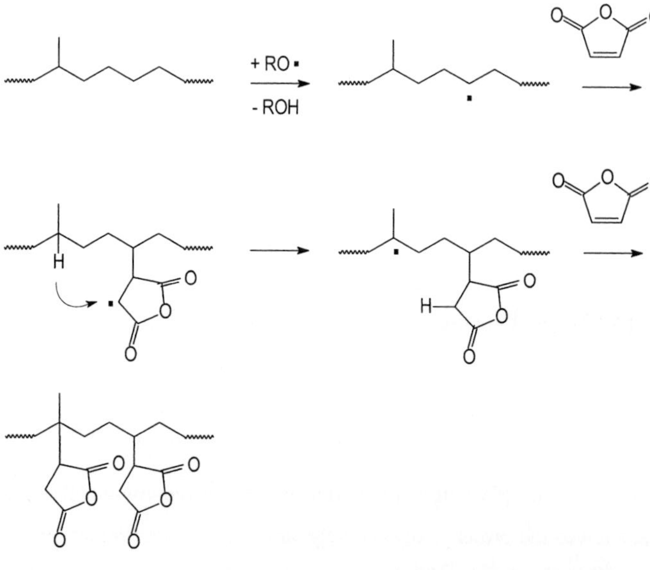

+ RO•

- ROH

H

H

Fig. 5. Formation of multiple MA graft structures via intramolecular H-transfer.

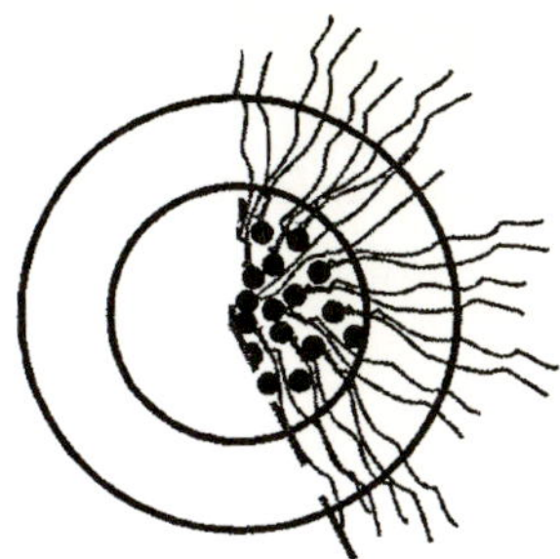

Fig. 6. Schematic representation of rigid MA-graft-rich domains in EPM-g-MA.

Applications

One of the major applications of EPM-g-MA is that as impact modifier of polyamides (PA).[16] Tough PA/EPM blends are usually produced via reactive blending, i.e. the compatibiliser is formed in-situ during blending via a reaction of the PA amine end groups with the grafted MA moieties. Using the sampling devices already mentioned, the changes in MA graft content, PA graft content and EPM particle size have been monitored along the extruder screw axis.[17] It was shown for a variety of PA-6/EPM-g-MA/EPM blends that all physico-chemical phenomena occur very fast upon melting of the PA-6 granulate in the first kneading zone. The residual MA graft content decreases to below 0.1 wt.%, the PA-6 graft content goes to a plateau value and the EPM particle size is reduced from mm to μm level (Figure 7). Further along the screw axis these quantities are not affected anymore. However, there is a continuous decrease of the PA-6 viscosity along the screw axis as a result of degradation.

Another typical application of MA-grafted polyolefins is that of adhesive between apolar and non-reactive polyolefins and polar substrates. In a recent study the adhesion between PA-6 and PP-g-MA was studied by determining the adhesion energy (E_{adh}) of sandwiches with a asymmetric double cantilever beam.[18] It was shown that E_{adh} is governed by diffusion of the MA-grafts towards the interface. XPS after removal of the non-grafted PA allowed the determination of the areal graft copolymer density (Σ). It was shown that E_{adh} is a linear function of $\log\Sigma$. Finally, it was demonstrated that the toughness of the interface is determined first by plastic deformation in the PP phase and ultimately by chain scission of PP blocks.

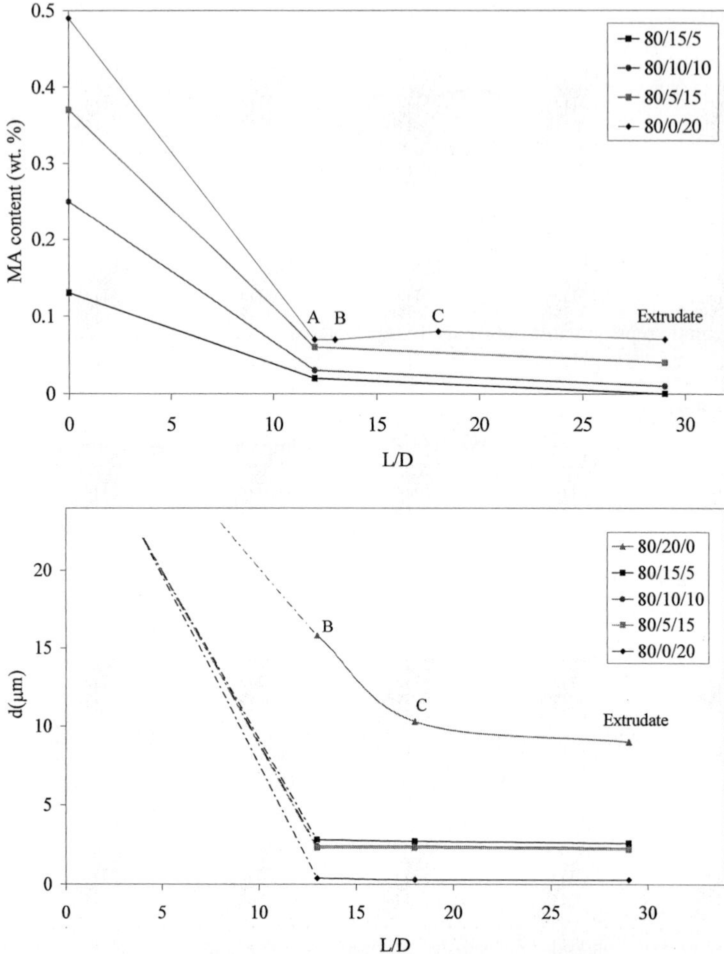

Fig. 7. Residual MA graft content of EPM-g-MA and average EPM particle size along screw axis for a variety of PA-6/EPM-g-MA/EPM blends (co-rotating twin-screw extruder: 6 kg/hr, 230 °C, 200 rpm).[17]

Concluding Remarks

From the results reviewed here it is clear that impressive progress has been made with respect to the production, the structure and the application of MA-grafted polyolefins. However, there is still a large number of questions to be addressed. An overview is given in Table 1, which is probably not exhaustive. Only, when more progress has been made in these areas, grafting of

MA onto polyolefins will be converted from alchemy to a true technology. Eventually, this should allow us to take maximum advantage of MA-grafted polyolefins.

Table 1. Questions related to MA-grafted polyolefins still to be addressed.

- Chemical structure/mechanism
  - identification/quantification of free-radical intermediates
  - intra/intermolecular H-transfer and termination reactions
  - kinetics
  - effect of polyolefin composition/tacticity ?
  - multiple graft structures in polymer system ?
  - clustering in melt ?
- Production
  - solubility/vapour pressure of MA in/above melt
  - effect of grafting recipe (MA dosage, type/amount of peroxide)
  - effect of residence time (distribution), temperature, rpm, screw lay out
  - process model, including chemical kinetics and mass/heat balances
  - low peroxide efficiency ?
  - strange maxima in MA graft content as function of recipe and conditions ?
- Application
  - study of interface
  - effect of (supra)molecular structure of MA-grafted polyolefin
  - relationships between interfacial structure and macroscopic properties
  - effectivity of monomeric MA graft vs. multiple MA graft vs. high-molecular-weight SMA co-polymer graft

Acknowledgements

Ana Vera Machado (University of Minho) and Marielle Wouters (Technical University of Eindhoven, currently at TNO Industry) are acknowledged for figures from their PhD theses.

[1] „*Reactive Extrusion: Principles and Practice*", M. Xanthos (ed.), Hanser Publishers, Munich 1992.
[2] "*Reactive Modifiers for Polymers*", S. Al-Malaika (Ed.), Blackie Academic & Professional, London 1997.
[3] A. V. Machado, J. Covas, M. van Duin, *J. Appl. Polym. Sci.* **1999**, *71*, 135.
[4] A. V. Machado, J. Covas, M. van Duin, *J. Polym. Sci., Part A, Polym. Chem.* **2000**, *38*, 3919.
[5] A. V. Machado, J. Covas, M. van Duin, *Polymer* **2001**, *42*, 3649.
[6] W. Heinen, C. H. Rosenmöller, C. B. Wenzel, H. J. M. de Groot, J. Lugtenburg, M. van Duin, *Macromolecules* **1996**, *26*, 1151.
[7] W. Heinen, M. van Duin, C. H. Rosenmöller, C. B. Wenzel, H. J. M. de Groot, J. Lugtenburg, *Marcromol. Symp.* **1998**, *129*, 119.
[8] W. Heinen, S. W. Erkens, M. van Duin, J. Lugtenburg, *J. Polym. Sci., Polym. Chem.* **1999**, *37A*, 4368.
[9] K. E. Russell, E. C. Kelusky, *J. Polym. Sci., Part A, Polym. Chem.* **1988**, *26*, 2273.
[10] S. Ranganathan, W. E. Baker, K. E. Russell, R. A. Whitney, *J. Polym. Sci., Part A, Polym. Chem.* **1999**, *??*, ???.
[11] S. Nemeth, T. Jao, J. H. Fendler, *Macromolecules* **1994**, *27*, 5449.
[12] V. Vangani, J. Duhamel, S. Nemeth, T. Jao, *Macromolecules* **1999**, *32*, 2845.
[13] V. Vangani, J. Drage, J. Mehta, A. K. Mathew, J. Duhamel, *J. Phys. Chem. B*, **2001**, *105*, 4827.
[14] M. E. L. Wouters, V. M. Litvinov, F. L. Binsbergen, J. G. P. Goossens, M. van Duin, H. Dikland, accepted by *Macromolecules*.
[15] J. Yang, T. Laurion, T. Jao, J. H. Fendler, *J. Phys. Chem.* **1994**, *98*, 9391.
[16] M. van Duin, R. Borggreve, "*Blends of Polyamides with Maleic Anhydride containing Polymers: Interfacial Chemistry and Properties*", in "*Reactive Modifiers for Polymers*", S. Al-Malaika (Ed.), Blackie Academic & Professional, London 1997, ch.3.
[17] A. V. Machado, J. Covas, M. van Duin, *J. Polym. Sci., Part A, Polym. Chem*, **1999**, *37*, 1311.
[18] E. Boucher, J. P. Folkers, H. Hervet, L. Leger, *Macromolecules*, **1996**, *29*, 774.

Macromol. Symp. **2003**, *202,* 11–23

Novel Copolymers via Nitroxide Mediated Controlled Free Radical Polymerization of Vinyl Chloride

Thomas Wannemacher,[1] *Dietrich Braun,*[1,*] *Rudolf Pfaendner*[2]

[1] Deutsches Kunststoff-Institut, 64289 Darmstadt, Germany
[2] Ciba Spezialitätenchemie Lampertheim GmbH, 68623 Lampertheim, Germany

Summary: Controlled free radical polymerization (CFRP) of vinyl chloride (VCM) and copolymerization with several comonomers have been studied in aqueous suspension. Therefore di-tert-butylnitroxide and three novel nitroxyl radicals were used as mediating agents. Copolymerization of VCM with styrene, partly combined with acrylonitrile, maleic acid anhydride and maleic acid imide as well as methyl methacrylate, n-butyl methacrylate, butyl acrylate and butadiene have been achieved, demonstrating an efficient route for novel vinyl chloride copolymer architecture.

Keywords: block copolymers; CFRP; living polymerization; nitroxides; vinyl chloride

Introduction

In the last years living or controlled free radical polymerization (CFRP) processes have achieved a high level of industrial and academic interest. Besides Werrington and Tobolsky, who performed first attempts of controlling a radical polymerization reaction with specific disulfides in the mid fifties,[1] Otsu et al. introduced the IniFerTer-concept[2] as a true CFRP-concept thirty years later. First studies with nitroxide radicals or ethers were carried out and published in an US-Patent by Solomon[3] and coworkers. Finally atom transfer radical polymerization, ATRP[4] by Matyjaszewski and Sawamoto in 1995 and two years later Moad's RAFT-concept[5] (reversible addition fragmentation chain transfer) were introduced. Today the CFRP techniques are widely spread due to versatility with regard to monomers and functional groups, due to access to a wide range of processes and the possibility to build new polymer architectures.

As PVC is one of the most important plastics with a number of applications, investigations of the polymerization of VCM in the presence of controlling agents and the synthesis of defined copolymers are of great interest. However, more than eighty percent of PVC is produced by suspension polymerization.[6] Therefore achievement of water based CFRP is a must.

 DOI: 10.1002/masy.200351202

Due to the fact, that the ATRP technique is not applicable to some monomers such as vinyl chloride and RAFT is complicated in water based processes, other methods of controlling polymerization reactions are gaining relevance. Therefore, the nitroxide mediated CFRP shows many advantages in controlling the polymerization of different monomers and in building up new polymer architectures like block and graft copolymers which cannot be achieved by standard reaction methods.[7]

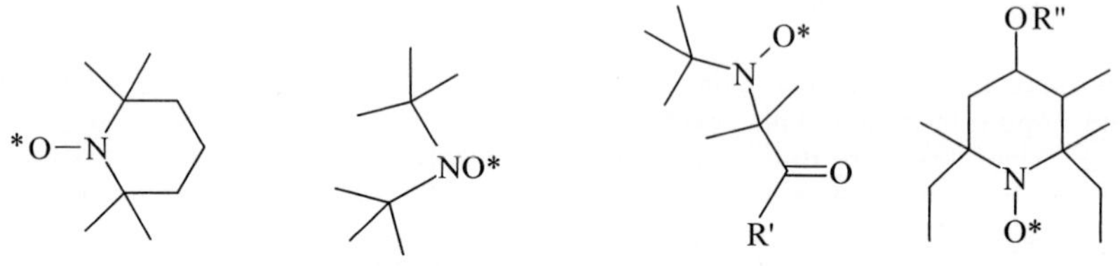

(1) (2) (3) (4)

The well-known 2,2,6,6-terамethyl-piperidine-1-oxyl (TEMPO) (1), which is successfully used in CFRP processes with common monomers such as styrene,[8],[9] is not efficient in vinyl chloride polymerization. First, the VCM-polymerization temperature is at least about thirty degrees less than in styrene polymerization. Second, due to a certain water solubility TEMPO is not very attractive in suspension processes.[10] The novel, somewhat more sophisticated nitroxides used in this study are open chain structures, such as di-tert-butylnitroxide (2) with two symmetrical R's and a structure with different R's (3). Furthermore highly sterically hindered piperidine-like structures (4), comparable to TEMPO, have been investigated. These nitroxyl radicals have been selected due to their reduced water solubility and higher polymerization efficiency at lower temperature.

The main characteristic of the nitroxyl mediated polymerization is a thermal equilibrium as explained in scheme 1. At lower temperatures, the equilibrium is strongly shifted to the side with the attached nitroxide at the growing chain. The reversible termination of the growing chain is the key step for controlling the polymerization, resulting in a very low concentration of free radical chain ends. Due to this, irreversible termination reactions, such as chain-combination and disproportionation are minimized. Neglecting any side reaction, the polymer growth occurs in a living type manner with a high degree of control. After complete monomer consumption the polymer appears as a dormant chain with the nitroxide end group.

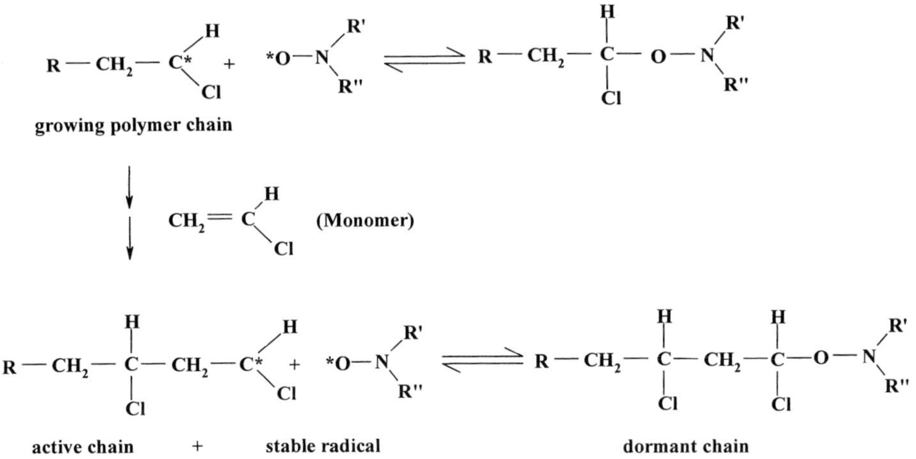

$$R-CH_2-\dot{C}^{*}(H)(Cl) + {}^{*}O-N(R')(R'') \rightleftharpoons R-CH_2-C(H)(Cl)-O-N(R')(R'')$$

growing polymer chain

$$\downarrow \; CH_2=C(H)(Cl) \quad \textbf{(Monomer)}$$

$$R-CH_2-C(H)(Cl)-CH_2-C^{*}(H)(Cl) + {}^{*}O-N(R')(R'') \rightleftharpoons R-CH_2-C(H)(Cl)-CH_2-C(H)(Cl)-O-N(R')(R'')$$

active chain + **stable radical** **dormant chain**

Scheme 1. Mechanism of nitroxide mediated controlled free radical polymerization.

A basic point of view is the so called persistent radical effect (PRE).[11] As shown in scheme 2 the initiator radical (X*) can undergo a reversible coupling with the nitroxide radical. A small amount of initiator radicals (transient radicals) irreversibly reacts with other radicals (initiator radicals or growing chain), which leads to a loss of initiator activity. By its nature, the mediating or persistent radical (the nitroxide) does not undergo any radical-radical coupling, so that a small amount of excess-radicals occurs. Therefore the ratio initiator : nitroxide plays an important role. The higher the concentration of excess free nitroxide, the higher is the amount of dormant chains. The reaction time increases, the polymerization rate and monomer conversion decreases. The overall results depend on temperature, monomer and the activity of the nitroxide.

Transient and persistent radicals are formed during the polymerization at the same extent, the persistent radicals do not interact with other persistent radicals but couple reversibly with transient radicals. The transient radicals, however, react among themselves irreversibly in self termination, the excess amount of persistent radicals grows in course of the reaction by simple stoechiometry. Because of this, the cross reaction gains on importance, the self terminating reactions are strongly inhibited, but never disappear completely.

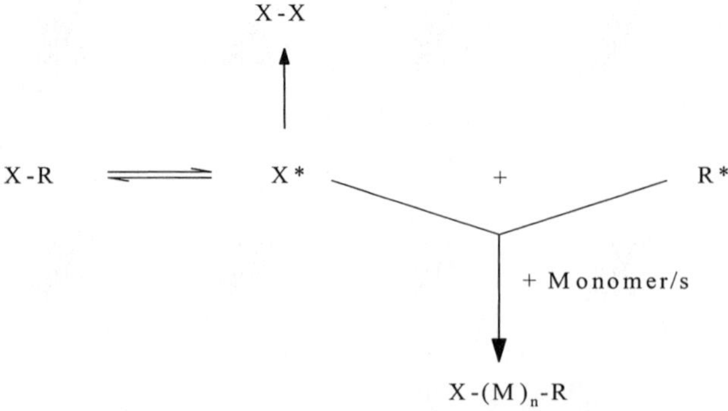

Scheme 2. Description of the self-accelerating persistent radical effect (PRE).

Due to the fact that the polymerization of vinyl chloride bears some specific characteristics, the nitroxide mediated CFRP of VCM depends on the choice of the nitroxide compound. To our knowledge no literature data on nitroxide mediated CFRP of VCM have been published until now. Because of the chlorine, VCM has a very high resonance stability, and hence no cationic initiation is possible. Due to a high transfer constant, chain transfer reactions become very important. Molecular weight and polydispersity (PD) are mainly determined by the polymerization temperature. The lower the temperature, the higher the molecular weight, whereas reaction time, conversion, as well as concentration and type of the initiator have only a minor influence on the resulting polymer. Another point of view is, that with rising temperatures not only a decrease of the molecular weight appears, but a great reduction of the thermal stability of the polymer, which is undesirable.

Experimental Part

Materials

Vinyl chloride monomer (VCM) 3.7, 99,97 %, stabilized (supplier: Messer Griesheim) and butadiene 99,5 % (supplier: Linde) were used without purification. Styrene (supplier: BASF), methyl methacrylate (supplier: Röhm), n-butyl methacrylate (supplier: Röhm), butyl acrylate (supplier: BASF) and acrylonitrile (supplier: Fluka) were destabilized by shaking with NaOH solution or with a dehibit 200 column (polyscience inc.) and distilled under reduced pressure before use. Poly (vinyl alcohol) 7200, degree of hydrolysis 98 % (supplier: Merck) and citric acid (supplier: Merck) were used as suspending agents and maleic acid anhydride and maleic acid imide (supplier: Deutsche Texaco AG) as comonomers were used as received. 1,1-Dimethyl-2-ethylhexaneperoxoate (Luperox 26, supplier: Atofina) was used as initiator. All nitroxides were supplied by Ciba Spezialitätenchemie Lampertheim GmbH.

Homopolymerization of Vinyl Chloride Monomer

The polymerization was carried out according to the suspension process in batch operation. A double jacketed autoclave reactor (Buechi) of a volume of 500 ml was operated at a temperature between 70 °C and 90 °C with a stirrer velocity of 1000 rpm. The pressure amounts to 12-18 bar depending on the temperature used for the polymerization. Temperatures and pressure were recorded via pc during the reaction. Initiator concentration was 0.1 mole-% of acylic peroxide based on the vinyl chloride monomer. In order to avoid side reactions and to reduce undesired interactions with other compounds, the recipe was consciously reduced to the minimum of necessary substances. As suspending agent 0.3 g poly (vinyl alcohol) with a molecular weight of 40000, partly combined with a small amount of citric acid (15 mg) was used. The water : monomer ratio was 1 : 2.7, which is in the range of technical processes. Dosage of the VCM was accomplished by a Swagelock sample cylinder, connected with quick coupler. This also allowed a dosage of substances against the reactor pressure during the reaction. After a reaction time of 21 hours which was shown to be reasonable time for a monomer conversion of 50 % or more the obtained polymer was isolated by filtration resp. centrifugation. The crude polymer was washed with water, filtered, washed with ethanol, and finally dried under vacuum at 40 °C to constant weight.

Copolymer Formation Using Nitroxide Containing PVC as a Macroinitiator

Re-initiation experiments were carried out in a 100 ml Schlenk-tube. The solution consisting of PVC, monomer and solvent (chlorobenzene, 20 ml) was degassed and heated for 15 h or 21 h at the appropriate temperature under stirring. The mass ratio monomer : macroinitiator was 4 : 1; the ratio monomer/macroinitiator to solvent was 0.3. After cooling in an ice bath, the polymer was precipitated into 1500 ml methanol, filtered, washed with ethanol and dried under vacuum as indicated before.

Analytical Techniques

Molecular weights and polydispersities were determined in THF using a Waters modular GPC system equipped with 3 columns (PL, Polymer Laboratories $1x10^5$, $2x10^4$, $2x10^3$ and $1x10^2$ nm) thermostated at 30 °C and differential refractometer R401 (Waters) and UV detector at 254 nm (Waters). Calibration was done with narrow distributed polystyrene standards (supplied by Polymer Standard Service, PSS).

Thermal stability analysis was carried out according DIN 53 381, part 1. Therefore a small sample (100 mg) was filled in a pyrolysis vessel heated up to 180 °C in an oil bath. A pre-heated nitrogen stream (6 l/h) purged the developed HCl into a conductivity cell with demin. water (25 ml). The conductivity change was recorded with a Starna CDC1068 electrode and the HCl-conversion was calculated. As a measure of the thermal stability of the PVC samples the induction period $t_{ind.}$ is introduced as the section of the tangent of conductivity curve and the x-axis. Furthermore, stabilization time $t_{stab.}$ in our case is determined as the time when conductivity of the water in the sample cell reaches 50 μS cm^{-1} due to absorption of the evolved HCl in analogy to DIN 53 381, part 1, method B.

DSC investigations have been carried out with a differential scanning calorimeter DSC7 from Perkin Elmer under nitrogen atmosphere. Temperature range was from –200 °C to +110 °C with a heating and cooling rate of 20 °C min^{-1}. The sample weight was 10 mg.

Results and Discussion

Nitroxide Mediated Controlled Free Radical Polymerization of Vinyl Chloride Monomer

Nitroxide containing poly (vinyl chloride) was obtained via suspension polymerization process. The presence of the chemically bonded nitroxyl group in the PVC polymer could be proven indirectly by the reinitiation experiments described later. As expected, the resultant polymers revealed in all cases a narrower polydispersity (PD) than compared with S-PVC polymerized without nitroxide mediation (table 1-4). However, the polydispersities are not as low as in common CFRP processes with other monomers. This may be due to the occurrence of side reactions like disproportionation, which is a well-known fact in PVC manufacturing and has its cause in the very high transfer constant of the VCM (C_M=12 – 24 * 10^4 at 70 °C). Nevertheless, high molecular weights are obtained even at elevated temperatures above 70 °C, which is very high for VCM polymerization. Variation of polymerization temperatures, nitroxide type and concentration has been investigated.

Table 1. Influence of the temperature on the molecular weight and the polydispersity.

	nitroxide	T (°C)	yield (%)	M_n (g mole^{-1})	M_w (g mole^{-1})	PD M_w/M_n	$t_{ind.}$ (min)	$t_{stab.}$ (min)
1	- [a)]	70	94	28000	51000	2.6	14	43
2	- [a)]	75	89	31000	72000	2.3	14	38
3	- [a)]	80	88	25000	58000	2.3	11	32
4	- [a)]	85	71	18000	47000	2.7	12	30
5	(2)	70	35	20000	41000	2.0	15	27
6	(2)	75	62	26000	56000	2.1	17	26
7	(2)	80	63	29000	56000	1.9	10	26
8	(2)	85	57	24000	50000	2.1	12	22
9	(2)	90	33	18000	38000	2.1	11	23

a) comparison experiments without nitroxide content
conditions: 0.1 mole-% Initiator, 0.05 mole-% nitroxide based on the vinyl chloride monomer, t=21h

As expected the addition of nitroxides leads to a reduction of the conversion rates. A narrower molecular weight distribution of the experiments with nitroxides indicates control of the polymerization. A direct influence on the thermal stability is not detectable in all cases except the experiments with TEMPO (cf. table 3), resulting in lower stability. However, as shown below TEMPO does not act as polymerization regulator (controlling agent) under the conditions of PVC polymerization.

Varying the nitroxide concentration (Table 2) also results in a variation of the initiator : nitroxide ratio which has a strong effect on the conversion. A decrease in its ratio corresponding to an increase in free nitroxyl leads to an increased reaction time and lower yield while the molecular weight and the thermal stability are not affected significantly.

Table 2. Influence of the nitroxide concentration at a polymerization temperature of 85 °C.

	nitroxide	concentration (mole-%)	yield (%)	M_n (g mole^{-1})	M_w (g mole^{-1})	PD M_w/M_n	$t_{ind.}$ (min)	$t_{stab.}$ (min)
10	-	-	76	19000	51000	2.7	11	30
11	(2)	0.025	56	20000	45000	2.2	15	28
12	(2)	0.030	65	20000	44000	2.2	9	19
13	(2)	0.050	42	19000	42000	2.3	9	19
14	(2)	0.075	38	17000	40000	2.3	10	20

conditions: 0.1 mole-% Initiator based on the vinyl chloride monomer, T=80 °C, t=21h.

Table 3. Influence of the nitroxide type.

	nitroxide	reaction time (h)	yield (%)	M_n (g mole^{-1})	M_w (g mole^{-1})	PD M_w/M_n	$t_{ind.}$ (min)	$t_{stab.}$ (min)
15	(1)	21	38	9000	22000	2.5	4.5	8
16	(2)	21	62	19000	46000	2.4	9	31
17	(3)	21	64	22000	51000	2.3	8	23
18	(4)	21	74	20000	46000	2.2	9	25

conditions: 0.1 mole-% Initiator, 0.05 mole-% nitroxide based on the vinyl chloride monomer, T=80 °C.

After 21h reaction time all different nitroxide types (with the exception of TEMPO) reveal higher molecular weights, lower polydispersities and much higher conversions, however lower than compared with the reaction without nitroxide addition. In all cases, the reaction time was prolonged up to 21 h. This is in contrast to conventional free radical suspension polymerization of VCM, where polymerization times of 4-7 h are usual. The highest conversion rate is achieved with a nitroxide concentration of 0.03 mole-%. For achieving control in PVC polymerization the use of nitroxyl structures with high steric hinderance is essential.

As clearly seen in Table 3, TEMPO (1) reduces conversion and molecular weight in contrast to other nitroxide types. As depicted before, TEMPO is not efficient for CFRP-processes at temperatures below 100 °C. Also TEMPO mediated PVC shows a relatively low thermal stability. This may be interpreted as an increase of side reactions leading to non-ideal (linear) PVC chain structures.

Random Copolymerization of VCM and Vinyl Acetate

As one of the most popular comonomer for VCM, vinyl acetate (9) is largely used in technical copolymerizations. Therefore it was obvious to use this comonomer in the CFRP process. The copolymerization reactivity rates of vinyl chloride / vinyl acetate are $r_1 = 1.68$ and $r_2 = 0.23$[12] at 68 °C, which implies that vinyl chloride preferentially polymerizes compared to the vinyl acetate. As a consequence, in normal radical polymerizations the initially formed copolymer is lower in vinyl acetate content relative to the comonomer composition. This ratio shifts during the copolymerization towards an increasing content of vinyl acetate. Due to the very low concentration of free radicals and the fact, that all chains are growing approximately with the same rate, the copolymers have a constant composition during the entire polymerization process. This has been shown by Hawker et al.[13] with the copolymerization of styrene/maleic acid anhydride mixtures.

In our experiments VCM / vinyl acetate random copolymers with a vinyl acetate content of 10-50 % (w/w) have been successfully obtained in suspension in the same way as described for the homopolymerization (Table 4).

Table 4. Conditions and results of random copolymerization of VCM and vinyl acetate.

	VCM	vinyl acetate	T (°C)	yield (%)	M_n (g mole^{-1})	M_w (g mole^{-1})	PD M_w/M_n
19	0.5	0.5	85	64	22000	49000	2.2
20	0.85	0.15	72	92	18000	42000	2.3
21[a)]	0.9	0.1	72	60	27000	57000	2.1
22	0.9	0.1	80	67	27000	54000	1.9

conditions: 0.1 mole-% Initiator, 0.05 mole-% nitroxide (2) based on monomers; t=21 h
a) t=45 h

The experiments with a high content of vinyl acetate mostly showed a coagulation of the suspension, but with a content of 10 % comonomer similar molecular weights compared to the pure PVC polymerization were obtained. At the temperature of 80 °C high molecular weight and the narrowest molecular weight distribution were reached. The copolymer composition and its regularity are still under investigation. Samples with a higher content of vinyl acetate revealed a strong coloring towards brown. This discoloration disappeared after re-precipitating.

Re-initiation Experiments

To demonstrate the capability of reinitiation and formation of block copolymers from the PVC obtained via nitroxyl mediated polymerization, these nitroxide containing dormant chains were re-awakened by heating up in the presence of a second monomer.

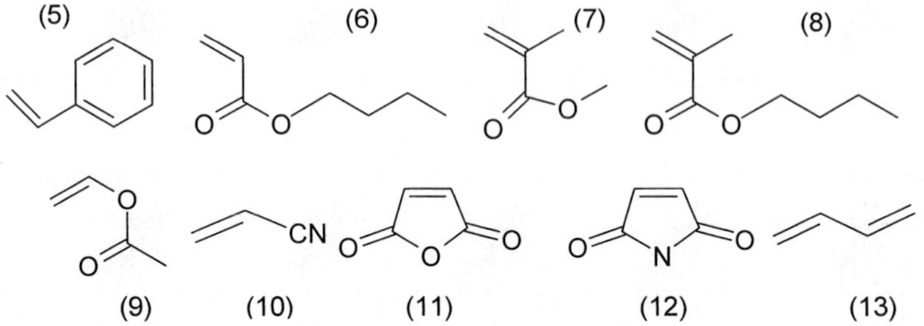

We investigated a number of common monomers such as styrene (5), butyl acrylate (6), methyl methacrylate (7) and butyl methacrylate (8), acrylonitrile (10), MAA (11), MAI (12) (in combination with styrene) and even butadiene (13) in terms of the possibility of re-initiation.
At temperatures of 100 – 130 °C, the nitroxyl containing PVC acts as a macro-initiator and starts a new polymerization reaction. Due to this elevated temperatures, the controlled reaction should proceed without any problem especially with monomers like styrene, which are known to undergo very easily the nitroxide mediated CFRP. The resulting polymer is an A-B-type two-block copolymer as proved by GPC measurements (cf. Fig. 1). The incorporation of styrene can be shown by an increase in molecular weight and by uv-activity of the sample.
Addition of campher sulphonic acid (CSA) (examples 27, 28) accelerates the polymerization rate as known from[14] and prevents the auto-initiation of styrene, which is a significant reaction at the chosen polymerization temperatures. With reference experiments it could be shown, that the auto-initiation process is largely suppressed. Furthermore the addition of stearyl-benzoyl-methane as a well-known thermostabilizer for PVC[15] to the reaction solution (example 29 and 30) results in reduced discoloration of the sample according to improved thermal stability of the PVC polymer resulting in a pure colorless block copolymer.

Table 5. Results of the re-initiation experiments.

	second / third monomer (+ optionally additive)	temp (° C)	yield[a)] (%)	M_n (g mole^{-1})	M_w (g mole^{-1})	PD M_w/M_n
23	NO*-containing pure PVC (13)	130	91 [b)]	20000	36000	1.8
24	styrene	100	30	17000	53000	3.5
25	butyl methacrylate	130	41	24000	60000	2.5
26	styrene + acrylonitrile (1:3)	110	15	24000	72000	3.0
27	styrene + MAA (1:1) + CSA 10 mg	130	98	Not determined		
28	styrene + MAI (1:1) + CSA 10 mg	130	68	30000	79000	2.7
29	methyl methacrylate + Stab 10 mg	130	54	29000	86000	3.0
30	butyl acrylate + Stab 10 mg	120	22	19000	39000	2.0

a) yield: conversion 2nd monomer
b) recovery of the original PVC
CSA = campher-10-sulfonic acid, MAA = maleic acid anhydride, MAI = Maleic acid imide, Stab = Stearyl-benzoyl-methane.

The successful reactivation reaction can be proved by GPC (Figure 1). There is no transfer to a new starting polymer chain of the second monomer. The initial nitroxide-containing PVC shows no UV-activity. After the re-initiation reaction, first an UV-absorption due to the aromatic styrene function can be seen. But there is also a shift to higher molecular weight and no second signal at the position of the initial PVC or any signal from other polymerization products is detected.

Furthermore butadiene (13) was used as second monomer. This monomer was easily dosed into the reactor after complete consumption of the VCM without purifying the original PVC. The reactor just was degassed and filled with gaseous butadiene monomer in the same manner as described for VCM. While butadiene is known as a good inhibitor of the VCM polymerization it can be proposed, that also an AB-type copolymer is resulting; however it was not confirmed by any analytical technique. The obtained polymer with the overall conversion of 13% shows a molecular weight of 39000 (M_w) and a polydispersity of 2.0. This material exhibits the expected rubber-like appearance and a soft touch contrary to the rigid hard PVC. Indication of copolymer formation is shown by DSC and by appearance of a mixed glass transition temperature T_g at 22 °C in addition to the typical T_g of polybutadiene at – 86 °C and PVC at + 87 °C. A complete analytical investigation in terms of copolymer composition and block structure will follow.[16]

a)

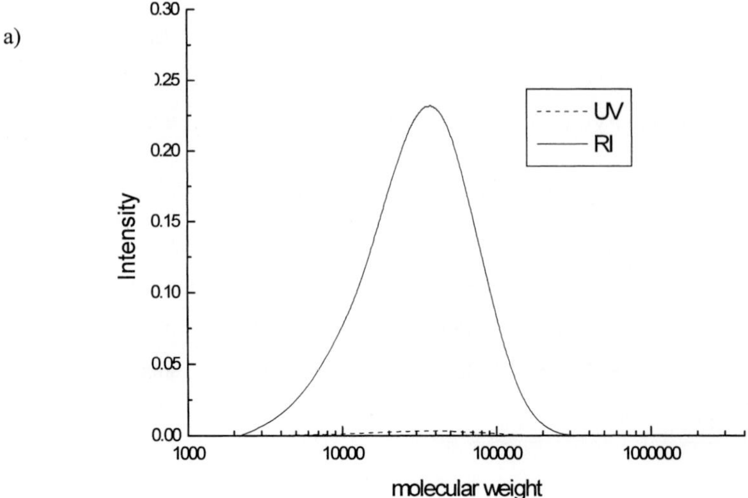

b)

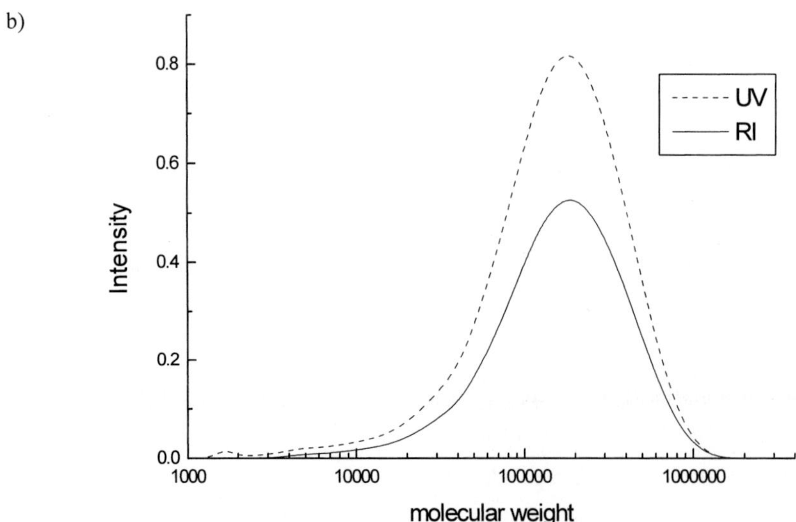

Fig. 1. a) Pure PVC polymerized in the presence of (4) is not absorbing in UV (only RI = Refractive Index).
b) Re-initiation of the PVC with styrene results in intense UV-absorption due to incorporated polystyrene and at the same time increase in molecular weight.

Conclusion

Nitroxide-mediated CFRP of vinyl chloride monomer was performed in suspension process, leading to PVC-homopolymers with a narrower molecular weight distribution. Contrary to classical PVC polymerization the molecular weight can be kept high at higher polymerization temperatures. The resulting nitroxide-containing PVC acts as a macroinitiator and forms different types of copolymers. Random copolymers of VCM with vinyl acetate, AB-type two-block copolymers with styrene, methyl methacrylate, n-butyl methacrylate, n-butyl acrylate and butadiene and terpolymers with one PVC block and two monomer(s) (styrene/methyl methacrylate, styrene/acrylonitrile, styrene/maleic acid anhydride and styrene/maleic acid imide) in the second block have been synthesized. Therefore, the living-type nitroxyl mediated polymerization of VCM opens new ways for the design of complex polymer architectures on the basis of PVC structures.

Acknowledgements

Ciba Spezialitätenchemie Lampertheim GmbH, Lampertheim, Germany is gratefully acknowledged for financial support, and for providing the nitroxide samples.

[1] T.E. Werrington, A.V. Tobolsky, *J.Am.Chem.Soc.* **1955**, *77*, 4510
[2] T. Otsu, M. Yoshida, *Makromole.Chem. Rapid Commun.* **1982**, *3*, 124
[3] D.H. Solomon, E. Rizzardo, P. Cacioli, *U.S. Patent 4,581,429*, **1986**
[4] a) J.-S. Wang, K. Matyjaszewski, *J.Am.Chem.Soc.* **1995**, *117* 5614; b) M. Kato, M. Kamigaito, M. Sawamoto, T. Higashimura, *Macromolecules*, **1995**, *28*, 1721
[5] J. Chiefari, Y.K. Chong, F. Ercole, J. Krstina, J. Jeffery, T.P.T. Le, R.T.A. Mayadunna, G.F. Meijs, C.L. Moad, G. Moad, E. Rizzardo, S.H. Thang, *Macromolecules* **1998**, *31*, 5559
[6] Verband Kunststofferzeugende Industrie **2002**
[7] C.J. Hawker, A.W. Bosman, E. Harth, *Chem.Rev.* **2001**, *101*, 3661
[8] M.K. Georges, R.P.N. Veregin, P.M. Kazmaier, G.K. Hamer, *Macromolecules* **1993**, *26*, 2987
[9] M.K. Georges, R.P.N. Veregin, P.M. Kazmaier, G.K. Hamer, *Trends Polym. Sci.* **1993**, *2*, 66
[10] C.J. Hawker, G.G. Barclay, A. Orellana, J. Dao, W. Devonport, *Macromolecules* **1996**, *29*, 5245
[11] a) H. Fischer, *Macromolecules* **1997**, *30*, 5666 b) H. Fischer, *Chem.Rev.* **2001**, *101*, 3581
[12] F.R. Mayo, C.Walling, F.M. Lewis, W.F. Halse, *J.Am.Chem.Soc.* **1948**, *70*, 1523
[13] D. Benoit, C.J. Hawker, E.E. Huang, Z. Lin, T.P. Russell, *Macromolecules* **2000**, *16*, 2143
[14] P. G. Odell, P. N. Veregin, G. K. Hamer, M. K.Georges, *U.S. Patent 5,608,023*, **1995**
[15] A. Michel, T.V. Hoang, B. Perrin, M.F. Llauro, *Polym.Degrad.Stab.* **1980**, *3*, 107
[16] T.Wannemacher, Dissertation, in press
[17] T.Wannemacher, Dissertation, in press

Macromol. Symp. **2003**, *202*, 25–35

Reactive Processing of Syndiotactic Polystyrene with an Epoxy/Amine Solvent System

Jaap Schut,[1,2] *Manfred Stamm,*[2] *Michel Dumon,*[1] *Jocelyne Galy,*[1] *Jean-François Gérard**[1]

[1] National Institute of Applied Sciences, Laboratory of Macromolecular Materials (IMP/LMM), UMR CNRS 5627; 17, Avenue Jean Capelle, 69621 Villeurbanne Cedex, France
E-mail: jfgerard@insa-lyon.fr
[2] Institute of Polymer Research Dresden, Physical Chemistry and Physics of Polymers; Hohe Strasse 6, 01069 Dresden, Germany

Summary: Syndiotactic polystyrene (sPS) is a new semi-crystalline thermoplastic which is believed to fill the price-performance gap between engineering and commodity plastics. In order to reduce the high processing temperature of sPS (>290°C), an epoxy-amine model system was used as a reactive solvent. Such a processing aid can be used to achieve a 50 to 500 fold lowering of the melt viscosity. When initially homogeneous solutions of sPS in a stoechiometric epoxy-amine mixture are thermally cured, Reaction Induced Phase Separation (RIPS) takes place, leading to phase separated thermoplastic-thermoset polymer blends. We focus our study on low (wt% sPS < 20%) and high concentration blends (wt% sPS > 60%) prepared by two processing techniques (mechanical stirring in a laboratory reactor or internal mixer/ reactive extrusion respectively). These blends have different potential interests. Low concentration blends (sPS domains in an epoxy-amine matrix) are prepared to create new, tunable blend morphologies by choosing the nature of the phase separation process, i.e. either crystallisation followed by polymerization or polymerization followed crystallisation. High concentration blends (sPS matrix containing dispersed epoxy-amine particles after RIPS) are prepared to facilitate the extrusion of sPS. In this case, the epoxy amine model system served as a reactive solvent. The time to the onset of RIPS is in the order of 7-9 min for low concentration blends, while it increases to 20-45 min for high concentration samples, as the reaction rates are substantially slowed down due to lower epoxy and amine concentrations. During the curing reaction the melting temperature of sPS in the reactive solvent mixture evolves back from a depressed value to the level of pure sPS. This indicates a change in the composition of the sPS phase, caused by (complete) phase separation upon reaction. We conclude that our epoxy amine system is suited for reactive processing of sPS, where final properties depend strongly on composition and processing conditions.

Keywords: reaction induced phase separation (RIPS); reactive processing; syndiotactic polystyrene; thermoset-thermoplastic blends

DOI: 10.1002/masy.200351203

Introduction

Syndiotactic polystyrene (sPS) is a relatively new and promising semi-crystalline engineering thermoplastic with a melting point around 270°C, which was first synthesized in 1985 by Ishihara et al. with the use of a titanium metallocene catalyst.[1] In addition to the advantages of atactic polystyrene (attractive price, low specific gravity, low dielectric constant, and good processability), sPS has a very good solvent, chemical and steam resistance, a high modulus of elasticity, high heat resistance and good dimensional stability. These properties originate from its semi-crystallinity, which can be between 15 to 60%. As it crystallises about two orders of magnitude faster than iPS (which has not been commercialised due to its low rate of crystallisation), it is therefore expected to fill the price-performance gap that currently exists between performance and commodity plastics. sPS has been commercialised by Dow Chemical Co. (as Questra™) and Idemitsu (as Xarec™). However, sPS has to be processed at temperatures exceeding 290°C, due to its high melting point, causing degradation of the polymer. This could be prevented by lowering the process temperature.

The goal of this study is therefore twofold: Firstly to facilitate the processing of sPS by lowering its melt temperature and viscosity with a curable epoxy/amine model system as reactive solvent, which will result in a thermoplastic-thermoset polymer blend. Secondly we aim to generate new blend morphologies by using the crystallisation of sPS as a variable parameter during the crosslinking of the epoxy/amine monomers. As the sPS-Epoxy/Amine system is as yet unexplored we also turned our attention to blends containing less than 20wt% sPS (yielding an epoxy resin continuous phase). Here the concept of reactive processing is of course no longer applicable, but one can obtain useful additional information about the crystallisation and phase separation behaviour of the system by studying these blends.

Pursuing the first objective will lead to an sPS matrix filled with epoxy-amine particles as the dispersed phase, while the second objective will give an epoxy-amine matrix filled with sPS particles as the dispersed phase.

When a homogeneous solution of a polymer in a mixture of monomers is thermally cured, the polymer will phase separate during the curing reaction as a consequence of the changing solvent nature (cause primarily by the increasing molecular weight but also by a changing Flory-Huggins interaction parameter [χ]). This so called "Reaction Induced Phase Separation" (RIPS)[2,3] leads to a phase separated thermoplastic-thermoset polymer blend. In a previous study[4] we found a depression of the crystallisation temperature of syndiotactic polystyrene by some 20-50 K in non-reactive blends of sPS with the epoxy monomer Diglycidylether of Bisphenol-A (DGEBA). From comparison with literature data on the shape of the phase diagram, we concluded that DGEBA was considered to be a poor solvent for sPS (where the word poor is used in its usual polymer physical meaning of a solvent that dissolves a polymer, however without the extensive swelling of the coiled chains that a good solvent would cause). The present study is concentrating on the experimental methods used for processing sPS with a DGEBA/MCDEA reactive solvent, and the characterisation of some important processing parameters such as melt viscosity, onset of phase separation, as well as the final morphology of the cured blend.

Experimental

Syndiotactic polystyrene (Questra QA 101™, kindly donated by Dow Chemical Co.), denoted sPS, of more than 99% syndiotacticity (^{1}H and ^{13}C-NMR), with a number average molar mass of 94,100 g mol^{-1}, and weight average molar mass of 192,000 g mol^{-1} was used. The epoxy resin reactive system was based on a stoichiometric mixture of diglycidylether of bisphenol-A (Dow DER 330™, n = 0.15, M = 383.1 g mol^{-1}, purchased from Dow Chemical Co.), denoted DGEBA, and the aromatic diamine 4,4'-methylenebis(3-chloro-2,6-diethylaniline) (Lonzacure M-CDEA™, 98%, M = 379.4 g mol^{-1}, purchased from Lonza) denoted MCDEA. sPS-DGEBA blends at concentrations below 30 wt% sPS were prepared in a large test tube equipped with a mechanical stirrer. Powdered sPS was added to vacuum degassed DGEBA and heated in a Wood's metal bath to 290°C to ensure complete dissolution within 10 min. This reactive mixture was then quickly transferred (minimising heat loss to prevent premature crystallisation) to small, preheated microtubes for turbidity measurements.

Blends containing 60 to 100 wt% sPS were prepared in a Haake internal mixer or a Clextral co-rotating twin-screw extruder, due to their high viscosities. Figure 1 shows the reactive extrusion set-up. The screw dimensions were as follows: diameter: 25 mm, length: 900 mm (l/d = 36), distance between screws: 21 mm. The temperature of the first zone was deliberately chosen very high (290°C, the advised extrusion temperature of pure sPS) in order to ensure fast and complete dissolution of the sPS in the DGEBA monomer. DGEBA at 80°C was injected just after the first counter flow element to ensure low counter pressure from the melt in the extruder barrel. At the same point the zone temperature was lowered to 280°C. The sPS/DGEBA mixture passed through a kneader zone for intense mixing followed by a transport

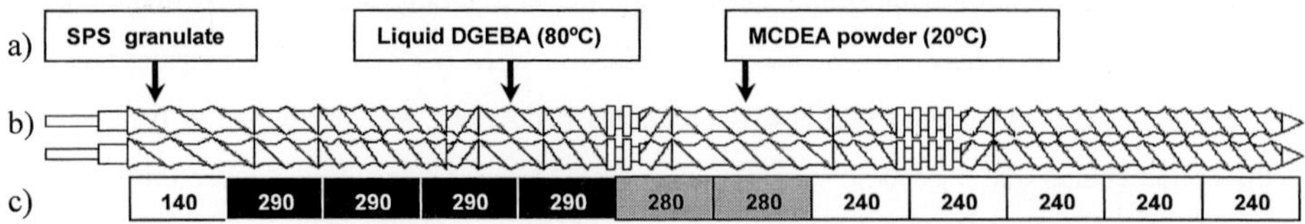

Fig. 1. The reactive extrusion process. a) The injection points of the 3 components; b) Transport, counterflow and kneading elements on the screws; c) Temperatures of heating zones.

zone. After this the temperature was further lowered to 240°C. MCDEA powder (at 20°C) was then added into the melt by a computer controlled vibrating funnel (feeder). Both the DGEBA pump container and the MCDEA funnel/container were placed on a balance in order to control and regulate the material flux into the extruder screw. The cure reaction started when MCDEA liquified and was mixed intimately with the melt by a second kneader followed by a counterflow element in order to ensure an appropriate residence time in the kneader zone. Upon exiting the extruder die the reacting melt (at low reaction conversion) was either captured in the hot, homogeneous state in a 35 mm test tube (which was kept at 240°C in a Wood's metal bath) *and subsequently cured in this tube* (giving RIPS morphology), or it was captured in a cold test tube and allowed to cool to room temperature causing thermal precrystallisation before being cured at 220°C (i.e. the sample underwent CIPS, Crystallisation Induced Phase Separation prior to curing), without remelting the sPS. Cloud point measurements were performed on an in house built apparatus, which consists of a temperature regulated and insulated oven in which a small test tube (8 mm internal diameter) is placed. Light emitted from a white lamp is guided through the sample by glass fibres and then to a photomultiplier and a photocell. The measured voltage is a measure of the relative turbidity of the sample. Rheological measurements were made on a Rheometrics Dynamic Analyser RDA II using a parallel plate geometry (2cm diameter plates, 3mm gap). Frequency scan rate was 1 Hz, temperature scan rate 2.2K min^{-1}. DSC measurements were performed on a water-cooled Perkin-Elmer DSC-7 under nitrogen atmosphere at a temperature scan rate of 10 K min^{-1}. All samples were held for 5 min at 300°C to erase thermal history prior to crystallisation scans.

Results

Non-reactive Systems

The phase diagram (crystallisation temperature vs. concentration) of the non-reactive sPS-DGEBA system as established by DSC is reported elsewhere.[4] It was found that 220°C is the minimum temperature at which low concentration blends (<40 wt% sPS) could be processed in homogeneous conditions, i.e. without premature crystallisation. On the other hand the minimum temperature needed for the reactive extrusion of high concentration blends (>60 wt% sPS) is 230-250°C. To investigate the effect of DGEBA on the melt viscosity of sPS, a number of blends were analysed by rheometry. Figure 2 shows the viscosity of several sPS/DGEBA blends in their molten states. More specifically, the plateau value of the viscosity in the melt of the mixture in question, i.e. the constant value of the viscosity from $T = T_c + 10$ K to $T = T_c + 30$ K (where T_c is the crystallisation temperature) during a crystallisation scan from 300 down to 80°C at -2.2 K min^{-1}. One can observe a large decrease in the melt viscosity upon addition of DGEBA: compared to pure sPS a 20-fold decrease from 100 to 4.8Pa s can be seen upon adding 50 wt% DGEBA, while adding 70 wt% DGEBA even causes a 500 fold decrease down to 0.2 Pa s. The latter viscosity is however too low to be processed with an extruder.

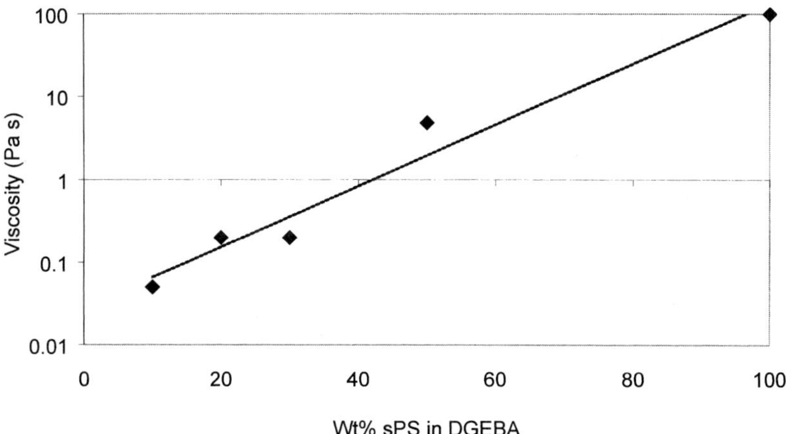

Fig. 2. Melt viscosity vs composition of syndiotactic polystyrene-DGEBA solutions above their crystallisation temperature ($T = T_c + 10$ K, scan frequency 1 Hz).

From the afore mentioned phase diagram and these rheological data, we conclude that DGEBA is a suitable reactive solvent for syndiotactic polystyrene.

Reactive Systems

The next step was to cure sPS/DGEBA blends with MCDEA diamine curing agent and follow the cure reaction by monitoring the Reaction Induced Phase Separation (RIPS). Figure 3 shows the phase separation process by following the turbidity of several low concentration, reactive sPS-DGEBA/MCDEA solution during isothermal curing at 220°C. Blends with concentrations above 25 wt% sPS were almost impossible to prepare in test tubes (due to their high viscosity) and contained too many air bubbles to be analysed by turbidity measurements.

After introduction of the sample in the device (after about 2 min reaction time), light transmission was stable at a plateau value for about 5-7 min (i.e. 7-9 min reaction time). After 7 to 9 min one can observe a decrease of the transmitted light intensity in all samples due to the appearance of a dispersed phase (the sample becomes opaque). This is the onset of RIPS, which takes place quite fast due to the high reaction temperature and high epoxy-amine concentrations. After a certain time the intensity falls to almost zero, indicating the end of the RIPS.

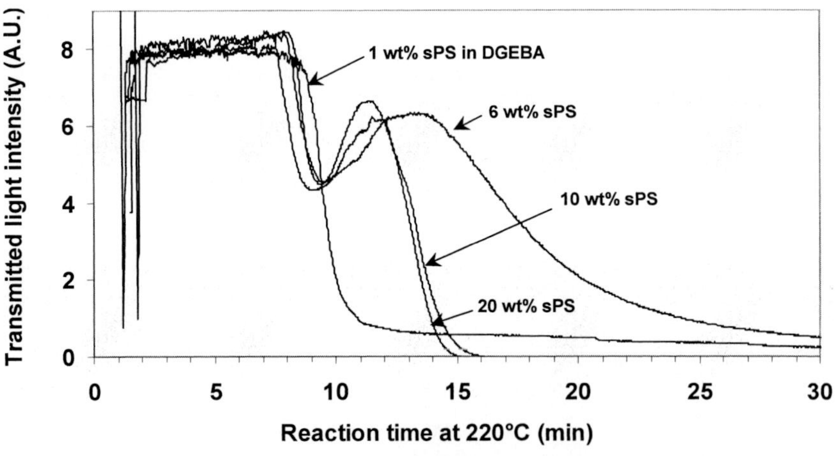

Fig. 3. Turbidity of low concentration sPS-DGEBA/MCDEA blends during cure at 220°C as a function of reaction time.

The cure reaction itself continues and samples are left to cure for 2 h.

The unexpected peak in the intensity after 11 to 15 min for the 6, 10 and 20 wt% samples is most probably caused by a crossing of the refractive index values of the two phases due to their continuous composition change during curing. The phase separation process ends after 12, 15, 16, and 30 min for the 1, 20, 10, and 6wt% sPS samples respectively, observed as a levelling off to a constant value of the observed intensities.

At the time of writing it is yet unresolved if the phase separation of sPS from the epoxy mixture consists of an initial liquid-liquid phase separation followed by crystallisation (liquid to crystalline phase transition) of the sPS component, ór if it is a transition of the dissolved sPS directly into its crystalline state. Literature on the system sPS-decalin suggests that crystallisation takes places directly out of a homogeneous solution, while liquid-liquid phase separation occurs at much lower temperatures.[5] More information on this topic is however scarce.

Although the time to RIPS is very short in the case of these low sPS content blends, the cure kinetics of the epoxy-amine reaction are substantially slowed down in samples with 60 wt% sPS or more. This dilution effect can be observed visually in the mixing chamber as an opacity of the melt in samples prepared in an internal mixer. Figure 4 shows the time to onset of turbidity vs blend composition for both high and low concentration blends.

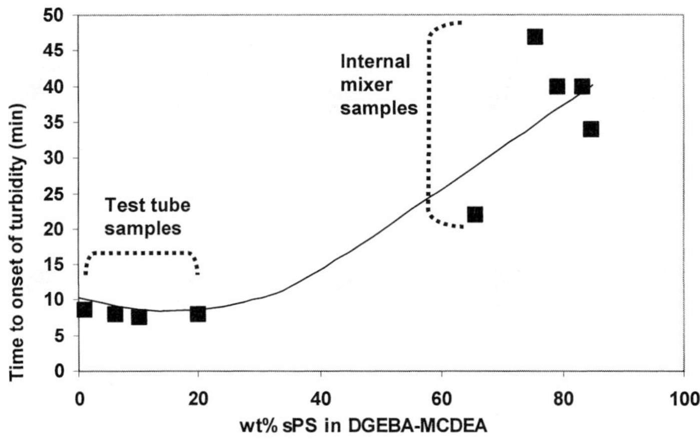

Fig. 4. Time to onset of turbidity in sPS/DGEBA-MCDEA blends during isothermal curing at 220°C as a function of their composition.

It can be seen that time to RIPS for the high concentration blends is now of the order of 20-45 min. DGEBA/MCDEA is therefore well suited for reactive extrusion (without premature phase separation), where the typical mean residence times is 3 to 5 min. Fig 4 reports the cloud point *times* while the correct parameter to characterize the extent of reaction at phase separation would be the cloud point *conversions*. Reaction conversion can generally be obtained by the determination of the residual heat of polymerization, which unfortunately proved to be impossible as explained later in the text We thus used time instead of conversion, which leads to cloud points having the same time corresponding to different reaction extents.

The next step was to investigate the feasibility of processing syndiotactic polystyrene in a co-rotating twin-screw extruder with DGEBA-MCDEA as a reactive solvent. Blends of sPS in DGEBA-MCDEA with compositions ranging from 60 to 90 wt% sPS were reactively extruded. The obtained extrudate was either captured directly from the extruder die into a preheated test tube at 220°C and subsequently cured in this tube (giving RIPS morphology), or it was captured into a cold test tube and allow to cool to room temperature to be precrystallised prior to curing at 220°C (CIPS morphology). The two routes have been found to give rise to very different morphologies. Both the blend morphology at different compositions as well as the tailoring of these morphologies by using the crystallisation ability of sPS as a parameter – yielding a range of interesting morphologies – will be discussed in more detail elsewhere.[6] Figure 5 shows the morphology of pure (beta) sPS and a 64wt% sPS/DGEBA-MCDEA reactively extruded sample. Pure sPS (***5a***) shows a rough fracture surface with no specific morphology, while a 64 wt% sPS/DGEBA-MCDEA blend (***5b***) shows 1-2 μm spherical particles of cured epoxy resin in an sPS matrix. This morphology is well known and characteristic for immiscible polymer blends[7,8] it therefore only serves to give a first impression of the morphologies of our system.

To check the advancement of the cure reaction, we intended to determine the residual heat of polymerisation of samples from the internal mixer taken at regular intervals during curing. To our surprise no residual reaction exotherm could be detected in DSC, which is probably caused by shielding of this exotherm by the melting endotherm of the sPS. We therefore turned our attention to the evolution of the melting temperature of sPS during the curing reaction.

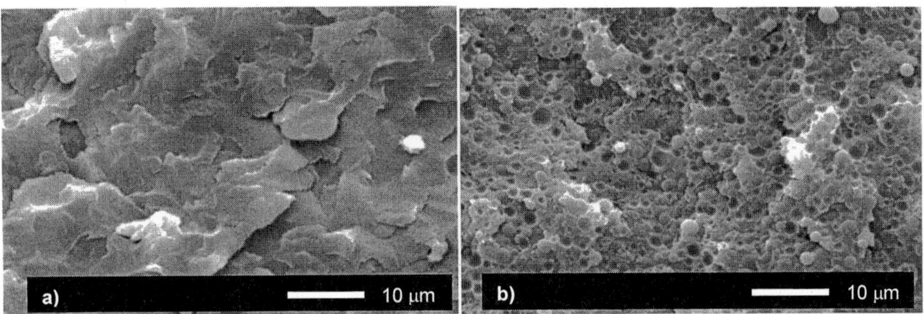

Fig. 5. SEM micrographs showing the morphology of: a) pure sPS (β form), b) 64 wt% sPS reactively extruded with DGEBA/MCDEA, cured at 250°C.

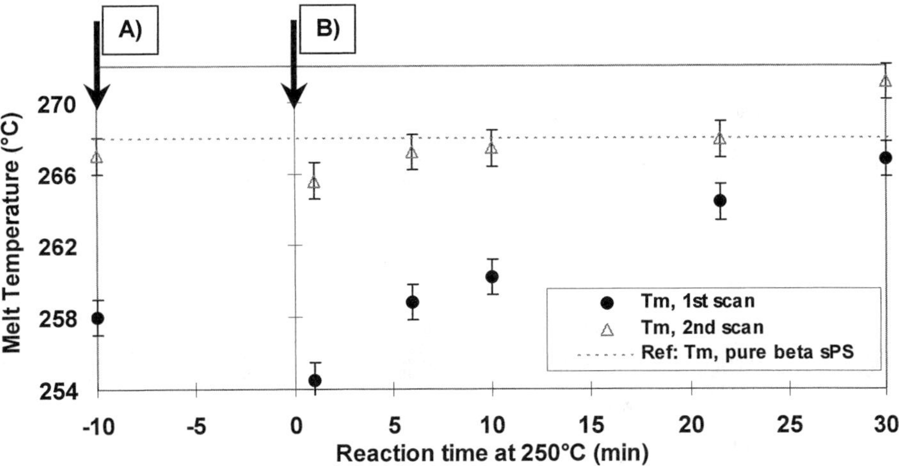

Fig. 6. The evolution of the melting temperature of sPS in sPS/DGEBA-MCDEA blends during reactive processing in an internal mixer at 250°C. A) Non reactive 85 wt% sPS/DGEBA blend. B) Cure reaction starts by adding MCDEA (resulting in a 80 wt% sPS/DGEBA-MCDEA blend).

Figure 6 shows the melting point evolution of an sPS/Epoxy-amine blend during cure, reflecting the effects of Reaction Induced Phase Separation. At point A) a non-reactive 85 wt% sPS/DGEBA blend is homogenised in the mixer for 10 min. At point B) (t=0) the MCDEA diamine curing agent is added to the melt, starting the reaction and changing the overall composition to 85 wt% sPS in DGEBA-MCDEA. One can observe an additional 3-4 °C drop of the melting temperature due to a further decrease of the sPS concentration. Once the cure reaction has started one sees a levelling off with time of the sPS melting point back to the value of pure sPS (268°C) beyond 30 min, indicating that complete phase separation into a pure sPS and a pure epoxy-amine phase has taken place during reaction.

Conclusions

Despite the poor solvent nature of DGEBA epoxy monomer for syndiotactic polystyrene, it decreases the melt viscosity of sPS 50 to 500-fold. When initially homogeneous blends of sPS and DGEBA-MCDEA are thermally cured, Reaction Induced Phase Separation (RIPS) takes place, leading to completely phase separated blends. The time to the onset of RIPS is in the order of 7-9 min for the 1 to 20 wt% sPS samples. The reaction kinetics of higher concentrated samples prepared in an internal mixer are substantially slowed down due to the lower epoxy and amine concentrations, time to RIPS is now in the order of 20-45 min. Together with a previously found melting point depression this leads us to conclude that DGEBA-MCDEA is a suitable reactive solvent to be used in the reactive processing of sPS. SEM shows 1-2μm cured epoxy resin spherical particles in an sPS matrix. Using DSC no residual cure exotherm could be detected in partially cured samples. We therefore monitored the evolution of the melting temperature during the course of the reaction. After the curing reaction started we observed a levelling off of the sPS melting point back to the value of pure sPS (268°C) with reaction time, indicating complete phase separation. We are currently studying the phase separation and crystallisation by WAXS/SAXS, and measuring the mechanical properties of the cured blends at different compositions and thermal histories. This will lead to an improved insight into the structure-properties relationship for these blends.

Acknowledgments

We gratefully acknowledge the funding of this research by the Research Training Networks (RTN) program under auspices of the Directorate-General for Research of the European Commission, under the name « PolyNetSet »; contract number HPRN-CT-2000-00146. Further thanks go to Hervé Perier-Camby (INSA, Villeurbanne) for his help with the reactive extrusion, and Dr. Nick Zafeiropoulos (IPF, Dresden) for the fruitful discussions.

[1] N. Ishihara, T. Seimiya, M. Kuramoto, M. Uoi, *Macromolecules*, **1986**, 19, 2464-2465.
[2] Girard-Reydet, H. Sautereau, J.P. Pascault, P. Keates, P. Navard, G. Thollet, G. Vigier, *Polymer*, **1998**, 39, 2269-2280.
[3] R.J.J. Williams, B.A. Rozenberg, J.P. Pascault, *Advances in Polymer Science*, **1997**, 128, 97-153.
[4] J.A. Schut, M. Stamm, M. Dumon, J.-F. Gérard, *Macromolecular Symposia, in press.*
[5] F. Deberdt, H. Berghmans, *Polymer*, **1993**, 34(*10*), 2192-2201.
[6] J.A. Schut, N.E. Zafeiropoulos, M. Stamm, M. Dumon, J.-F. Gérard, *Manuscript in preparation*
[7] R. Venderbosch, H.E.H. Meijer, P.J. Lemstra, *Polymer*, **1994**, 35(*20*), 4349-4357.
[8] J.G.P. Goossens, S. Rastogi, H.E.H. Meijer, P.J. Lemstra, *Polymer*, **1998**, 39(*25*), 6577-6588.

Controlled Synthesis of Block Polyesters by Reactive Extrusion

Graeme Moad, Andrew Groth, Michael S. O'Shea, Julian Rosalie, Ramon D. Tozer, Gary Peeters*

CRC for Polymers, CSIRO Molecular Science, Bag 10, Clayton South, Victoria 3169, Australia

Summary: The synthesis of block copolyesters by combining transesterification resistant oligoesters with poly(ethylene terephthalate) is described. Critical factors in the synthesis include processing conditions (temperature), end groups and catalysts used in oligoester synthesis. It is necessary to achieve a balance between ease of incorporation/synthesis and transesterification resistance. The process is discussed with reference to poly(neopentyl isophthalate) as a representative transesterification resistant oligoester. The primary motivation for the work has been the development of higher barrier poly(ethylene terephthalate) useful for making monolayer beverage containers.

Keywords: block copolyester; oligoester synthesis; PET; poly(ethylene terephthalate); poly(neopentyl isophthalate); reactive extrusion; transesterification resistance

Introduction

Conventional single layer poly(ethylene terephthalate) (PET) bottles do not have sufficient gas (CO_2, O_2) barrier for many applications (*e.g.* beer bottles, baby food containers). Other properties (*e.g.* heat distortion temperature) may also be less than optimal.

PET is a semicrystalline polymer with barrier properties that are largely determined by the rate of gas transmission through the amorphous phase.[1,2] PET copolymers may have markedly improved barrier performance and a large number of patents and papers relate to this area.[1,3,4] Other properties (*e.g.* % crystallinity, T_g, and T_m) are strongly influenced by the length of PET homopolymer segments.[5] The incorporation of comonomers into PET, which necessarily reduces the PET segment length, is therefore anticipated to have a strong influence over both these properties and the materials properties that depend on these parameters. These factors limit both the type and level of comonomer than can be introduced in an effort to reduce gas barrier.

It is also known that various properties (*e.g.* barrier properties, impact, heat distortion temperature) for block copolyesters or polyester blends can be superior to those of random copolyesters of the same nominal composition.[6] However, block copolyester and polyester

 DOI: 10.1002/masy.200351204

blends have a propensity to undergo rapid transesterification during processing (*e.g.* extrusion, solid stating, injection molding) leading to a deterioration of the abovementioned properties. To address this, various approaches have been described and applied to reduce the extent of transesterification; these include using the use of mild processing conditions and transesterification inhibitors.[7,8] Much of this effort is described in the patent literature.

The main thrust of our work has been to design and synthesize block polyesters that have both desirable properties (in particular, high gas barrier and PET-like processing) and a reduced tendency to undergo structural randomization and consequent property degradation by transesterification. This has been achieved though the use of blocks made from transesterification resistant segments.[9] In designing the transesterification resistant segment, it is necessary to achieve a balance between ease of incorporation/synthesis and transesterification resistance.

In this paper, we examine the synthesis of polyesters containing segments based on neopentyl glycol (NPG) by reactive extrusion and determine their resistance of the segments to transesterification during extrusion and subsequent solid stating.

Experimental Part

^{1}H NMR spectra were obtained with a Bruker Avance DRX500 on samples dissolved in 1:1 deuterochloroform:deuterotrifluroacetic acid. Chemical shifts are reported in ppm from tetramethylsilane. Spectra were recorded immediately after sample preparation to avoid unwanted trifluoroacetylation.[10]

Oligoester molecular weights quoted were determined on a Waters Associates GPC with THF as eluent at 1 mL/min at 22 °C. The columns were calibrated with narrow polydispersity polystyrene standards (Polymer Laboratories) and molecular weights are reported as polystyrene equivalents. PET molecular weights were determined by GPC on a Hewlett Packard 1090 high performance liquid chromatograph equipped with a diode array detector (270 nm) and four Waters Ultrastyragel columns with 10% HFIP in chloroform containing 1% tetrabutylammonium acetate as eluent at 1 mL/min at 30 °C.

Differential Scanning Calorimetry was carried out with a Metler Toledo DSC 821 equipped with a TSO801RO sample robot under nitrogen. For T_m measurements samples were weighed into 40μL aluminum sample pans, crimp sealed and pierced. A heating rate of 20 °C/min was used.

The PET used was Eastman 9663; a PET homopolymer with an intrinsic viscosity of 0.82. Two samples of oligo (NPG-IPA) were obtained from UCB: Crylcoat 2988 had M_n = 7900 (acid value from data sheet of 30 mg KOH/g), Crylcoat 690 had M_n = 8100 (hydroxyl value from data sheet of 30 mg KOH/g). The PET was dried in a desiccant drier system to give a moisture level of 40 ppm. The oligoester was cryoground to a powder and dried in a vacuum oven at 50 °C for 16 h.

Oligoester Synthesis

Oligoesters were prepared by polycondensation and end group functionality controlled by adjusting the stoichiometry of the reaction and confirmed by ^{1}H NMR. The following procedure is typical.

NPG (380 g) and catalyst (butylhydroxyoxostannane) (2.5 g) were heated to 80°C under nitrogen in a 500 mL flanged flask equipped with mechanical stirrer, insulated fractionating column and stillhead, and thermocouple. Isophthalic acid (251 g) was added portionwise with stirring (approximately 150 rpm). The temperature of the mixture was slowly ramped up to 240°C. After 1 h. the temperature had risen to about 170°C and the condensate had started to distil (b.p. 100°C). When 56 ml of condensate had distilled and the temperature at the stillhead had begun to fall (*ca* 6 h), the molten polyester was poured from the reaction flask and allowed to cool. The polyester had M_n = 4200.

Reactive Extrusion

Experiments were performed on a JSW TEX 30 twin screw extruder having a 30 mm screw diameter and an overall L/D of 42 [comprising ten temperature controlled barrel sections of L/D 3.5, three unheated sampling monitoring blocks of L/D 1.167 and a cooled feed block of L/D 3.5]. The PET and the oligoester were fed into the extruder using a JSW TTF20 gravimetric feeder and a K-Tron KQX gravimetric feeder respectively. The extruder was operated in co-rotating (intermeshing self wiping) modes with throughputs of between 1 and 5 kg/h. The screw design consisted of kneading, conveying and reversing elements as shown in Fig. 1. The screw speed was routinely set at 155 rpm (40% of motor output). The residence times were 37 minutes for a throughput rate of 1 kg/h, 17 minutes for 2 kg/h and 5 minutes for 5 kg/h. The barrel temperature profile was as shown in Fig. 1. The melt temperatures and pressures were monitored at three points along the barrel as well as in the die. The extrudate was air-cooled by passage along a conveyor belt and pelletized.

Co-Rotating Screw SP- 23

(Barrel 6)

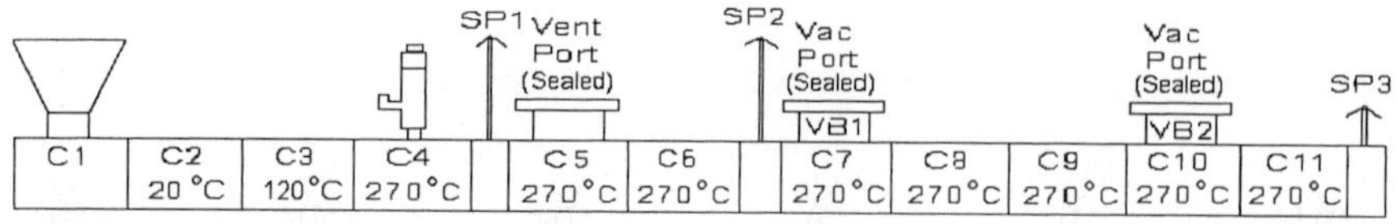

Fig. 1. Screw and barrel profile employed in processing oligomeric polyesters with Eastman 9663 PET homopolymer on JSW Tex 30 twin-screw extruder.

A selection of the extrudate samples were taken, cryoground and extracted in a Soxhlet extractor with chloroform for 2 days. The extract was evaporated to dryness, the residue weighed and then analyzed by ^{1}H NMR. The results of this analysis indicated that, in the case of the IPA-end capped oligoester, all of the oligoester (>95%) was chemically incorporated since no oligoester was identified in the extract by ^{1}H NMR analysis.

Solid Stating

A Buchi GKR-50 flash distillation apparatus was used to heat samples *in vacuo*. A sample of the block polyester was crystallized by heating to *ca* 70°C/30 mmHg for 16 h. The apparatus was then attached to a vacuum pump equipped with a liquid nitrogen trap. After 0.5 hours the vacuum was stabilized to 0.35-0.5 mmHg. and the temperature was raised to 220°C and maintained for 6.5 h.

The product was characterized by GPC, which showed that the molecular weight had increased (from number average molecular weight *ca* 16000 to 25000) the 'H NMR (Fig. 2) was unchanged showing that the polymer had not undergone transesterification during solid stating.

Results and Discussion

Oligoesters of the desired block length were prepared by conventional polycondensation (Scheme 1). End group control was achieved by adjusting the stoichiometry of the reaction. Oligo(neopentyl isophthalate) can also be obtained commercially with predominantly -OH or CO_2H chain ends. The oligoesters were melt blended at levels of 5-20 mole% of repeat units with PET, ideally to produce segmented block copolymers according to Scheme 2, where the segment length of the transesterification resistant block *m* is retained. Processing was carried out in a twin-screw extruder. The average melt temperature was 270 °C and residence times between 5-37 minutes (corresponding to throughput rates of between 5 and 1 kg/h) were explored.

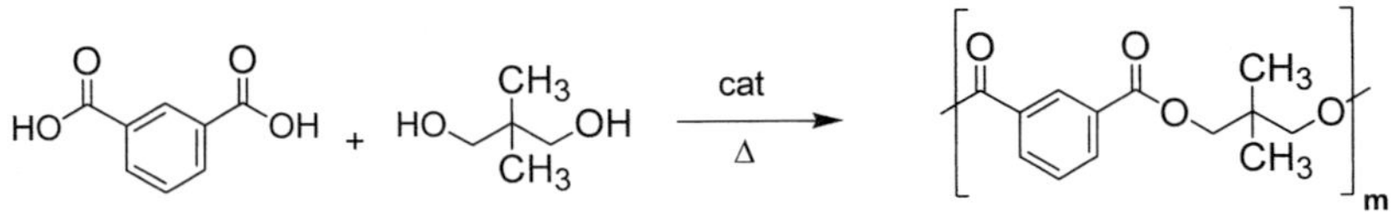

Scheme 1

Evidence for oligoester incorporation came from Soxhlet extraction (amount of residual oligomer), GPC (the expected MW decrease, disappearance of oligomer) and NMR (disappearance of end group signals). Evidence for retention of block integrity was provided by DSC (similar T_m, % crystallinity to PET) and NMR (number of EG-IPA-NPG sequences = block junctures).

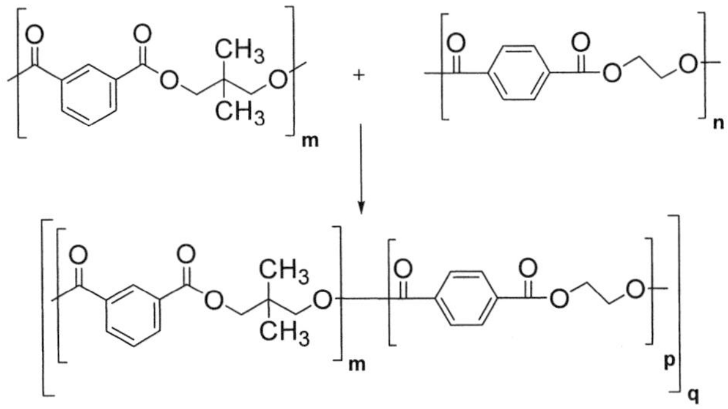

Scheme 2

This block copolyester was subsequently also subjected to solid stating (6 h at 220 °C/1 mm Hg) to increase the molecular weight of the copolyester and again the product examined by ^{1}H NMR to establish the block integrity.

The ^{1}H NMR spectra shown in Fig. 2 provide evidence for the retention of block integrity during processing and subsequent solid stating. The spectrum (a) of the oligoester shows signals due to H2 (refer Fig. 3) of the repeat unit (NPG-IPA-NPG sequence) at 8.70 ppm and a smaller signal, which we attribute to the H2 of NPG-IPA-H chain ends, at 8.73 ppm. This region is almost unchanged in spectrum (b) of the oligoester processed with PET and (c) of the solid stated polyester. In spectra (b) and (c) the signal at 8.72 ppm is due to the H2 of the NPG-IPA-EG sequence of the block juncture. In the spectrum (d) of random poly(ethylene isophthalate-*co*-ethylene terephthalate) the H2 signal for the repeat unit (attributable to a EG-IPA-EG sequence) appears significantly shifted at 8.77 ppm. That no signal at 8.77 ppm appears in spectra b and c shows that there are no EG-IPA-EG sequences and thus no substantial structural randomization during melt processing or solid stating.

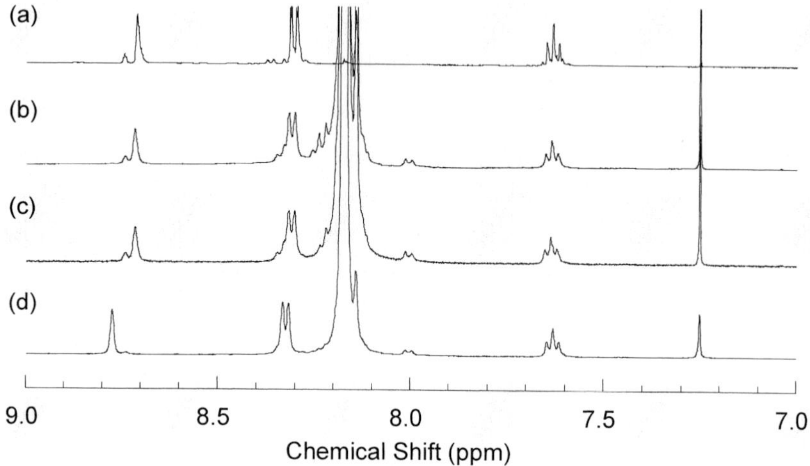

Fig. 2. Region 7-9 ppm of ^{1}H NMR spectra ($CF_3COOH/CDCl_3$ solvent) of (a) IPA capped oligo(neopentyl isophthalate) (Crylcoat 2988, **1**), (b) this oligoester (10% wt%) processed with PET (270 °C, 37 min), (c) this product solid stated (220 °C/1 mmHg, 6 h), (d) conventional poly(ethylene isophthalate-co-ethylene terephthalate).

Fig. 3. Designation of isophthalate hydrogens.

As a control, analogous ethylene glycol (EG) based oligomers were prepared and processed with PET under similar conditions. The extent of transesterification of oligo(ethylene isophthalate) poly(ethylene terephthalate) blends was also determined by ^{1}H NMR by determining triad fractions (ratio TPA-EG-IPA-EG-TPA : IPA-EG-IPA-EG-TPA : IPA-EG-IPA-EG-IPA) from the appearance of the triplet(s) at 7.62 ppm due to the H4 of the isophthalate repeat unit as described by Ha *et al.*[11] This analysis showed that there was substantial loss of block integrity during extrusion under conditions where poly(ethylene isophthalate-*co*-ethylene terephthalate) remains unchanged.

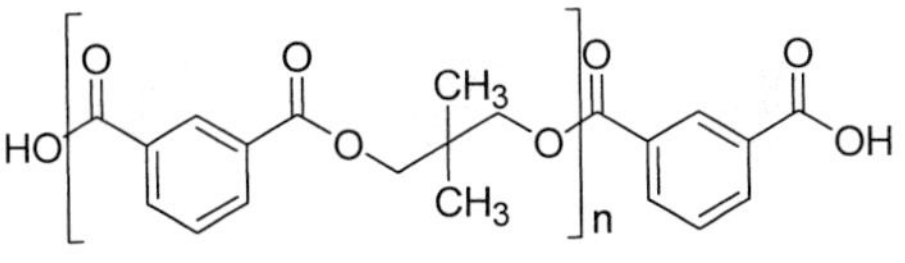

1 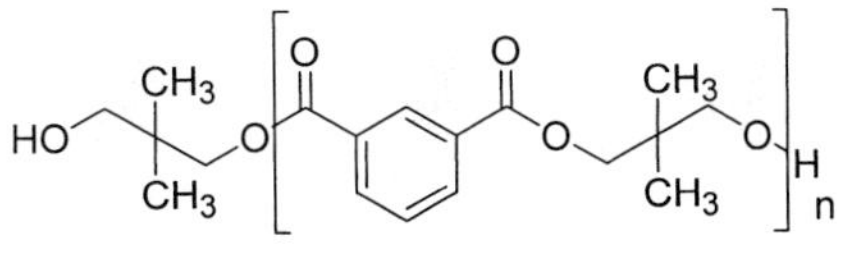 **2**

In the synthesis of transesterification resistant segments, attention must be paid to end groups since factors that increase resistance to transesterification can also create a resistance to incorporation. We found that the oligoester with IPA groups (**1**) showed a greater level of incorporation (>95%) than that with NPG chain ends (**2**) (*ca* 80% for an extrusion experiment with 30 min residence time). It is also known that end groups can have a significant effect on the rate of transesterification of PET polyester blends and that, for example, EG end capped PET is more prone to transesterification than TPA end capped PET.[12,13] It is also possible to facilitate incorporation of transesterification resistant segments either through the use of chain coupling agents (*e.g.* pyromellitic dianhydride) or through the use of an oligoester with reactive chain ends.[2]

Further evidence for resistance of oligomer (1) to transesterification comes from measurement of the melting transition (T_m) and % crystallinity of the polyester products by DSC (Fig. 4). Incorporation of up to 10 mole% of IPA-NPG repeat units from NPG or IPA end capped oligo(NPG-IPA) (M_n~8000) left the T_m and % crystallinity (36% after crystallization) unchanged. PET processed with IPA end capped oligo(EG-IPA) of similar molecular weight (M_n~7000) had substantially lower (T_m) and % crystallinity. It is known[5,14,15] that random PET-PEI copolyesters have markedly reduced crystallinity and T_m (for PET T_m=257 °C, with 10% IPA T_m=226 °C, with 20% IPA T_m=204 °C).[15]

Reasons for the transesterification resistance of NPG based polyester segments still need to be defined. NPG based polyesters are well known to be more resistant to hydrolysis than analogous EG based copolymers and are widely used in the coatings industry for this property.[16] The hydrolysis resistance is often attributed to steric factors but other factors such as relative hydrophobicity must also be considered. It is also known that the rate of alcoholisis of polyesters by neopentyl glycol is slow relative to that by ethylene glycol or propylene glycol.[17] However, a number of factors will influence the rate of transesterification including the ester group reactivity, end group reactivity, miscibility of the polyester segments, and the presence of residual polycondensation catalysts. The extent of miscibility of poly(IPA-NPG) and PET is not established. Qualitative evidence for miscibility is the finding that films with up to 20% poly(IPA-NPG) are highly transparent and that multiple T_g's are not evident in the DSC thermograms. More detailed kinetic studies on transesterification involving these and other oligoesters are in progress.

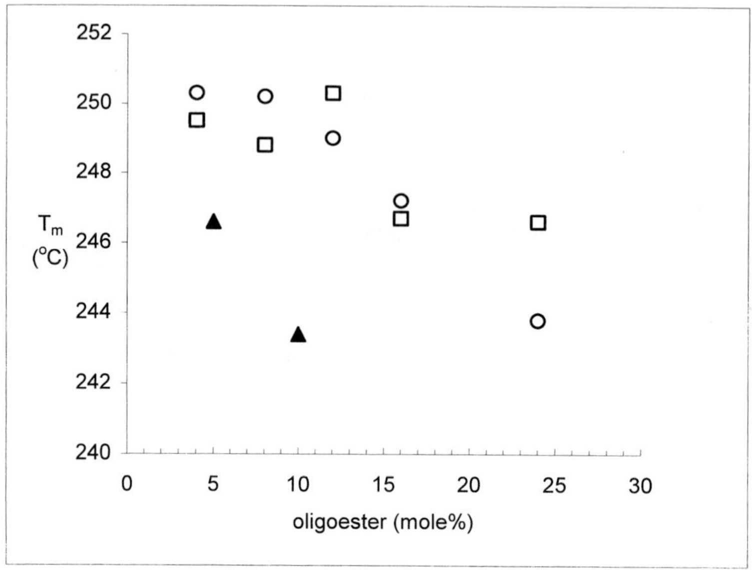

Fig. 4. Variation of T_m with on incorporation of (a) NPG end capped oligo(NPG-IPA) = □, (b) IPA end capped oligo(NPG-IPA) = ○ and (c) oligo(EG-IPA) = ▲ incorporated by reactive extrusion (set temperature 270 °C, throughput rate of 1 kg/h, residence time 37 minutes - see Experimental). X axis is mole % of oligoester repeat units *vs* PET repeat units.

The barrier properties of amorphous films prepared from copolyesters containing NPG based blocks appear similar or better than those of EG based copolyesters of similar overall composition.[9] Those prepared from the poly(IPA-NPG) based block copolyesters had oxygen barrier slightly better than random PET-PEI copolyesters of similar IPA content. The replacement of EG by NPG does not cause deterioration of barrier properties. Barrier properties are anticipated to improve on biaxial orientation of the films.

Conclusions

The controlled synthesis of block copolyesters by reactive extrusion has been achieved. Oligoester segments based on NPG, in particular Oligo(IPA-NPG), have been demonstrated to have resistance to transesterification such that the block integrity can be maintained during melt processing with PET and subsequent solid stating allowing the synthesis of block copolyester products with defined structure. This has enabled the synthesis of block polyesters with crystallinity, melting point and processing properties similar to those of PET but with significantly improved barrier properties.

[1] A. Polyakova, E.V. Stepanov, D. Sekelik, D.A. Schiraldi, A. Hiltner, E. Baer *J. Polym. Sci. Part B: Polym. Phys.* **2001**, *39*, 1911-9.
[2] R.Y.F. Liu, D.A. Schiraldi, A. Hiltner, E. Baer *J. Polym. Sci. Part B: Polym. Phys.* **2002**, *40*, 862 - 77.
[3] A. Polyakova, D.M. Connor, D.M. Collard, D.A. Schiraldi, A. Hiltner, E. Baer *J. Polym. Sci. Part B: Polym. Phys.* **2001**, *39*, 1900-10.
[4] A. Polyakova, R.Y.F. Liu, D.A. Schiraldi, A. Hiltner, E. Baer *J. Polym. Sci. Part B: Polym. Phys.* **2001**, *39*, 1889 - 99.
[5] P. Lodefier, A.M. Jonas, R. Legras *Macromolecules* **1999**, *32*, 7135-9.
[6] V.M. Nadkarni, A.K. Rath In *Handbook of Thermoplastic Polyesters*; Fakirov, S. Ed.; Wiley-VCH: Weinheim, 2002; vol 1, p 319-89.
[7] N.R. James, S.S. Mahajan, S. Sivaram In *Transreactions in Condensation Polymers*; Fakirov, S. Ed.; Wiley-VCH: Weinheim, 1999, p 220-65.
[8] Z. Denchev In *Blends of Thermoplastic Polyesters*; Fakirov, S. Ed.; Wiley-VCH: Weinheim, 2002; vol 2, p 757-813.
[9] G. Moad, A.M. Groth, M.S. O'Shea, R.D. Tozer WO0222705 *Chem. Abstr.* **2002**, **136**:263636
[10] B. Fox, G. Moad, G. Van Diepen, R.I. Willing, W.D. Cook *Polymer* **1997**, *38*, 3035-43.
[11] W.S. Ha, K.C. Yong, S.S. Jang, D.M. Rhee, C.R. Park *J. Polym. Sci., Part B: Polym. Phys.* **1997**, *35*, 309-15.
[12] S. Collins, S.K. Peace, R.W. Richards, W.A. MacDonald, P. Mills, S.M. King *Macromolecules* **2000**, *33*, 2981-8.
[13] A.M. Kenwright, S.K. Peace, R.W. Richards, A. Bunn, W.A. MacDonald *Polymer* **1999**, *40*, 5851-6.
[14] Z.M. Li, M.B. Yang, A. Lu, B.H. Xie, J.M. Feng, R. Huang *J. Mat. Sci. Lett.* **2002**, *21*, 1063-7.
[15] G.P. Karayannidis, I.D. Sideridou, D.N. Zamboulis, D.N. Bikiaris, A.J. Sakalis *J. App. Polym. Sci.* **2000**, *78*, 200-7.
[16] F. Belan, V. Bellenger, B. Mortaigne, J. Verdu *Polym. Degr. Stab.*, **1997**, *56*, 301-9.
[17] A.K. Bulai, V.S. Kalinina, B.M. Arshava, Y.G. Urman, R.S. Barshtein, I.Y. Slonim *Polymer Science USSR* **1976**, *18*, 2821-9.

Synthesis and Properties of Poly(*m*-carborane-siloxane) Elastomers

Mogon Patel, Anthony C. Swain, Anthony R. Skinner, Leslie G. Mallinson, Gerard F. Hayes*

Atomic Weapons Establishment, Aldermaston, Reading, RG7 4PR, UK
Email: mogon.patel@awe.co.uk

Summary: Organic boron polymers are of interest to AWE, and a significant improvement would be to make components from a single phase material that enables the boron content to be maximised and the elastomeric properties to be tuned to requirement. This paper is a summary of the work carried out towards the synthesis and characterisation of poly(*m*-carborane-siloxane) rubbers. These rubbers have been synthesised by the Ferric Chloride-catalysed condensation reaction between dimethoxy-*m*-carborane terminated monomer and dichlorodimethylsilane. Phenyl modified and Phenyl-vinyl modified variations have been synthesised in addition to the unmodified poly(*m*-carborane-siloxane) polymers. The unmodified poly(*m*-carborane-siloxane) shows some crystallinity but the modified versions are amorphous and have good elastomeric properties. Overall, the measured boron content is in close agreement to that expected from the repeat unit, suggesting the synthesised rubbers have been successfully prepared. It is also demonstrated that the carborane unit offers increased thermal stability and retention of 'rubber-like' properties even after ageing at high temperature.

Keywords: condensation polymerisation; elastomeric properties; phenyl, vinyl modified; poly(*m*-carborane-siloxane); thermal stability

Introduction

The term "carborane" is commonly used in a generic sense to describe compounds composed of boron, hydrogen, and carbon, whose molecular geometry consists of polyhedral fragments.[1] Several families of carboranes exist with general formulae, CB_nH_{n+2}, $C_2B_nH_{n+2}$, for example. Of these, the neutral closed polyhedral $C_2B_nH_{n+2}$ (n=10) species have been used in the preparation of poly(*m*-carborane-siloxane) polymers.[2] The carboranyl fragment is usually denoted by the formula [-$CB_{10}H_{10}C$-], here we use the cage structure shown in Figure 1 to denote the *m*-substituted carborane. Poly(m-carborane-siloxane) polymers were discovered in the early 1960s and demonstrated useful properties under extreme conditions.[2] These polymers are prepared by the ferric chloride-catalysed copolymerisation of dichloro and dimethoxy-terminated

 DOI: 10.1002/masy.200351205

monomers. The generation of a network structure is readily achieved through the crosslinking of introduced vinyl groups.[3]

Our aim is to synthesise polymers where the *m*-carborane is chemically bound to the network and the following advantageous properties are required;

- elastomeric (stress absorbing) in nature
- relatively high *m*-carborane content
- solid or foam components made using our in-house technology
- enhanced ageing characteristics & thermal stability compared to current materials

We report on the synthesis, chemical and physical properties of poly(*m*-carborane-siloxane) elastomers. They offer the potential to prepare materials combining useful mechanical properties with good thermal stability. Our approach also offers the capability of fine-tuning the elastomeric properties to requirement by changing functional/pendant groups. The Lewis acid (Ferric Chloride) catalysed condensation between dimethoxy-*m*-carborane and a dichlorodimethylsilane was utilised in this study. A cross-linked foamed poly(dimethylsiloxane) rubber, not containing *m*-carborane, was also prepared in this study and used as a reference to which the properties of the carborane polymers could be assessed.

Materials

1,7-bis-(dimethylmethoxysilyl)-*m*-carborane was purchased from Katchem limited. Dichlorodimethylsilane, dichloromethylphenylsilane, dichloromethylvinylsilane and anhydrous Ferric Chloride, were purchased from Aldrich Chemicals and used as received.

The polydimethylsiloxane resin used in this work was purchased from Rhodia silicones, under the code RTV5370. It is a polysiloxane silanol containing small amounts of hydrogen-methyl polysiloxane silanol and tetraalkoxy silane, together with 24 wt% fumed silica filler. The curing catalyst (XY-70, supplied by Rhodia and manufactured by Rhone Poulenc) contains stannous 2-ethylhexanoate, supplied as a 77% w/w solution in 2-ethyl hexanoic acid.

Experimental Procedures

Synthesis of Unmodified Rubber

1,7-bis-(dimethylmethoxysilyl)-*m*-carborane (40g), dichlorodimethylsilane (16g), and anhydrous Ferric Chloride (1mol % of the carborane) were mixed in a three neck reaction vessel (500 ml). The vessel was purged with dry nitrogen, placed in an oil bath on a magnetic hot plate and heated to 140°C. The equimolar condensation reaction between chloro- and methoxy- (on the *m*-carborane precursor) groups which takes place generates chloromethane gas (see Fig. 1). The resulting product was a brownish-black polymer, which was washed with acetone to remove residual catalyst.

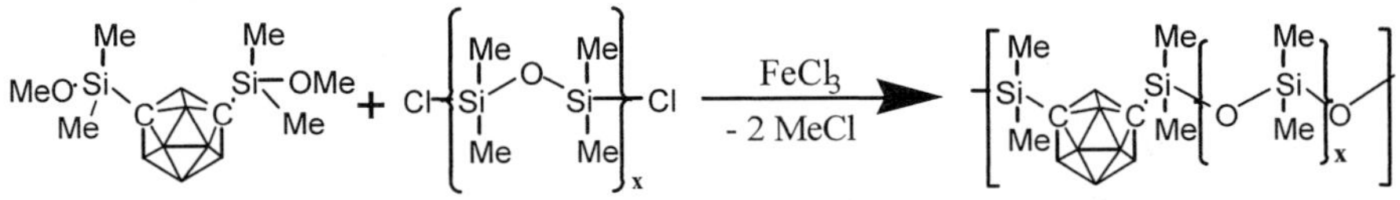

Fig. 1. Lewis acid catalysed condensation reaction.

Synthesis of Phenyl Modified Rubber

The procedure used was similar to that described above however, dichloromethylphenylsilane was used as an additional silicon source. A typical reaction comprised the following: *m*-carborane precursor (18g); dichlorodimethylsilane (6.12g, 85 mol%); dichloromethylphenylsilane, (1.60g, 15 mol%) and Ferric Chloride (0.1g). Fig. 2 shows the repeat unit of the synthesised rubber.

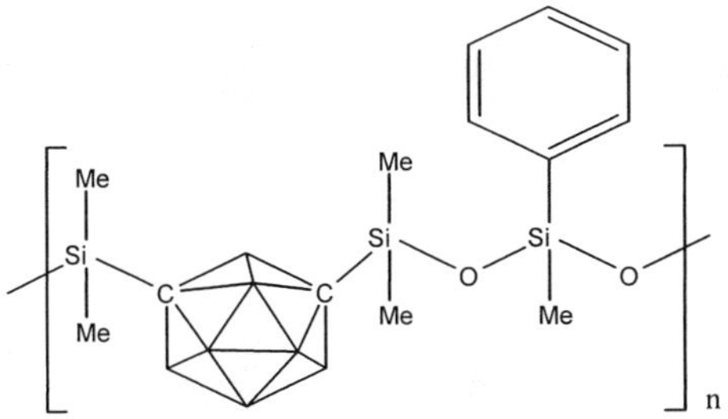

Fig. 2. Phenyl modified rubber (idealised).

Synthesis of Phenyl Vinyl Modified Rubber

The reagents were as follows: m-carborane precursor (24g,); dichlorodimethylsilane (7.2g); dichloromethylvinylsilane (0.52g); dichloromethylphenylsilane (2.85g) and Ferric Chloride (0.12g). Fig. 3 shows the repeat unit of the synthesised rubber.

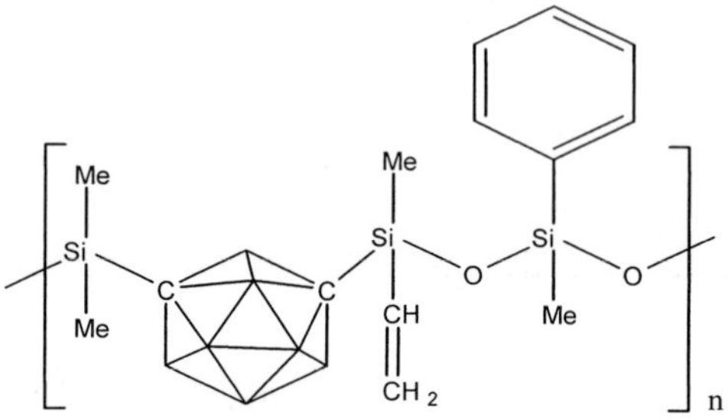

Fig. 3. Phenyl-vinyl modified rubber (idealised).

Fabrication of Shaped Components

After synthesis the modified siloxane gums were fabricated into shaped components using standard siloxane vulcanisation and fabrication technology. Dichlorobenzylperoxide (1% by wt) was used as the curing agent and mixed into the soft dough. Shaped components from the modified poly(*m*-carborane-siloxane) rubbers were prepared by compression moulding at 150°C. Shaped components from the unmodified poly(*m*-carborane-siloxane) rubber were readily made by compression moulding at 70 °C.

Synthesis of Poly(dimethylsiloxane) Rubber

A known amount (5% wt) of Stannous 2-ethylhexanoate catalyst (XY70) was added to 30g of RTV 5370 polysiloxane resin and containing the curing agent tetrapropylorthosilicate. The polymer (see Fig. 4) was cured under ambient conditions for 20 minutes followed by a subsequent post-cure at 70°C for 16 hours. This rubber does not contain *m*-carborane and was used as a reference to which the properties of the poly(m-carborane-siloxane) elastomers could be assessed against.

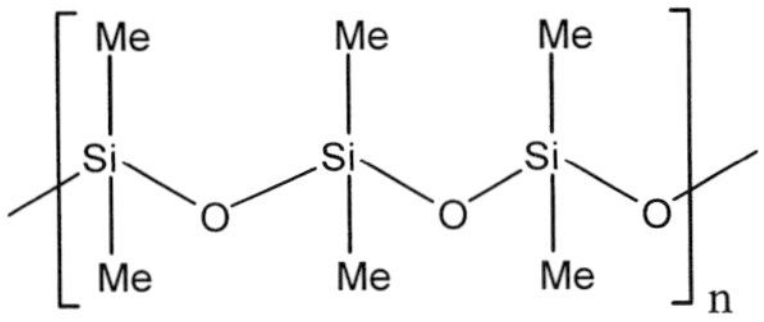

Fig. 4. Structure of Polydimethylsiloxane (PDMS) rubber.

Results and Discussion

Gel Permeation Chromatography

The unmodified poly(*m*-carborane-siloxane) rubber was readily soluble in THF and molecular mass characterisation was conducted using GPC (using a flowrate of 1ml/min and styrogel 10μm column with refractive index detector). A typical GPC chromatogram (see Fig. 5) shows a single broad molecular mass distribution with a polydispersity (M_w/M_n) value of 3.9. A weight average molecular weight (M_w) of 154,000 g/mol was determined relative to linear polystyrene standards, demonstrating the synthetic route generates relatively high molar mass polymer.

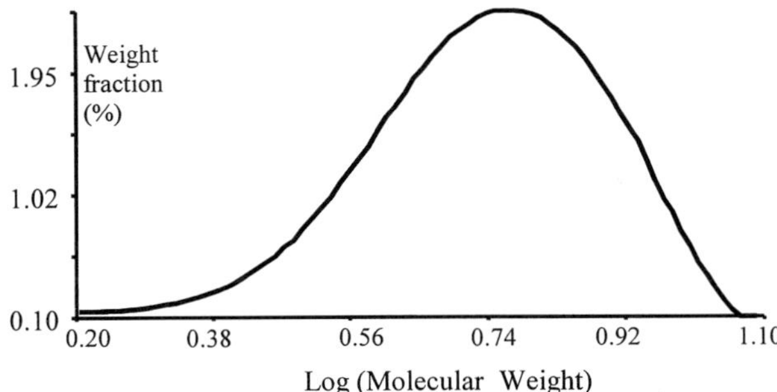

Fig. 5. GPC spectrum of the unmodified poly(*m*-carborane-siloxane).

The corresponding Mark-Houwink parameter determined from this analysis was 0.6, indicating that the polymer exists as a lightly perturbed coil in THF solution. No evidence of branching was observed (which might occur during the polymerisation) suggesting the polymer is essentially linear.

Nuclear Magnetic Resonance

$^{11}B\{^{1}H\}$ NMR of the unmodified poly(*m*-carborane-siloxane) was conducted to assess the boron species within the rubber. Fig. 6 shows a spectrum typical for that of a carborane fragment[4], showing that the cage structure remains intact during the synthesis. The carborane cage clearly exhibits several different boron environments including B-B, B-C, B-H bonds and a characteristic broad line due to unresolved ^{11}B-^{11}B couplings.

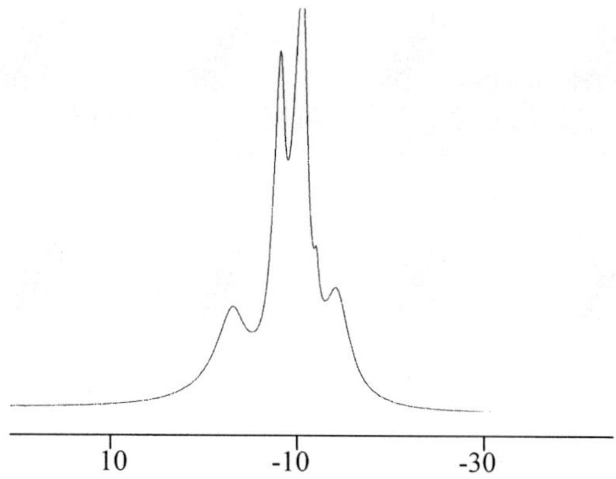

Fig. 6. $^{11}B\{^{1}H\}$ NMR of the unmodified poly(*m*-carborane-siloxane).

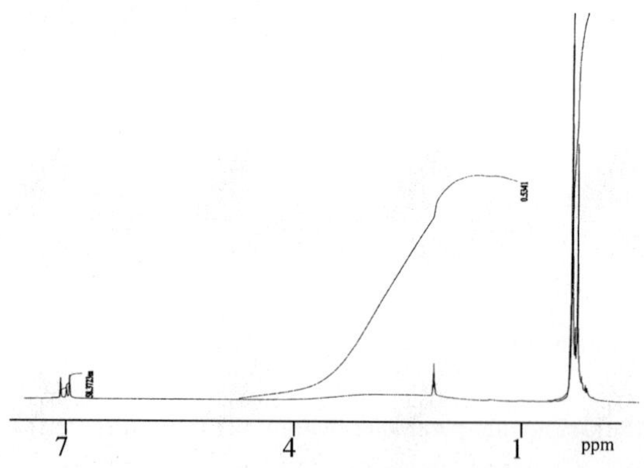

Fig. 7. ^{1}H NMR of the phenyl modified poly(*m*-carborane-siloxane).

^{1}H NMR of the toluene soluble fraction from the phenyl modified poly(*m*-carborane-siloxane) rubber was carried out to assess the composition of the material, see Fig. 7. Two main proton environments are observed i.e. those of the carborane (2.1ppm) and the sily methyl groups (0.5ppm), in an integral ratio of approx. 54%. This is close to the theoretical value of approximately 60% expected from the repeat unit (see Fig. 2). Overall, the ^{1}H NMR data provides further evidence that the carborane unit is an integral part of the polymer backbone and not simply acting as a filler within the material.

These data are supported by a broad singlet centred on 68.5 ppm in the $^{13}C\{^1H\}$ NMR spectrum (see Fig. 8) which is indicative of carbons associated with the carborane-siloxane linkages. Signals ascribed to carbons associated with methyl (0 to 20ppm) and phenyl (127-137 ppm) groups were also observed. $^{29}Si\{^1H\}$ NMR of the unmodified poly(*m*-carborane-siloxane) rubber showed the characteristic D (i.e -O-Si-O-) and M (i.e Si-O) silicon environments, consistent with the proposed structure.

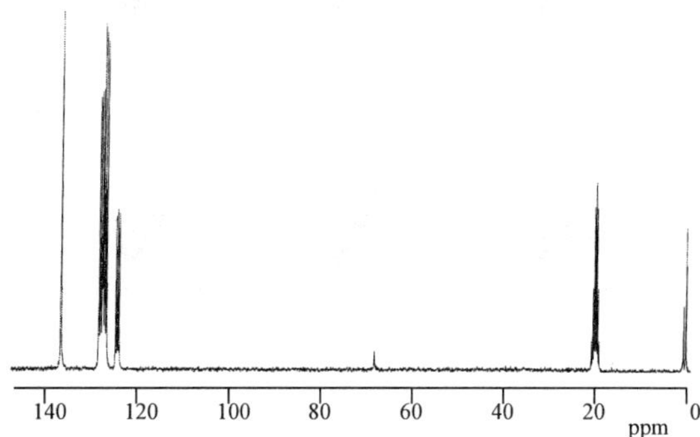

Fig. 8. Carbon NMR of the phenyl modified rubber.

InfraRed Analysis

Infrared analysis of these polymers after toluene extraction shows the typical Si-O-Si asymmetric stretching vibration in the 1090 - 1020 cm^{-1} region (see Fig. 9). The region 2500-3000 cm^{-1} displays absorbency peaks characteristic of the carborane cage (B-H absorbance is shown at 2600 cm^{-1}). Since the samples examined were extracted in toluene prior to testing, the fact that the B-H absorption is detected strongly suggests it originates from carborane cages attached to the chain. The weak absorbance at 1430cm^{-1} is characteristic of the aromatic C-H stretch in the phenyl-containing elastomers. There was no evidence of Si-OCH_3 resonances (a sharp intense band at 2840 cm^{-1} and a medium intense band 1190 cm^{-1}) in any of the spectra indicating complete reaction of methoxy groups from the *m*-carborane precursor. There was no clear evidence of Si-Cl absorptions (which normally appear within the range 625-425 cm^{-1}) suggesting the absence of chlorosilane residues.

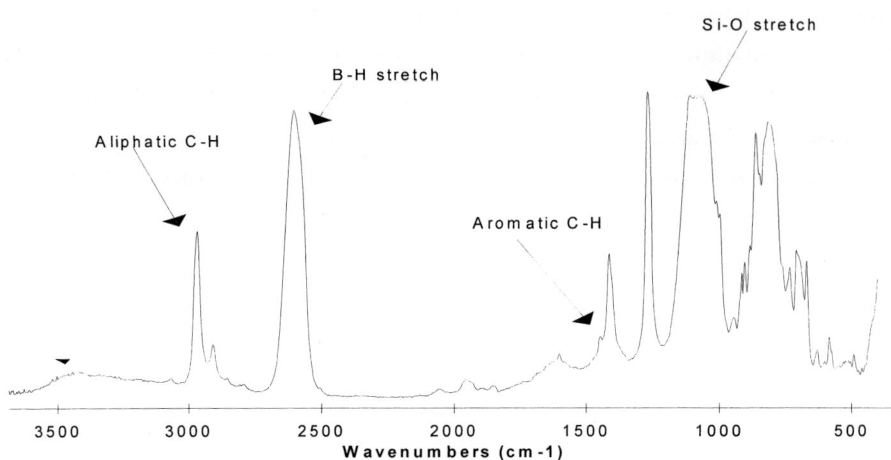

Fig. 9. Infrared spectrum of the phenyl vinyl modified poly(*m*-carborane-siloxane).

The absence of the ubiquitous vinyl resonances at 1600 cm^{-1} and 1410 cm^{-1} provides evidence for the participation of vinyl groups during the gelation and fabrication process. The spectrum obtained for the unmodified elastomer is in close agreement with that reported in the literature.[5]

Other Physical and Chemical Properties

Table 1 shows the results of swell tests in toluene. The phenyl vinyl modified poly(*m*-carborane-siloxane) rubber showed the lowest degree of swell, indicative of the greater cross-link density of this material. This is due to the additional cross-links introduced by the vinyl groups during gelation/fabrication. This is in agreement with the Infrared data discussed earlier, suggesting the depletion of vinyl groups in this material through reaction during the gelation and fabrication process. The phenyl-modified poly(*m*-carborane-siloxane) rubber was found to swell to a far greater extent than the phenyl-vinyl modified rubber indicating a lower cross-link density. The unmodified rubber was found to swell considerably in toluene and was found to dissolve partially in the solvent.

The unmodified poly(*m*-carborane-siloxane) rubber shows a Boron content of 28.8% and the phenyl vinyl modified poly(*m*-carborane-siloxane) polymer shows the lowest boron content. Surprisingly, the phenyl-modified poly(*m*-carborane-siloxane) polymer showed a higher Boron content than the unmodified rubber (29.5% compared with 28.8%). This difference could be because the synthesis of the unmodified material resulted in some moisture deactivation of the carborane precursor (thus a lowing of carborane incorporated in the polymer). Moisture hydrolysis of the chlorosilane in the preparation of the phenyl modified rubber is also possible and would lead to higher than expected Boron in the polymer network. Overall, the measured % Boron is in close agreement to that expected/calculated from the repeat unit suggesting that the synthesised rubbers have been successfully prepared.

Table 1. Some physical and chemical properties.

	PDMS	Unmodified Carborane rubber	Phenyl modified rubber	Phenyl-vinyl modified rubber
Linear toluene swell (% of thickness) per unit density	70	dissolves	40	19
Coefficient of Thermal Expansion (μm/m°C)	319	128	259	264
% B from Atomic Emission spectroscopy (Cal. repeat unit)	0	28.8 ± 0.4 (31)	29.5 ± 0.2 (30)	26.4 ± 0.2 (30)

Coefficients of thermal expansion (determined using a TA Instruments 2910 thermomechanical analyser operating at 5°C/minute with a static weight of 0.1N) show that the introduction of carborane units into a polysiloxane decreases thermal expansion (see Table 1). The unmodified poly(*m*-carborane-siloxane) rubber has the smallest expansion coefficient due to the crystalline nature of the material, in contrast to poly(dimethylsiloxane)s, which have extremely low cohesive energies and therefore do not generally show significant crystallinity at room temperature. The poly(*m*-carborane-siloxane)s possess both Lewis Acidic (carborane) and Lewis Basic (oxygen atoms) regions within the back-bone, thereby allowing strong inter-chain interactions. These attractions increase the tendency of carborane modified polymers to adopt crystalline conformations thereby inhibiting elastomeric characteristics.[6]

Thermal Transitions

A TA Instruments 2940 Differential Scanning Calorimeter (DSC) was used to assess thermal phase transitions within the polymers, Table 2 shows typical results. The introduction of the bulky carborane unit into the siloxane backbone clearly reduces the flexibility of the chains as seen by the increase in the glass transition temperature. The unmodified poly(*m*-carborane-siloxane) rubber exhibits crystallinity due to the strong attraction of the carborane cage for the oxygen on the siloxane backbone. The phenyl modified poly(*m*-carborane-siloxane) shows no melting transition suggesting the introduction of phenyl units into the polymer backbone prevents the chains aligning into crystalline conformations.

Table 2. Thermal properties using differential scanning calorimetry.

	PDMS	Unmodified Carborane rubber	Phenyl modified rubber	Phenyl-vinyl modified rubber
Glass transition (°C)	-115	-39	-35	-35
Melting Point (°C) Enthalpy of melting (Jg^{-1})	-45 22.7	60 9.9	No melting	No melting

Thermomechanical Properties

The Thermomechanical Analyser (TMA) was also used to assess the load-bearing properties of the samples. The synthesised rubbers have good elastomeric characteristics (see Fig. 10 and Fig. 11) with compression set values equal to or lower than that shown by a poly(dimethylsiloxane) foam (although, of course, the carborane rubbers are not foamed). It should however, be noted that the comparison here is of foam material and solid material, and that foams can show much more compression set. Solid RTV5370 poly(dimethylsiloxane) rubber was not available for this study as it can only be produced in the foamed state. The data shown in Fig. 10 and Fig. 11 have been normalised for density to allow for this difference. The poly(dimethylsiloxane) rubber shows signs of damage at around 250°C. Relative to poly(dimethylsiloxane), the poly(*m*-carborane-siloxane) rubbers show less change in mechanical properties at temperatures up to 400°C. It is evident that, with increasing temperature, the loss of elastomeric properties (increase in compression set and deformation rate) in the poly(dimethylsiloxane) rubber is much greater than that in the poly(*m*-carborane-siloxane) rubbers. The *m*-carborane unit clearly offers increased thermal stability and retention of 'rubber-like' properties after ageing at high temperature. However, residual $FeCl_3$ catalyst impurities within the rubbers (from the synthesis process) would be expected to accelerate the thermal ageing in these rubbers by co-ordination with water and breaking of Si-O-Si linkages. It is therefore important to minimise the residual catalyst content in these rubbers in order to maximise the thermal stability properties at high temperatures.

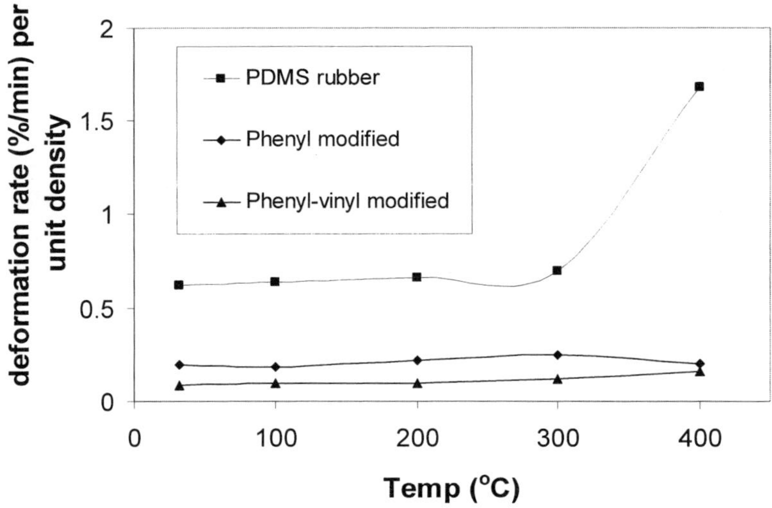

Fig. 10. Mechanical response at different temperatures using 0.001 $Nmin^{-1}$ force ramp.

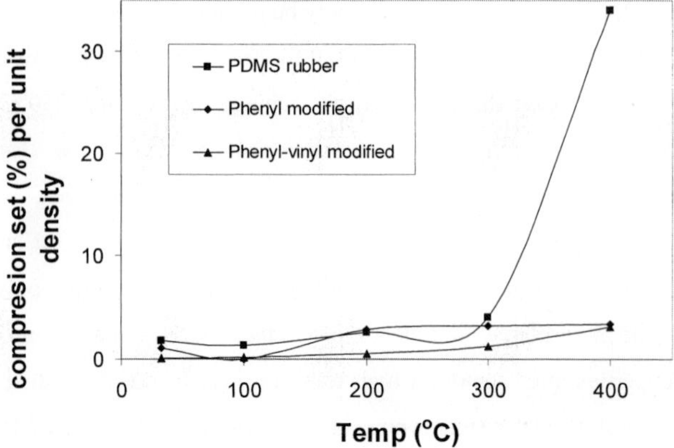

Fig. 11. Growth of compression set with temperature (0.02N loading for 2 hours followed by 30 minute recovery).

Conclusions

Poly(*m*-carborane-siloxane) elastomers have been successfully prepared by the Ferric Chloride-catalysed co-polymerisation between dichlorodimethylsilanes and a di-methoxy-*m*-carborane. The phenyl-modified poly(*m*-carborane-siloxane) polymers show enhanced elastomeric characteristics due to the disruption of crystallinity found in the unmodified poly(*m*-carborane-siloxane) rubber. The poly(*m*-carborane-siloxane) polymers show enhanced thermal stability relative to conventional poly(siloxane) rubbers because the carborane units act as an energy trap. Overall, the measured boron content is in close agreement to that expected from the repeat unit, suggesting the synthesised rubbers have been successfully prepared

Acknowledgement

Mr. A.F. McDonald is thanked for the GPC results. The authors would also like to thank Dr. J.J. Murphy and Mr. P.R. Morrell for useful discussions on carborane chemistry.

[1] R.N Grimes, *"Carborane"*, Academic Press, New York **1970**.
[2] H. Schroeder, O.G. Schaffling, T. B Larcher, F. F. Frulla and T.L. Heying, *Rubber Chem. Technol*, **1960**, *39*, 1184.
[3] K. O. Knollmueller, R. N. Scott, et. al., *J. Polym. Sci*, **1971**, *Part A-1*, 1071.
[4] G.R. Eaton, *J.Chem.Educ.*, **1969**, *46*, 547.
[5] R.Adams, *Inorg.Chem.*, **1963**, *2*, 1087.
[6] R.E.Kesting, K.F.Jackson, E.B.Klusman and F.J.Gerhart, *J. Appl. Polym. Sci.*, **1970**, *14*, 2525.

Impact Modification of PA-6 and PBT by Epoxy-Functionalized Rubbers

*Emiliano Lievana, József Karger-Kocsis**

Institut für Verbundwerkstoffe GmbH (Institute for Composite Materials), Kaiserslautern University of Technology, POBox 3049, D-67653 Kaiserslautern, Germany
E-mail: karger@ivw.uni-kl.de

Summary: Glycidylmethacrylate (GMA) grafted apolar (ethylene/propylene/diene - EPDM) and polar (acrylonitrile/butadiene - NBR) rubbers were melt blended with polyamide–6 (PA-6) and polybutylene terephthalate (PBT). The toughness of the blends containing 5, 10 and 50 wt.% epoxy functionalized rubbers was assessed by various methods (notched Charpy, perforation impact) as a function of temperature (T=23 and –40 °C). The notched Charpy tests served to deduce the fracture toughness (K_c) and energy (G_c) data. It was established that EPDM-g-GMA is a slightly better impact modifier than is NBR-g-GMA albeit the latter polar rubber is more compatible with both matrices than the less polar EPDM-g-GMA. This finding was traced to the difference in the glass transition temperature (T_g of NBR is higher than that of EPDM) and to the dispersion of the epoxy funtionalized rubbers.

Keywords: EPDM; epoxy functionalization; grafting; NBR; PA-6; PBT; toughness

Introduction

Funtionalized rubbers are widely used for the impact modification of thermoplastic polycondensates, such as polyamides (PA) and polyesters. Maleated (maleic anhydride grafted) polymers work as efficient tougheners in PAs,[1-4] which is mostly due to the high reactivity between the related functional groups. Maleated polymers are, however, far less efficient in linear polyesters, such as polyethylene (PET) and polybutylene terephthalate (PBT).[5] To improve the interfacial compatibility of polar polyesters and apolar additives during reactive blending most straightforward is to exploit the reactivity of epoxy groups towards the –OH and –COOH functions of the polyesters. This concept has been approved recently by several researchers.[5-9] It is worth of noting that the epoxy group is highly reactive also with the functional groups of PAs (i.e., $-NH_2$, -COOH, -NH-CO-). So, epoxy functionalized rubbers may improve the toughness of PAs, as well. It was recently demonstrated that the initial polarity of the grafted rubbers strongly affects the blend compatibility with thermoplastic polyesters.[5,7] This work was aimed at studying the toughness response of PA-6- and PBT-based blends

 DOI: 10.1002/masy.200351206

containing epoxy funtionalized apolar (ethylene/propylene/diene - EPDM) and polar (acrylonitrile/butadiene - NBR) rubbers. The related rubbers, denoted further on as EPDM-g-GMA and NBR-g-GMA, were home made by melt grafting with glycidylmethacrylate (GMA).

Experimental

Materials and Blending

The EPDM (component ratio: 71/24/5 wt.%, Type: Buna® EPG 6470) and NBR (acrylonitrile content: 40 wt.%, Type: Perbunan® NT 3946) rubbers of Bayer were melt grafted by GMA in the presence of peroxide. The grafted GMA content of EPDM-g-GMA and NBR-g-GMA was 8.7 and 5 wt % respectively. EPDM-g-GMA contained also 0.7 wt% homopolymerized GMA according to the analytics used.[7,10] PA-6 and PBT were Ultramid® B3 and Ultradur® B 4520 grades of BASF. Blends were produced on twin-screw extruders: the PA-6 based one on Brabender DSE 25, whereas the PBT-based blends on Werner-Pfleiderer ZSE II. The maximum melt temperature was set for both blends at 250°C.

Specimens Preparation and Testing

The granulated blends were injection molded on an Arburg Allrounder following the recommendations of the polymer producer. Film-gated plaques (80x80 mm^2) in 1 mm thickness along with dumbbells (according to ISO 3167) were produced.

Specimens for dynamic mechanical thermal analysis (DMTA) and for instrumented notched Charpy tests were taken from the gauge section of the dumbbells. This sampling along with the determination of the fracture toughness (K_c) and fracture energy (G_c) at 1.2 m/s deformation rate are described in our former paper.[11]

Plaques of 1 mm thickness were used for instrumented falling weight impact (IFWI) performed on Ceast Dartvis® at v = 3.2 m/s. Further details to this technique are disclosed in Ref. [12]. Both instrumented impact tests (viz. Charpy high-speed flexural and falling dart) were performed at T=23 and –40 °C, respectively.

Results and Discussion

Phase Structure

Figure 1 compares the DMTA spectra of the PBT blends with EPDM-g-GMA and NBR-g-GMA.

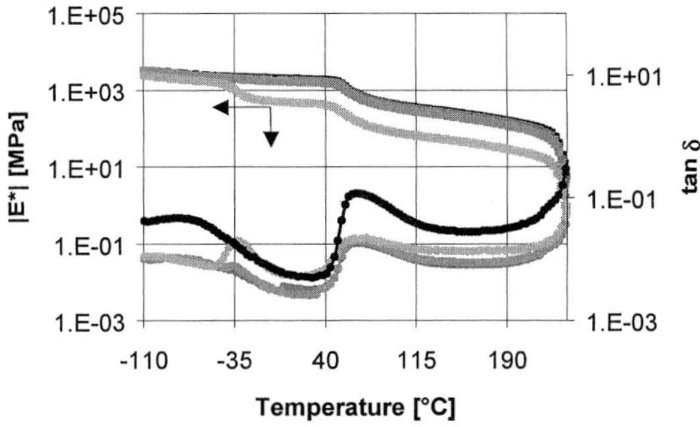

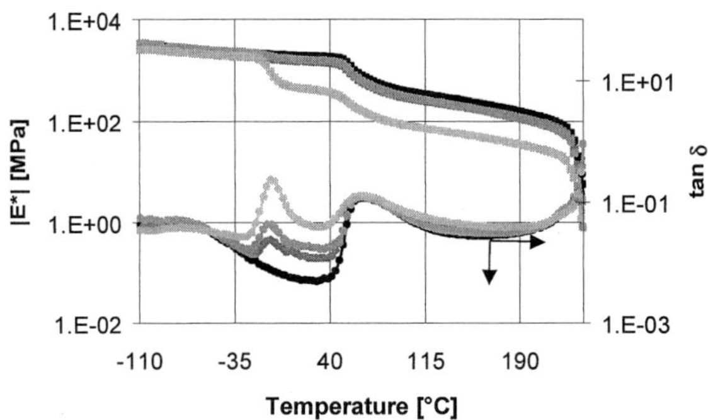

Fig. 1. DMTA spectra of PBT blends with different amounts of EPDM-g-GMA (top) and NBR-g-GMA (bottom). Designations: • 0 %, • 5%, • 10%, • 50% rubber (i.e. with increasing rubber content the related lines become less black).

As expected, the complex moduli decrease with increasing amount of funcionalized rubber. Based on the fact that the glass transition temperatures (T_g) of the blend constituents are well discernible and their position did no change in the blends one can presume the presence of a two-phase structure. Note that the T_g of the EPDM-g-GMA is far less pronounced compared to the NBR-g-GMA. This is related to the T_g peak position of the related rubber. NBR-g-GMA has a higher T_g (–5 °C) than EPDM-g-GMA (T_g ~ -30 °C) and thus there is no overlapping with the β-relaxation transition of the PBT matrix. The |E*| vs T curves (cf. Figure 1) of the blends with 50 wt.% rubbers suggest the onset of a co-continuous structure.[5,7] On the other hand, the blends with ≤ 10 wt% rubber blend show a disperse-type morphology. Similar results, as discussed above, were received also for the PA-based blends.

Tensile Mechanical Results

Results listed in Table 1 demonstrate that with increasing rubber content stiffness, strength and ductility of the blends decrease. Based on the ultimate elongation one can claim that EPDM-g-GMA is a more suitable toughener than NBR-g-GMA. Attention should be paid to the fact that the ethylene/GMA (E/GMA) copolymer outperforms both epoxy functionalized rubbers in respect with the 'wet' tensile performance. The property retention with the PBT blends was similar: EPDM-g-GMA yielded higher values than NBR-g-GMA both at T=23 and -40 °C. It is noteworthy that the blends with 50 wt.% rubber showed ultimate elongations (≥ 130 % except the NBR-g-GMA containing blend at T= -40°C) which are characteristics for thermoplastic elastomers. This is a further hint for the formation of a co-continuous structure in the related blends.

Table 1. Tensile mechanical properties of the PA-based blends after conditioning. Notes: for comparison purpose results achieved on blends with ethylene/GMA (Lotader® AX 8840 of Atofina) are also indicated.

Blend Composition	E-Modulus [MPa]	Tensile Strength [MPa]	Elongation at Break [%]
PA-6	892 ± 23	42 ± 0.6	200 ± 5
PA-6 + EPDM-g-GMA			
5 wt%	640 ± 18	36 ± 1.0	120 ± 5
10 wt%	545 ± 10	32 ± 0.7	110 ± 10
PA-6 + NBR-g-GMA			
5 wt%	659 ± 21	35 ± 0.7	59 ± 8
10 wt%	566 ± 17	32 ± 0.2	54 ± 5
PA-6 + E/GMA			
5 wt%	680 ± 16	44 ± 3.3	230 ± 26
10 wt%	605 ± 20	41 ± 2.2	240 ± 15

Fracture Toughness and Energy

Table 2 shows that both functionalized rubbers toughen the PBT. Their efficency is comparable at both testing temperatures. One can also notice that incorporation of more than 5 wt.% grafted rubber has no effect on the linear elastic fracture mechanical parameters. Contrary to the predictions based on the DMTA and tensile mechanical results NBR-g-GMA yields slightly better values than EPDM-g-GMA. Note that similar fracture results as given in Table 2 were reported for the neat PBT in earlier works.[13-14] Similar tendency, as deduced for the PBT-blends, was observed also for the PA-6 blends. However, in this case E/GMA outperformed both EPDM-g-GMA and NBR-g-GMA which was in line with the prediction based on the tensile mechanical performance (cf. Table 1).

Table 2. Fracture toughness (K_c) and energy (G_c) as a function of blend composition and testing temperature. Notes: K_c was determined by the method described in Ref. [15]. G_c was computed via $G_c = K_c^2/E$, where E was taken from the DMTA spectra.

Blend Composition	K_c [MPa.m$^{1/2}$]		G_c [kJ/m^2]	
	23 °C	- 40°C	23 °C	-40 °C
PBT	5.5	5.3	15.6	12.5
PBT + EPDM-g-GMA				
5 wt%	6.3	5.9	22.1	16.0
10 wt%	6.1	5.9	23.1	16.6
PBT + NBR-g-GMA				
5 wt%	6.1	5.9	24.6	18.1
10 wt%	5.7	6.8	20.0	19.6

Scanning electron microscopic (SEM) pictures taken on the fracture surface of PBT blends containing 5 wt.% grafted rubber are depicted in Figure 2.

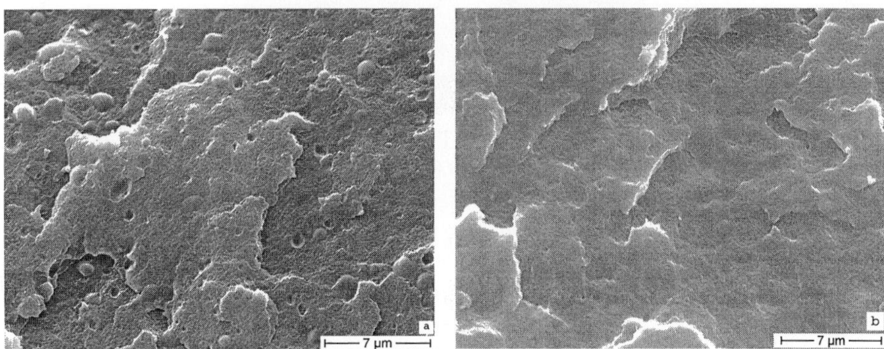

Fig. 2. SEM pictures taken on the fracture surface (T= -40°C) of PBT blends with 5 wt% EPDM-g-GMA (a) and NBR-g-GMA (b), respectively.

According to Figure 2a EPDM-g-GMA is present in a coarse dispersion (particle size <2 µm). On the other hand, NBR-g-GMA produced likely a much finer dispersion (particle size <0.5 µm). This is, however, an artifact. Etching the particle surface by xylene a bimodal, coarse dispersion became visible - cf. Figure 3.

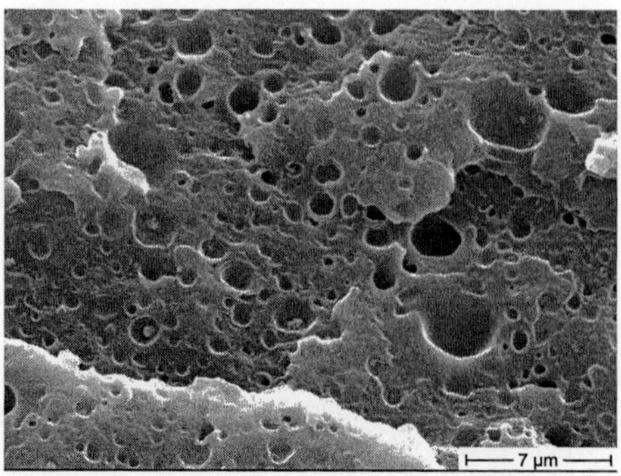

Fig. 3. SEM picture taken on the xylene etched fracture surface (T=23 °C) of PBT blended with 10 wt.% NBR-g-GMA.

In case of the blends with 50 wt.% functionalized rubbers a co-continuous morphology was found as shown on the example of PBT+50 wt.% EPDM-g-GMA (cf. Figure 4).

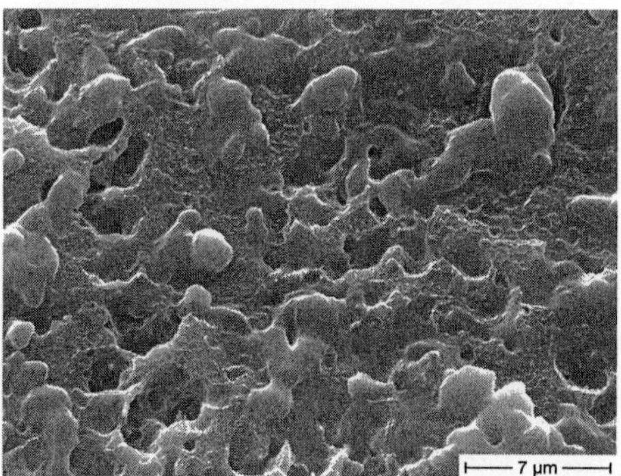

Fig. 4. Co-continuous phase structure in PBT with 50 wt.% EPDM-g-GMA (Note: SEM picture was taken from the fracture surface).

Instrumented Falling weight Impact (IFWI)

The effects of type and amount of the epoxy functionalized rubbers, as well as that of the testing temperature, are summarized in Figure 5. Figure 5 shows that the specific (i.e. thickness related) perforation energy of the blends with both rubbers is similar at least at ambient temperature. However, EPDM-g-GMA outperforms NBR-g-GMA, when tested at T = - 40°C. This behavior should be linked with the T_g of the corresponding rubber.

The fact that the specific perforation impact of PBT blended with EPDM-g-GMA does not depend on the testing temperature should be traced to the co-continuous phase structure.

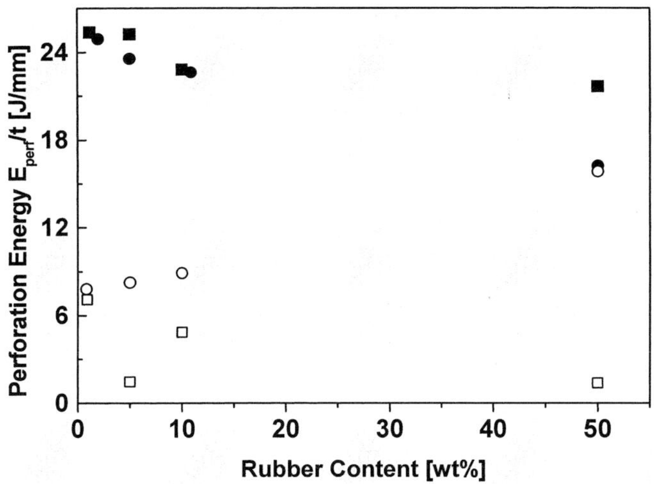

Fig. 5. Specific perforation impact energy as a function of rubber content and type when tested at T=23 and =-40 °C, respectively. ■NBR-g-GMA, T=23°C; • EPDM-g-GMA, T=23 °C; o EPDM-g-GMA, T= -40°C; □ NBR-g-GMA, T= -40°C.

Conclusions

Based on this work devoted to study the toughening efficiency of the epoxy functionalized apolar (EPDM-g-GMA) and polar (NBR-g-GMA) rubbers blended with PA-6 and PBT, respectively, the following conclusions can be drawn:

- The rubbers are dispersed at low (≤ 10 wt.%), whereas form a co-continuous structure at high (~ 50 wt.%) concentrations.
- The enhancement in the fracture toughness and energy is similar for both rubbers in the tested temperature range. Adding more than 5 wt.% epoxy functionalized rubber does not improve the linear elastic fracture mechanical response.
- The perforation impact (biaxial loading) of the blends were similar for both rubber at ambient temperature. At T= -40°C, however, EPDM-g-GMA was superior to NBR-g-GMA, which reflected the difference in the T_g of the rubbers.

[1] B. Majudmar and D. R. Paul in: "Polymer blends". Formulation, D. R. Paul, C. B. Bucknall, Eds., Wiley, New York, **2000**, 1, 539.
[2] O. Okada, H. Keskkula, D. R. Paul, Polymer **2000**, 41, 8061-8074.
[3] S. C. Tjong, Y. C. Ke, Plast. Rubb. Compos. Process. Appl. **1996**, 25, 319.
[4] Gy. Marosi, I. Csontos, Gy. Bertalan, Sz. Szabó, Polym. News **2000**, 25, 353.
[5] J. Karger-Kocsis in: "Handbook of Thermoplastic Polyesters", S. Fakirov, Ed., Wiley-VCH, Weinheim, **2002**, 2, 1291.
[6] N. Torres, J. J. Robin, B. Boutevin, J. Appl. Polym. Sci. **2001**, 81, 2377.
[7] N. Papke, Karger-Kocsis, Polymer **2001**, 42, 1109.
[8] M. Pracella, L. Rolla, D. Chionna, A. Galeski, Macromol. Chem.. Phys. **2002**, 203, 1473.
[9] M. Pracella, F. Pazzagli, A. Galeski, Polym. Bull. **2002**, 48, 67.
[10] N. Papke, J. Karger-Kocsis, J. Appl. Polym. Sci. **1999**, 74, 2616.
[11] J. Karger-Kocsis, J. Varga, G. W. Ehrenstein, J. Appl. Polym. Sci. **1997**, 64, 2057.
[12] O. Benevolenski, J. Karger-Kocsis, Macromol. Symp. **2001**, 170, 165.
[13] Z. A. Mohd Ishak, U. S. Ishiaku, J. Karger-Kocsis, J. Appl. Polym. Sci. **1999**, 74, 2470.
[14] Z. A. Mohd Ishak, U. S. Ishiaku, J. Karger-Kocsis, Compos. Sci. Technol. **2000**, 60, 803.
[15] M. Maspoch, A. Tafsi, H. E. Ferrando, J. I. Velasco, A. M. Benasat, Macromol. Symp. **2001**, 169, 159.

Compatibilization of PA6/Rubber Blends by Using an Oxazoline Functionalized Rubber

R. Scaffalo,[1] F.P. La Mantia,[1] R. Bertani,[2] A. Sassi[2]*

[1] University of Palermo, Department of Chemical Engineering and Materials, Viale delle Scienze, 90128, Palermo, Italy
[2] University of Padova, Department of Chemical Processes and C.N.R., Institute of Sciences and Molecular Technologies, Via F.Marzolo 9, 35131, Padova , Italy

Summary: The compatibilization of blends of polyamide 6 with a nitrile butadiene rubber has been investigated. The procedure consists of two steps: modification of the nitrile groups of the rubber into oxazoline in the melt through condensation of ethanolamine with formation of a molecule of ammonia, followed by use of the modified rubber as a compatibilizing precursor which is melt mixed with the polyamide to produce the compatibilized blend. The modification reaction has been detected by NMR analysis and a rheological, mechanical and thermomechanical characterization has been carried out on the all the blends. The results indicate that the modification reaction occurs but the conversion of nitrile into oxazoline is relatively low. Use of the modified rubber in the preparation of binary polyamide/rubber blends, leads to an increase in viscosity, which is typical of compatibilized systems, and to enhanced tensile, impact and thermomechanical properties. These phenomena can be explained by the formation of in situ rubber/polyamide copolymers that act as compatibilizers, due to the reaction between oxazoline and the end groups of the polyamide. The presence of residual low molecular compounds, from the modification or from the purification of the rubber worsens all of the properties and inhibits the compatibilizing effect of the modified rubber.

Keywords: compatibilization; nitrile rubber; oxazoline; polyamide; reactive processing

Introduction

One of the most popular topics and targets of both scientific and industrial research is the preparation of high performance polymer blends with the final aim of producing new materials with improved properties as compared to those of the original polymer. The polymer couples are generally immiscible and incompatible, the final properties of the materials are generally poor and a proper compatibilizing technique is required to improve the otherwise weak interfacial adhesion, the gross morphology and the poor stability of the blend.[1-5]

 DOI: 10.1002/masy.200351207

A highly desirable result could be achieved by coupling of the impact resistance of rubber with the good mechanical and thermomechanical properties of polyamides but an appropriate compatibilization system has to be developed.

One of the most versatile methods for the compatibilization of incompatible A/B polymer blends is to add a graft A-g-B copolymer that locates at the interface, thus allowing a reduction of the interfacial tension with an enhancement of the dispersion and stabilization of the morphology.[1-3] The preparation of these copolymers can be achieved in a separate step or in situ, during the preparation of the polymer blend.[1-3,6] Of course, the second method is particularly desirable for industrial purposes and is relatively easy to achieve if one or both of the polymers contain functional groups able to react, such as maleic anhydride, acrylic acid, glycydil methacrylate, oxazoline.[6-11]

Oxazoline has been found to be highly reactive toward a number of other functional groups e.g. carboxyl, amine, phenol and mercaptan.[12] To graft an oxazoline ring to the polymer thus in situ it is possible to use compounds already containing an oxazoline ring[13-22] or to form it through condensation of ethanolamine on nitrile groups either in solution or in the melt.[23-27]

In the current work the nitrile groups of nitrile butadiene rubber (NBR) have been converted into oxazoline through reaction with ethanolamine in the presence of a catalyst. The modification reaction is depicted in Scheme 1. The modified rubber has then been used for the compatibilization of polyamide6/NBR blends. The possible method of formation of the compatibilizer is illustrated in Scheme 2.

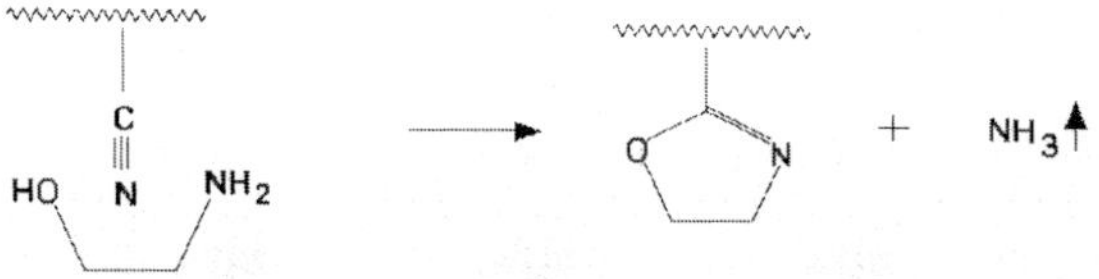

C
N
HO
NH_2
O
N
+
NH_3

Scheme 1. Reaction of conversion of nitrile into oxazoline.

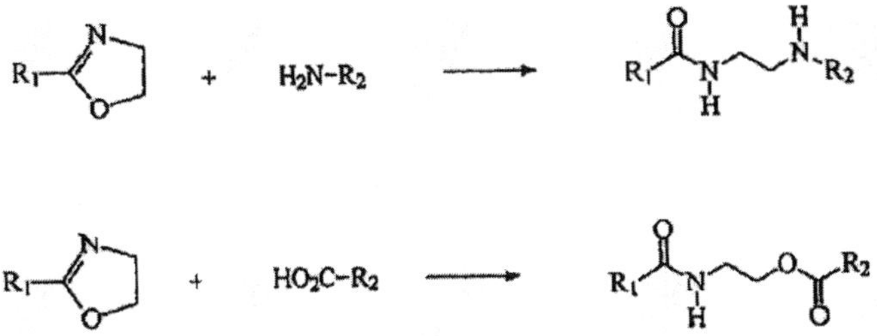

Scheme 2. Reaction of oxazoline with amine and carboxyl groups.

Experimental

Materials and Preparation

The rubber used in this work was kindly supplied by Enichem Elastomeri (a nitrile content of 30%). The polyamide 6 (PA6) was the ASN27 grade, supplied by Rhodia. The functionalizing agent employed was 2- ethanolamine (EA) and the catalyst used according to previous works[23–28] was zinc acetate, both purchased from Aldrich. All the materials were used as received without further purification.

The functionalization was performed in a Brabender (PLE 330) batch mixer between 160 °C and 180 °C and a speed of 64 rpm, requiring about 10 minutes. A solution of EA and catalyst (100:6 w/w) was added to molten rubber (rubber/solution 75/25 w/w) and the mixer was closed in order to avoid reactant loss through vaporisation. The rubber was then removed from the chamber and either used as it was or after further purification, consisted of solubilization it in tetrachloroethane followed by precipitation from methanol in order to remove any trace of reactants and low molecular weight compounds and drying under vacuum at 85 °C to remove any remaining solvent.

80/20 w/w binary blends of polyamide and rubber (modified and unmodified) were also prepared using the same apparatus as used for the functionalization process. The conditions employed were : 220 °C, a rotational speed of 64 rpm and about 10 minutes mixing time, until a constant value of torque was reached. Both the polyamide and the rubber were dried overnight in a vacuum oven before processing.

Characterization

The modification reaction was investigated by CP MAS NMR using a Bruker 200AC spectrometer operating at 4.7 Tesla and equipped for solid state analysis. Samples were spun at 5000 Hz in 7 mm diameter zirconia rotors with Kel-F caps. The 50.32 MHz ^{13}C CP MAS NMR spectra were obtained using the standard Bruker cross polarization pulse sequence with high power dipolar decoupling during acquisition, 3 ms Contact Time, 30 seconds Relaxing Delay, and were processed with 20 Hz exponential line broadening. ^{13}C chemical shifts were externally referenced to solid sodium 3-(trimethyl-silyl)-1-propane sulfonate at 0 ppm. The magic Angle condition was adjusted observing ^{79}Br resonance in a rotor containing 5% of KBr.[29]

Tensile tests were carried out using an Instron 1122 (United Kingdom) tensile testing machine on samples cut from sheets obtained by compression moulding (Carver laboratory press) at 220 °C.

Impact tests were carried out using a CEAST (Italy) impact test machine on notched samples at –20 °C. Thermomechanical properties were evaluated using a Rheometric DMTA V (USA) in the range 20-110 °C at a frequency of 1 Hz.

Results and Discussion

Modification

FTIR analysis, not reported here, is not useful for monitoring the conversion of the nitrile group of NBR into oxazoline. This is probably due to the poor degree of achieved conversion during the modification reaction. It is worth noting that this is not a negative result as, for compatibilization purposes, even very low conversions are sufficient to obtain a good performance of the modified polymer. Even if the effectiveness of the modified rubber as a compatibilizing agent can be used as indirect evidence of the modification reaction, direct investigations have been carried out by using ^{13}C NMR in the solid state analysis. These results are reported in Figs. 1a-b.

In Fig. 1a the spectrum of neat NBR is reported. The two signals characteristic of the rubber can be clearly seen due to the aliphatic and aromatic carbon atoms, respectively, centered at 31ppm and 130ppm.

The same peaks are also present in the spectrum of the modified rubber (Fig. 1b), but in this case additional weak peaks are also present. In particular the signal centered at 61 ppm can be attributed to the carbon atoms bonded to the nitrogen in the oxazoline ring[30] grafted onto the main backbone of the polymer. Their relative weakness confirms the very low conversion achieved in the modification reaction. Different reaction temperatures have also been employed, but the characteristics of the different materials obtained are exactly the same, at least in the range of conditions investigated in the present work. A separate study on the degradation of the rubber,[31] showed that a slightly higher reaction temperature led to less degradation, probably due to the lower melt stress resulting from the reduced melt viscosity,. Thus, all the functionalized rubber used for compatibilization purposes was prepared at 180 °C.

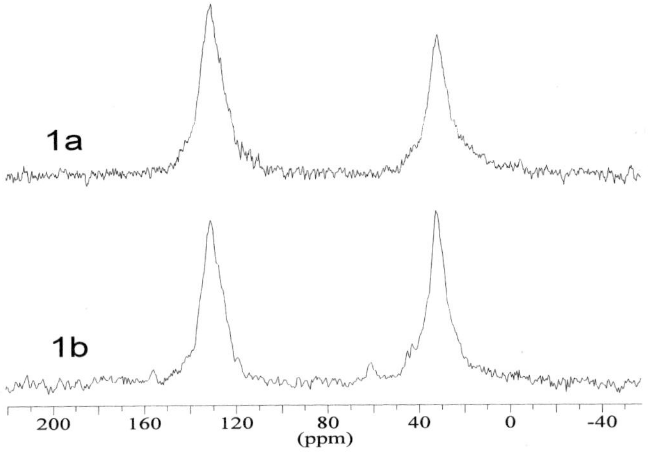

Fig. 1. Solid state ^{13}C NMR spectra of: (a) unmodified NBR, (b) modified and purified NBR.

Compatibilization

In Fig. 2 the torque obtained during the modification reaction is reported for the following blends: PA6 with modified and unpurified NBR (PA6/NBR mod.); PA6 with modified and purified NBR (PA6/NBR mod. pur.) and PA6 with pure NBR which has been previously processed under the same functionalization conditions (PA6/NBR lav.).

The mixing torque is a measure of the power absorbed during processing of the rubber and is strictly related to the melt viscosity of the material inside the mixing chamber. As can be clearly seen, the curve with the lowest torque values is that obtained for the binary blend containing modified and unpurified NBR. This can be explained by the presence of low molecular weight residual reactants that act as plasticizers lowering the melt viscosity of the blends to values even lower than that displayed by the uncompatibilized PA6/NBR lav. blend. The presence of these compounds has been also confirmed by the generation of smoke products in the chamber during the preparation of the blend.

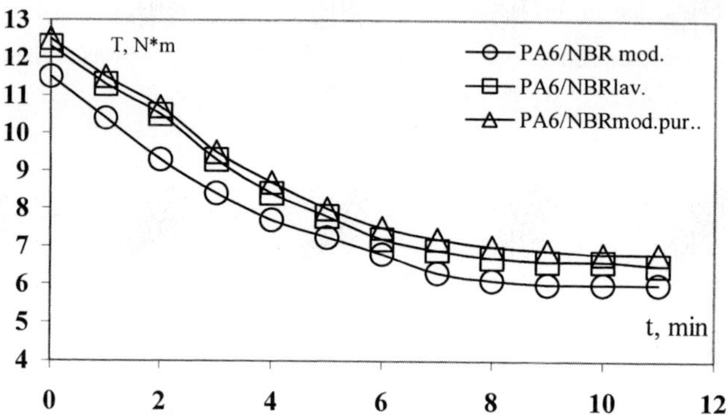

Fig. 2. Mixing torque as a function of time of compatibilized and uncompatibilized PA6/NBR.

The situation is different when the rubber is purified prior being added to the PA6. In this case no smoke products are detected during processing and the viscosity is higher than that of the uncompatibilized blend. This phenomenon is expected as the reaction between the oxazoline modified NBR and the PA6 leads to the formation of NBR-PA6 copolymers that, localizing at the interface of the two polymers, can act as compatibilizers for the blend. This, causes a greater interaction between the phases with the consequent increase of the melt viscosity. It is clear that this phenomenon also occurs when modified and unpurified NBR is used, but in this case the increase in the viscosity is overcome by the plasticizing effect caused by the presence of the low molecular weight compounds.

In Table 1 the tensile mechanical properties of the three binary blends are reported. The mechanical results confirm the results of the rheological analysis.

The mechanical properties of the blend containing unpurified modified NBR show the worst properties as compared with the other blends, especially for the uncompatibilized blend. Once again, this is due to the presence of residual low molecular weight compounds that act as defects in the solid state and induce premature rupture exhibiting low values of tensile stress, elongation at break and of the breaking energy. This effect is more intense than the plasticizing effect evidenced by the lowering of the elastic modulus.

Table 1. Mechanical properties of compatibilized and uncompatibilized PA6/NBR binary blends.

Material	Modulus MPa	Tensile Stress MPa	Elongation at Break %	Energy to Break J
PA6	300	60	335	-
PA6/NBR lav.	640	36.3	75	4.96
PA6//NBR mod.	574	36.0	70	3.91
PA6/NBR mod. pur.	625	38.6	101	5.72

The situation is completely different for the blend containing purified modified NBR. In this case the modulus is comparable with that of the uncompatibilized blend while all the other properties are enhanced. This improvement in the mechanical properties of the blend is be ascribed to the presence of copolymers that locate at the interface between PA6 and NBR and increase the interfacial adhesion and therefore the tensile properties.

To further verify the effectiveness of modified NBR as compatbilizer impact tests have been performed at –20 °C. This temperature is compatible with possible outdoor applications of rubber toughened PA6. The relative results are reported in Table 2.

Table 2. Impact strength measured at –20 °C of compatibilized and uncompatibilized PA6/NBR binary blends.

Blend	Impact strength, J/m^2
PA6/NBR lav.	3940
PA6//NBR mod.	2300
PA6/NBR mod. pur.	5250

The trend is the same as that displayed foe the tensile properties. Once again, the lowest value is obtained for the modified and unpurified NBR and the highest is obtained for purified modified NBR blend. As before, the presence of low molecular weight compounds is the responsible for the low value obtained . The conclusion is that the removal of reactants is crucial to the obtainment of good properties for these materials.

To confirm the presence of in situ formed compatibilizers, thermomechanical tests have been performed on neat polyamide, on the uncompatibilized binary blend and on the compatibilized blend with purified modified NBR. The curves of tan δ as a function of temperature for the above mentioned materials are reported in Fig. 3.

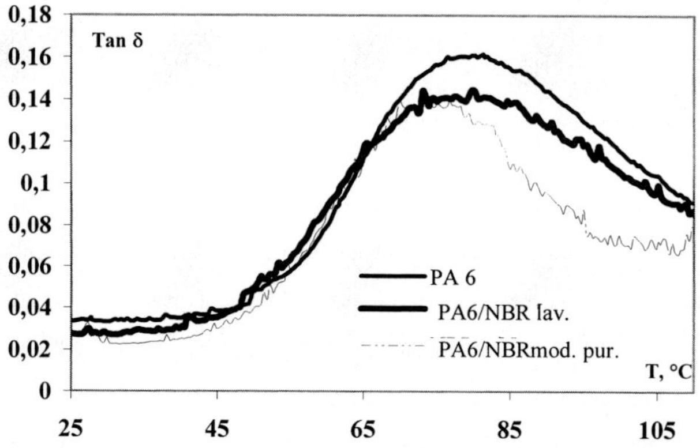

Fig. 3. Tan δ as a function of temperature for neat PA6, for the binary uncompatibilized blend PA6/NBR lav. and for the compatibilized blend with purified modified NBR.

Can be clearly seen in the figure, the curves for the neat polyamide and the uncompatibilized blend show approximately the same tan δ peak while the curve corresponding to the compatibilized blend shows a peak at a slightly lower temperature.

As the tan δ peak is associated with the glass transition temperature of the phase, a reduction of the peak temperature corresponds to a lowering of the glass transition temperature.

When no interactions between the polyamide and other phases are present, such as in the case of the uncompatibilized blend, there are no differences between the glass transition temperature of the polyamide phase in the blend and of the polyamide alone. However, in the compatibilized blend there is an interaction between the phases, resulting from the presence of in situ formed NBR-g-PA6 compatibilizers at the interface of the polymers. A qualitative explanation of this phenomenon is that the presence of these copolymers significantly reduces the crystallinity of the polyamide phase and weakens the connection between the crystalline and amorphous phase. As a consequence a reduction in the energy required for this transition, and therefore of the temperature of the transition, can be expected.

Conclusions

In this work compatibilization of NBR/PA6 blends has been investigated. The nitrile groups of the rubber have been modified with oxazoline by a condensation reaction that employes ethanolamine in the presence of a catalyst. Degree of conversion is very small and not easy to detect but it is nevertheless efficient for compatibilization purposes.

Binary blends of PA6 with modified and unmodified NBR have been prepared in a batch mixer. The results indicate that the modified rubber is able to improve the tensile and the impact properties of the blends but only if a proper purification stage is carried out in order to completely remove residual low molecular weight compounds. The formation of in situ NBR-g-PA6 copolymer at the interface can be reasonably invoked to explain the increased compatibility of the blends.

[1] L. A. Utracki, R. A. Weiss, Eds, *Multiphase Polymers: Blends and Ionomers*, ACS, Washington, D.C., 1989
[2] L. A. Utracki, *Polymer Alloys and Blends, Thermodynamics and Rheology*, Hanser Publ., New York, 1989
[3] L. A. Utracki, *Two-Phase Polymer Systems*, Hanser Publ., New York, 1991
[4] C. Koning, M. Van Duin, C. Pagnoulle, R. Jerome, *Progr. Polym. Sci.*, **1998**, 3/4, 707.
[5] M. Xanthos, S. S. Dagli, *Polym. Eng. Sci.*, **1991**, 31, 929.
[6] M. Xanthos, J. E. Biesenberger, *Reactive Extrusion, Principles and Practice*, Hanser Publ., New York, **1992**
[7] W. J. McKnight, R. W. Lenz, P. Musto, R. Somani, *Polym. Eng. Sci.*, **1985**, 25, 1124.
[8] G. Fairley, R. E. Prud'homme, *Polym. Eng. Sci.*, **(987**, 27, 1495.
[9] Z. Liang, H. L. Williams, *J. Appl. Polym. Sci.*, **1992**, 44, 699.
[10] J. Rösch, R. Mülhaupt, *Makromol. Chem., Rapid Commun.*, **1993**, 14, 503.
[11] K. Y. Park, S. H. Park, K.-D. Suh, *J. Appl. Polym. Sci.*, **1997**, 66, 2183.
[12] J. A. Frump, *Chem. Rev.*, **1971**, 71, 483.
[13] T. Vainio, G.-H. Hu, M. Lambla, J. Seppälä, *J. Appl. Polym. Sci.*, **1996**, 61, 843.
[14] P. Hietaoja, M. Heino, T. Vainio, J. Seppälä, *Polym. Bull.*, **1996**, 37, 353.
[15] T. Vainio, G.-H. Hu, M. Lambla, J. Seppälä, *J. Appl. Polym. Sci.*, **1997**, 63, 883.
[16] U. Anttila, C. Vocke, J. Seppälä, *J. Appl. Polym. Sci.*, **1999**, 72, 877.
[17] C. Vocke, U. Anttila, J. Seppälä, *J. Appl. Polym. Sci.*, **1999**, 72, 1443.
[18] N. C. Liu, H. Q. Xie, W. E. Baker, *Polymer* **1993**, 34, 4680.
[19] N. C. Liu, W. E. Baker, *Polymer* **1994**, **35**, 988.
[20] Y. Fujita, T. Sezume, K. Kitano, K. Narukawa, T. Mikami, T. Kawamura, S. Sato, T. Nishio, T. Yokai, T. Natura, EP308179 A2 **(1989)**
[21] G. H. Hu, R. Scaffaro, F. P. La Mantia, *J. Macr. Sci.-Pure Appl. Chem.*, 1998, **A35**, 457.
[22] F. P. La Mantia, R. Scaffaro, C. Colletti, T. Dimitrova, P. Magagnini, M. Paci, S. Filippi, *Macr. Symp*, **2001**, 176, 265.
[23] D. T. Hseih, D. N. Schulz, D. G. Pfeiffer, *J. Appl. Polym. Sci.*, **1995**, 56, 1667.
[24] D. T. Hseih, D. N. Schulz, D. G. Pfeiffer, *J. Appl. Polym. Sci.*, **1995**, 56, 1673.
[25] G. H. Hu, R. Scaffaro, F. P. La Mantia, *J. Mat Sci. Pure Appl. Chem.*, **1998**, A35, 457.
[26] R. Scaffaro, G. Carianni, F. .P. La Mantia, A. Zerroukhi, N. Mignard, R. Granger, A. Arsac, J. Guillet, *J. Polym. Sci., Part A: Polym. Chem.*, **2000**, 38, 1795.
[27] F. P. La Mantia, R. Scaffaro, X. Liu, *J. Appl. Polym. Sci.*, in press
[28] H. Witte, W. Seelinger, *Angew.Chem.Int.Ed.Engl.* **1972**, 11, 287.
[29] J. S. Frye, G. E. Maciel, J.Mag. Res. **1982**, **48**, 125
[30] C. J. Pouchet, J..Behnke, Eds., the Aldrich library of ^{13}C and ^{1}H FT-NMR Spectra **1992**
[31] A. Casale, R. S. Porter, *Polymer Stress Reactions*, Academic Press, London **1978**

Functionalization on Polyamide Complex with Iodide: A Review

Akio Kawaguchi,[*1] *Naoki Tsurutani*[2]

[1]Research Reactor Institute, Kyoto University, Kumatori, Osaka 590-0494, Japan
[2]Faculty of Science, Kyoto University, Kitashirakawa, Kyoto 606-8502, Japan
E-Mail: akawagch@rri.kyoto-u.ac.jp

Summary: Application with iodinated polyamide-6 (iodine/PA-6 complex) is introduced for modification of structure and functionality on the polyamide. Coordination between polyiodides (I_n^- , n = 3, 5, ...) and the polyamide advances rapidly and drastically within polyiodide solutions even in bulky samples of the polyamide, and responsiveness in the complex with other molecules such as water is enhanced by coordination. Such response and variation of the structure of the complex can introduce novel functionality with other molecules or ions or salts into the polyamide or hydrophilic polymers.

Keywords: complex; iodine; modification; polyamides; structure

Introduction

Doping of iodine into polyamide-6 (PA-6, nylon 6) can be introduced by an easy procedure with a I_2-KI aqueous solution, but a complex consequently prepared by doping ("nylon 6/iodine complex") shows various structures and some complicated activation. While the doping procedure and following reduction process (releasing iodine) were applied for conversion of hydrogen-bondings in PA-6 ($\alpha-\gamma$ transition), the intermediate state in which polyiodides ions (I_n^-; n = 3, 5, ...) are coordinated with the polyamide has not been clarified adequately for the complicated structures or chemical reactions.[1,2]

The doping process can be achieved by immersing PA-6 (bulk or oriented film) into aqueous solutions containing polyiodides (for example, I_2-MI aqueous solutions, where MI is an alkali-iodide), and the complex can be prepared easily and rapidly at a room temperature. However, the structures or reaction of the complex are complicated by various processes, such as diffusion of polyiodides or alkali cations (maybe, adsorbed simultaneously), adsorption/release of moisture, transition of the doped polyamide within crystalline region, etc.[3-6] Since the structures of the complex are attributed to coordination of polyiodides with amide groups on the PA-6 chains, it is frequently required on investigation for physical structures of the complex to consider chemistry. In previous researches, we have found some points to consider this complex; (1) PA-6 is a crystalline polymer with comparatively high crystallinity due to the hydrogen-bonding by amide

 DOI: 10.1002/masy.200351208

groups; (2) nevertheless, the original polyamide is also a hydrophilic polymer with high ability for adsorption of moisture; (3) on the other hand, adsorption of iodine from the aqueous solutions should be estimated considering coordination of polyiodide ions, not a mono-iodide ion (I^-; $n = 1$).[6-9] These points indicate that the structures of the complex should originate from coordination and hydrophilicity. Then, it is considered that, in the complex, coordination in the crystalline region (not only in the amorphous region) causes variation of intercalated structures and activated responses to environments in the ordered states. In following reports, we introduce some characteristic behaviours of the iodinated complex and views for application to functionalization with the complex.

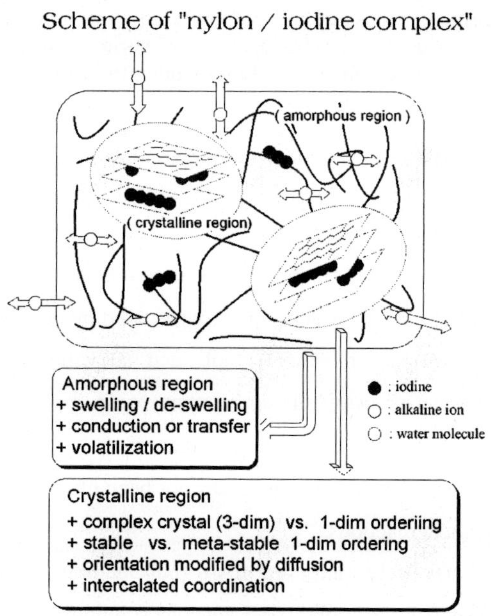

Fig. 1. Scheme of structures in iodinated PA-6 complex.

Preparation

Melt-crystallized plane samples (non-oriented or doubly oriented) or commercial films ("Rayfan™ NO", Toray Plastic Film Co.Ltd.) were used as starting materials. These non-doped samples were annealed in vacuum and their crystalline structure was estimated as the α-phase.

The cut pieces of the samples were immersed into the I_2-MI aqueous solutions (M^+: K^+, Na^+, NH_4^+,,, etc.) which contain the polyiodides of condensation of 0.2-1.0 N at 5-20 °C. If necessary, the doping process was interrupted by rinsing the samples in water. In investigation for response to humidity, doped samples were aged in vacuum to minimize influences by volatilized elements. Or, for the time-sliced investigation with a synchrotron radiation (SR), droplets of the solution were poured into the sample cells at room temperature.

Structures and Characteristic Behaviours

Modification of Diffracted Intensities with Swelling

At first, the term of the "hydrogen-bonded sheets" in the complex does not mean normal hydrogen-bondings between the anti-parallel polyamide chains but is used to support us to understand the layered structure; the structures in the complex are constructed by coordinate bonds. Whereas, the "crystalline region" is not equal to a complex crystal which is attributed to three dimensional symmetry; the "crystalline region" may be the crystallites of the α-phase of PA-6 before doping but, in the complex, it contains various structures (complex crystal with three dimensional symmetry, meta-stable structures, one dimensional ordered structures,,, etc.) and transition among them also exists.

The hydrogen-bondings in PA-6 construct layer structures of the α form in the crystalline region of the non-doped samples.[10] And, the layer structures in the iodine-doped complex attributed to the "hydrogen-bonded sheets" provide sites for coordination with the polyiodides in the crystalline region; the amide groups or hydrogen-bondings are introduced as the host sites with "long range order" which actively interact with the polyiodides, cations or water molecules. Such coordination between the amide groups and the polyiodides is also expected in the amorphous region and it is achieved actually. However, it is regarded that an amount of the doped iodine in the amorphous region is less than that in the crystalline region,[7] or, as static structures, stability of the amorphous complex is relatively lower than that for the crystalline complex. Functionality of the amorphous complex will not be mentioned in detail here while it will induce some interests in view of ionic conduction.[6]

The most elementary and stable coordination is that all layer sites between the "hydrogen-bonded sheets" are occupied with the dopants; this scheme reveals a spacing of 1.56 nm and it is indicated both in comparatively swollen samples and dried ones, i.e., in fresh samples just after rinsing with water or in ones aged in vacuum for a several months, though there is observed transition in intensities of the orders on the 1.56 nm-spacing between the swollen samples and dried ones.[4] Investigation with Raman spectroscopy and WAXD showed that the polyiodides doped in the complex are I_3^- and I_5^- and that they are oriented on the tilt to the chain axis of PA-6 and parallel to it, respectively.[5,6] Furthermore, each complex prepared with the I_2-MI aqueous solutions (M^+: K^+, Na^+, NH_4^+) shows the 1.56 nm-spacing while there is also deviation in intensities of the orders on the spacing depending on the cations; the cations are sure to be doped simultaneously and there should be deviation for their structures and functionality.[11] Therefore, we may require more discussion for modification of the intercalated polyiodides with ageing or swelling since the previous results of Raman spectroscopy indicate information only from the surface of the samples where drying and volatilization take the lead.

Modification of Orientation

On investigation with the "doubly oriented samples", the double orientation can be changed by diffusion of the dopants. It is that, while chain orientation is held parallel to drawn direction, there are two schemes of orientation of the "hydrogen-bonded sheets"; the sheets in the complex can be either parallel or normal to surface of the filmy samples even though the sheets were oriented parallel to the surface in the original non-doped doubly oriented PA-6 samples.(Fig. 2)[12] On the other hand, on investigation with the non-oriented samples, Murthy reported re-orientation of the chain axis normal to the surface of the samples.[3] The two cases may seem to make a conflict for each other, but we should consider that the results with the drawn samples indicate too strict orientation of the chain direction for the PA chains to reconstruct their orientation. On the other hand, a restraint for the chains would not be so strict around the crystallites in the non-oriented PA-6, and then it is possible to reconstruct the chain direction with doping. If so, considering the oriented structures with the polyiodides, we can expect deviation of ability of the polyiodides on diffusion. And the most fundamental point is that diffusion of the dopants into the polyamide can induce re-orientation of the chain direction or the "hydrogen-bonded sheets". Such modification of the orientation is confirmed by a time-sliced observation with the SR at SPring-8 in Nishi-harima, Japan; in the earliest stages on diffusion of the polyiodides, we can observe both creation of the complex structures and, simultaneously, re-orientation of the chains normal to surface of the non-oriented samples or modification of the "hydrogen-bonded sheets" in the doubly oriented samples.[13] Furthermore, it is interesting that the structures formed on the early stage of the doping seem to neither be the most stable intercalation of the spacing of 1.56 nm nor other intercalated structures induced by swelling.(following) Research is going on with the SR.

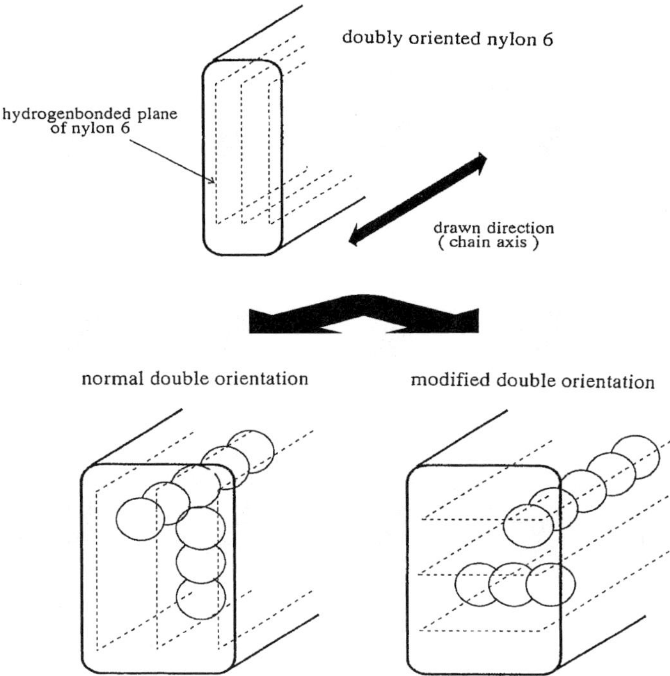

Fig. 2. Modified orientation in doubly oriented complex.

Effects by Moisture-activated Conduction and Modification of Intercalation

As a prior response to moisture, conduction in the amorphous region of the dried complex is activated in a few minutes after transfer into a moist environment. Conduction is ionic and has dependence on cations, M^+, which are doped simultaneously with the polyiodides from the I_2-MI solutions.[14,15] In the view of ionic conduction, however, composition of the polyiodides may not be proper as an application for an electric devices because of problems of volatilization of the iodine.[16]

In the crystalline region, as mentioned above, the most popular intercalation of the polyioidides between the "hydrogen-bonded sheets" shows the spacing of 1.56 nm, where all of the layer-sites between the sheets are occupied with polyiodides or ions. On the other hand, other spacing can be observed by additional procedures. One of them is immersion of the complex samples into an KI aqueous solution, and other is re-swelling by moisture. The former treatment leads "1.95 nm spacing" which is estimated as a one of the layer structures of intercalation with the dopants occupying the sites partially. That is applied with an "elimination effect" by monoiodide ions, I^-.[9] On the other hand, swelling by moisture induces slower and calmer elimination of the dopants from the crystalline region of the complex, and then the swelling process induces not only the "1.95 nm spacing" but also other intercalation with longer spacings.[17] Consequently, modified intercalated layer structures are achieved within the complex system swollen by moisture. And the modified structures vanish in dry condition and are reversed into the most stable intercalation with the spacing of 1.56 nm.

Application for Advanced Doping with other Ions or Salts

Generally, the crystalline region are not so active for conduction or transfer in polymeric systems; they are comparatively static in general swollen systems and occasionally behave as static cross-linking points in a gel. In the iodine complex as mentioned above, however, swelling induces not only the activated ionic conduction in the amorphous regions but also transition of the ordered structure in the crystalline region. It suggests that swelling or the "elimination effect" with the monoiodide ion can be available for "secondary doping" of other ions or molecules which can not be doped directly into the polymeric systems or their crystalline regions of them. And, if the "secondary doping" can be introduced, we can expect novel functionalities and behaviours.

For example, it is possible to import micro composites of AgI through the iodinated PA-6 complex. AgI is a hardly soluble salt and is hard to be dispersed on hydrophilic polymers. However, treatment with Ag^+ solutions to the iodinated complex can induce extraction of micro particles of AgI into it, and the nano-composite of AgI within PA-6 shows characteristic behaviours. For example, the composite shows response of structures to humidity in spite of hard-solubility of AgI for water.[18] And Gotoh et.al. have been researching condution of AgI crystallites and indicated interests in relation between orientation of the crystallites and the polymeric system.[19,20] Iodination may not be always available for modification of general hydrophilic polymers, but, if possible, it will induce novel structures and functionalities standing on coordination and activation with polyiodides.

Conclusion

Iodinated PA-6 complex should be investigated not only as the intermediate state on $\alpha-\gamma$ transition of PA-6 but also as the activated states complicated by multi-components. The procedure of iodination has a potential to introduce characteristic structures and activated functionality on the polyamide or other hydrophilic polymers.

Acknowledgements

The authors thank Prof. Miyaji of Kyoto University for very great advice and members in his laboratory for usage of WAXD instruments. We thank Toray Plastic Films Co.,Ltd. for a kind supply of "Rayfan™ NO" films. Some of the results were succeeded with an IP reader and an analyzing system which were purchased by a project at RIKEN, "The Development and Application of Neutron Optics (NOP)", and the project was approved by support of Special Coordination Funds for promoting Ministry of Education and Science of Japanese Government. Some results and continuous researches have been succeeded through subjects adopted at BL45XU, a beam line in SPring-8 (Nishi-harima, Japan).

[1] H. Arimoto, *J. Polym. Sci.*, **1964,** *A-3*, 2283.
[2] I. Abu-Isa , *J. Polym. Sci. Pt. A-1,* **1971**, *.9*, 199.
[3] N.S. Murthy, et.al., *J. Polym. Sci. Polym. Phys. ed.*, **1985**, *.23*, 2369.
[4] A. Kawaguchi, *Polymer* **33**, 3981 (1992).
[5] A. Kawaguchi, *Polymer* **1994**, *35*, 2665.
[6] A. Kawaguchi, *Sensors and Actuators B.*, **2001**, *73*, 174.
[7] N.S. Murthy, *Macromolecules*, **1987**, *20*, 309.
[8] W.R. Vieth, "*Diffusion In and Through Polymers*", Hanser Publishers, Munich:, 1991.
[9] A. Kawaguchi, *Polymer*, **1996**, *37*, 4877.
[10] D.R. Holmes, et.al., *J. Polym. Sci.*, **1955**, *42*, 159.
[11] A. Kawaguchi, *Polym. Prep.Jpn.*, **1999**,*.48*, 868.
[12] A. Kawaguchi, *Polymer* **1994**, *35*, 3797.
[13] A. Kawaguchi, et.al., *SPring-8 User Exp. Rep.* **2000**, *5*, (2000A), 354.
[14] A. Kawaguchi, *Sensors and Actuators B.*, **2000**, *63*, 10.
[15] A. Kawaguchi, *Sensors and Actuators B.*, **2001**,. *56*, 220.
[16] T. yamamoto, et.al., *J. Mat. Sci.*, **1986**, *21*, 604.
[17] A. Kawaguchi, et.al., *Polym. Prep.Jpn.*, **2000**, *.49*, 441.
[18] A. Kawaguchi, et.al., *Polym. Prep.Jpn.*, **2002**, *51*, 393.
[19] Y. Gotoh, et.al., *Polym. Prep.Jpn.*, **2002**, *51*, 742.
[20] Y. Gotoh, et.al., *Polym. Prep.Jpn.*, **2002**, *.51*, 2259.

Macromol. Symp. **2003**, *202*, 85–95

Influence of Processing Method and Components Molecular Structure on the Phase Behaviour of Polyethylenes/Dye Blends

Andrea Pucci,[1] Giacomo Ruggeri,[1,2] Camillo Cardelli,[3] Giovanni Conti[1]*

[1] Department of Chemistry and Industrial Chemistry, Via Risorgimento 35, 56100 Pisa, Italy
[2] INSTM, UdR Pisa, Via Risorgimento 35, 56100 Pisa, Italy
[3] IFAM-CNR, Via Alfieri 1, 56010 Pisa, Italy
E-mail: grugge@dcci.unipi.it

Summary: The phase dispersion of terthiophene alkyl derivatives on different polyethylene matrices was investigated. The PE affinity toward dichroic dyes with different structure, the effect of blending process and the influence of a polyolefinic compatibilizer on the homogeneity of host-guest blends were comparatively investigated by calorimetry, DSC and SEM analyses. For these purposes, polyethylenes with different molecular weights and densities and EVAc were used as host matrices. The polymeric compatibilizer was prepared by radical functionalization of a commercial low density polyethylene. The dichroic nature of the guest phase allowed to perform UV-Vis measurements in polarized light on oriented blend film samples. The dyes affinity toward PE is one of the key factor in obtaining oriented polyolefinic films with high optical performances for several applications.

Keywords: dichroic blend films properties; melt-processing; phase dispersions; polyolefinic blends; solution-casting

Introduction

Multicomponent polymeric materials based on conventional polyolefins and polar polymers or guest aromatic molecules present a great importance because they allow the optimization of their properties compared with the isolated components. For this purpose, the miscibility and the phase behaviour of polymer blends have received much attention due to the difficulty to obtain materials with good phase dispersion. It has been demonstrated that phase separations phenomena lead to materials with poor mechanical,[1] electronic[2] and optical[3] properties. In this work the phase and the dispersion behaviour of blends based on guest dichroic functionalized terthiophene dyes and host polyethylenes were studied by varying the structure

 DOI: 10.1002/masy.200351209

of the guest molecule and the type of blending process utilizing different processable polyolefinic matrices. In addition, in order to reduce the interfacial tension[4] between the components of the blend and to improve consequently their phase dispersion, the effect of a polyolefinic compatibilizer on the phase behaviour of these systems was investigated. The functionalization of polyolefins with polar monomers seems to be the favourite route to prepare polymeric compatibilizers due to the good availability of the materials and the extensive knowledge of the process. Indeed, it was demonstrated that functionalization degrees of about 2 mol-% of modified olefinic polymers positively affect the dispersion of the two phases of incompatible polymer blends leading to materials with better properties and performances.[5] The dispersion behaviour of host-guest materials were analyzed by calorimetry, differential scanning calorimetry (DSC) and microscopy. In addition, due to the dichroic nature of the guest chromophores,[6] the homogeneity of the blends prepared was also investigated by UV-Vis spectroscopy in polarized light, monitoring the anisotropic behaviour of the binary blend films after tensile deformation at high temperature. The key property to characterise the molecular orientation in such systems is the dichroic ratio[7] (R) defined by the equation $R = A_{//}/A_{\perp}$, where A is the respective absorbance and the subscripts // and $\perp$ denote, respectively, the direction parallel or perpendicular to the drawing direction. These results were discussed and used for the preparation of dye/polyolefin blends with specific thermo-mechanical properties and suitable optical performances for applications as linear polarizers.

Experimental

Apparatus and Methods

^{1}H- and ^{13}C-NMR spectra were recorded by a Varian Gemini-200 MHz spectrometer on 5-10% $CDCl_3$ (99.6+ atom % D, Aldrich) solutions. NMR spectra were recorded at 20°C and the chemical shifts were assigned in ppm using solvent signal as reference. FT-IR spectra were recorded by a Perkin-Elmer Spectrum One spectrophotometer by deposition of a drop of liquid between two KBr windows or on dispersions in KBr. Mass spectra were recorded with a Varian Saturn 2000 connected to a gas-chromatograph Varian 3800, on solutions of diethyl ether. The melting points were accomplished by a Reichert Polyvar optical microscope with crossed polarizers, equipped with a programmable Mettler FP 52 hot stage. Elementary analyses were made by microanalysis laboratories at the Faculty of Pharmacy, University of Pisa. Optical absorption studies were carried out in dioxane ($5 \cdot 10^{-5}$ M) solutions with a Jasco 7850 UV-Vis spectrophotometer or on polymer films in polarized light with the same instrument, fitted with Sterling Optics UV linear polarizer, or with a Perkin-Elmer Lambda 900, fitted with Glan-Thomson polarizers. Differential scanning calorimetry (DSC) analysis were performed by a Perkin-Elmer DSC7 calorimeter equipped with a CCA7 cooling device. The calibration was carried out by using Mercury (m.p. –38.4°C) and Indium (m.p. 156.2°C) standards for low-temperature scans and Indium and Zinc (m.p. 419.5°C) for high-temperature ones. Heating and cooling thermograms were carried out at standard rate of 10°C/min. The Scanning Electron Microscopy (SEM) analysis was performed with a Jeol 5600-LV microscope, equipped with Oxford X-rays EDS microprobe, instrument at the Chemical Engineering Department of Pisa University. Polymer processing were performed by a Brabender plastograph mixer (mod. OHG47055, 30cc) under nitrogen atmosphere. A Campana PM20/200 press was used for moulding the polymeric samples. Heat of solution measurements were performed by a Calvet type differential calorimeter Mod. BT-200 from SETARAM (France).

Materials

All reactions of air and water sensitive materials were performed under an atmosphere of argon or nitrogen, using glassware which was previously flame-dried at reduced pressure (0.05 mbar). The solvents used in the reactions (Aldrich, J.T. Baker, AnalaR, Fluka and Carlo Erba) were dried by conventional methods and freshly distilled under an inert atmosphere. Spectroscopic grade dioxane (Aldrich) was used for the absorption experiments in solution. 5,5”-bis-thiooctadecyl-2,2’: 5’,2”-terthiophene ($C_{18}S$-TT-SC_{18}) was synthesized as described in a previous work.[8] Very Low Density Polyethylene (VLDPE) supplied by EniChem (Italy) is characterized by 9 mol.-% of 1-butene as comonomer, melt flow index = 1.6 g/10min (190°C/2.16 kg, ISO 1133) and density = 0.9 g/cm^3. Ultra High Molecular Weight Polyethylene (UHMWPE), $\overline{M}_w = 3.6 \cdot 10^6$, density = 0.928 g/cm^3 (Stamylan UH210, DSM, The Netherlands) and Ethylene-Vinyl Acetate copolymer (EVAc, Greenflex FF35), supplied by Polimeri Europa (Italy) and characterized by 9 wt.-% of vinyl acetate and melt flow index = 1.6 g/10 min (190°C/2.16 kg, ISO 1133) were used as polymer matrices.

Film Preparation by Solution Casting

0.5 g of UHMWPE and the appropriate amount of chromophore were dissolved in 75 ml of *p*-xylene at 125°C and stirred until complete dissolution occurred; the solution was then cast on a cold glass and slowly evaporated, first at room temperature, then at 50°C. For the compatibilized sample, a 1:1 by weight respect to the dye amount of the polymeric additive was co-dissolved in xylene before adding UHMWPE.

Film Preparation by Mechanical Mixing

The blends and the polymer thin films were prepared according to a procedure previously reported.[7]

Polymer Orientation

Solid state drawings of the host-guest films was performed as reported in literature.[6]

Functionalization of VLDPE with Diethyl Maleate (VLDPE-g-DES)

The radical bulk functionalization of VLDPE was carried out according to procedures previously reported.[9] The determination of the functionalization degree (FD) was performed by proton nuclear magnetic resonance (^{1}H NMR) and infrared spectroscopy (IR). In the latter case, on the basis of Fodor et al.,[10] known amounts of PE and poly(diethyl fumarate) were mixed to give a composition close to that predictable for the functionalized polymer (from 1 to 6 –COOEt groups per 100 ethylene) in order to build a calibration curve. Poly(diethylfumarate) was previously prepared by radical polymerization of diethylfumarate according to a literature procedure.[11]

Results and Discussion

Three different strategies were accomplished in order to prepare host-guest polymeric systems with improved phase dispersion. In particular, the chemical structure of the guest dichroic molecule, the influence of the type of blending process and the use of VLDPE-g-DES as a polymeric compatibilizers were investigated.

Influence of the Structure of Terthiophene Based Dyes

The guest phase studied in our works is represented by dichroic terthiophene chromophores functionalized by long thio-alkyl lateral chain and different electron withdrawing groups.[6] In a recent work[12] it was demonstrated that the structure of the alkyl lateral chain of the dye plays a fundamental role on the molecule phase dispersion into UHMWPE films prepared by solution-casting. In particular, the favoured molecule's dispersion was achieved by functionalizing the aromatic terthiophene's core with a branched alkyl chain, 5"-thio-(3-butyl)-nonyl-2,2':5',2"-terthiophene (1), conferring the chromophore a very low tendency to crystallize. Actually, branched chromophore (1) is liquid at room temperature (m.p.= -5.6°C; ΔH_m= 21.4 J/g) whereas the linear one, 5"-thiooctadecyl-2,2':5',2"-terthiophene (2), obtained by functionalizing the terthiophene molecule with a long linear chain, presents higher melting temperature and crystallinity (m.p.= 82/91°C, ΔH_m= 122.3 J/g). As proposed in previous papers[7,13] the crystallinity of the dye plays an important role in controlling the thermodynamic of mixing. For this purpose, a group contribution procedure based on modified UNIQUAQ equation[14] was developed[15] to predict the enthalpy of mixing of all terthiophene molecules functionalized by an alkyl lateral chain. At a fixed entropy of mixing, the solution enthalpy has been considered as an index of the solute affinity for *n*-heptane, and then for linear PE, rather than a real miscibility indicator. As evidenced by the plotted curve reported in Figure 1, this quantity decreases increasing the number of carbon atoms of the alkyl chain.

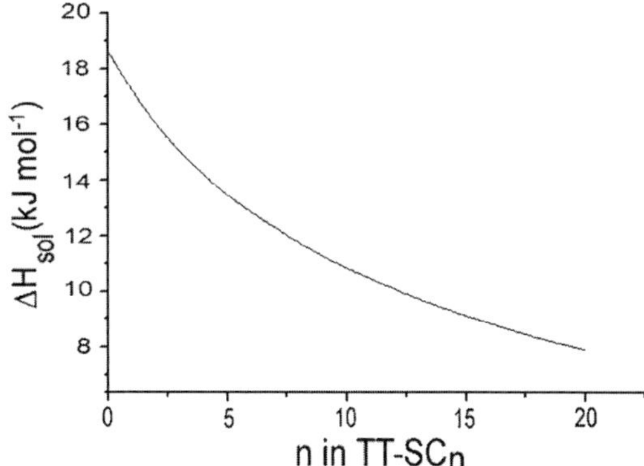

Fig. 1. Calculated ΔH_{sol} of terthiophene molecules as a function of the number of carbon atoms in the alkyl chain.

At the same time, heat of solution measurements between the alkyl terthiophene derivative and polyethylene were measured by calorimetry dissolving at 25°C the chromophore in *n*-heptane used as an *analogue* of linear PE. The $\Delta H_{sol} = 9.5\ kJ \cdot mol^{-1}$ measured for the branched dye (1) whose alkyl chain is composed by 13 carbon atoms, confirmed the effectiveness of the theoretical procedure. Thus, while the introduction of alkyl linear chains on the terthiophene nuclei increases its chemical affinity for PE, the whole dissolution process is made less favoured due to the contemporary increase of molecule's crystallinity. Hence, it is confirmed by calorimetry that the branched alkyl functionalization represents the best compromise for optimized chromophore dissolution into polyethylene matrices.

Influence of the Blending Process: Use of Processable Polyethylenes as Host Matrices

Different works have been performed in these years in order to increase the dispersion of the guest phase into the polymeric matrix by changing the type of blending process.[16,17] Respect to polymer blends obtained by casting of dilute solutions, characterized by molecular segregation of the guest phase from the host matrix, the melt blending technique leads to multicomponent materials with much better phase dispersions. It has been recently[8] demonstrated that the melt processing of dichroic chromophores and polyethylenes with different molecular weight and density (HDPE and LLDPE respectively) allowed to obtain homogeneous materials leading to devices with better optical performances. According to these results, new blends based on 0.2-0.5 wt.-% of the easy prepared terthiophene derivative ($C_{18}S$-TT-SC_{18}) functionalized by two long linear thio-alkyl lateral chains and EVAc as polymer matrix were prepared by melt-processing in a Brabender mixer. The scanning electron microscopy images of the sections of the films prepared by compression moulding of melt-processed blends based on EVAc or polyethylenes and $C_{18}S$-TT-SC_{18} revealed that the polymer matrix with a significant content of polar groups (EVAc) allowed a more effective distribution of the dye within the polymer bulk (Figure 2). Actually, the distribution of the chromophore along a section of LLDPE and HDPE films is just near the contact surface polymer-air, whereas it appears very similar on the polymer film surface of both three host matrices.

In order to evaluate how the chromophore dispersion could affect its orientational behaviour and the anisotropic properties of the host-guest films, UV-Vis spectra were performed exciting the oriented polymer samples with a linearly polarized radiation respectively parallel and perpendicular to the drawing direction.

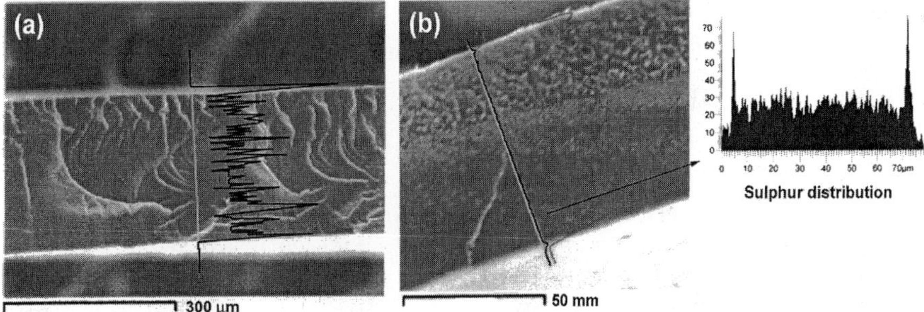

Fig. 2. Scanning electron micrographs and sulphur concentration profiles on the sections of EVAc (a) and LLDPE (b) based films (C_{18}S-TT-SC_{18} concentration = 0.5 wt.-%).

Table 1. Dichroic parameters of oriented films prepared by melt-processing.

Blend	Polymer	C_{18}S-TT-SC_{18} (wt.-%)	λ	R
LL-0.5	LLDPE	0.5	8	3.9
LL-0.5	LLDPE	0.5	12	11.4
LL-0.2	LLDPE	0.2	10	6.5
LL-0.2	LLDPE	0.2	12	10.6
HD-0.5	HDPE	0.5	8	5.3
HD-0.5	HDPE	0.5	12	10.8
HD-0.2	HDPE	0.2	9	5.7
HD-0.2	HDPE	0.2	12	10.0
EVAc-0.5	EVAc	0.5	3	3.2
EVAc-0.5	EVAc	0.5	8	4.4
EVAc-0.2	EVAc	0.2	3	3.5
EVAc-0.2	EVAc	0.2	8	7.3

The anisotropic behaviour of the films reported in Table 1 as a function of the drawing ratio (λ, defined as the ratio between the length of the sample after and before the stretching of the film respectively) revealed interesting dichroic ratio (R) values for films based on EVAc denoting a good tendency of the chromophore to align along the oriented fibres of the copolymer. Comparing the dichroic behaviour of all the oriented blends prepared by melt processing, EVAc-0.2 showed the highest dichroic ratio (R= 7.3) at λ maximum (λ= 8).

Influence of the Polymeric Compatibilizer on the Chromophore Dispersion into UHMWPE

The degree of functionalization (FD), defined as the number of functional groups grafted on the polyolefin backbones per 100 repeating units (FD in mol-%), evaluated by IR using a calibration curve and ^{1}H NMR spectroscopy calculating the integrals of the diagnostic signals, and the molecular weight of VLDPE-g-DES are reported in Table 2.

Table 2. VLDPE-g-DES features.

Polymer	FD (mol-%)[1]	FD (mol-%)[2]	$\overline{M}_n$	$\overline{M}_w$	$\overline{M}_w/\overline{M}_n$
VLDPE-g-DES	1.7	1.8	76400	363900	4.8

[1] evaluated by IR; [2] evaluated by ^{1}H NMR

The effect of VLDPE-g-DES on the phase dispersion of C_{18}S-TT-SC_{18} into UHMWPE films prepared by casting of xylene solutions (Table 3) was evaluated by SEM, DSC and by optical analyses.

Table 3. UHMWPE blends prepared in this work.

Blend	Polymer	C_{18}S-TT-SC_{18} (wt.-%)	VLDPE-g-DES (mg, [wt.-%])
UH3-3	UHMWPE	3	-
UH3-VL	UHMWPE	3	15, [3]

SEM images of UHMWPE films revealed an increased phase dispersion of the chromophore in host-guest systems containing the polymeric compatibilizer (Figure 3).

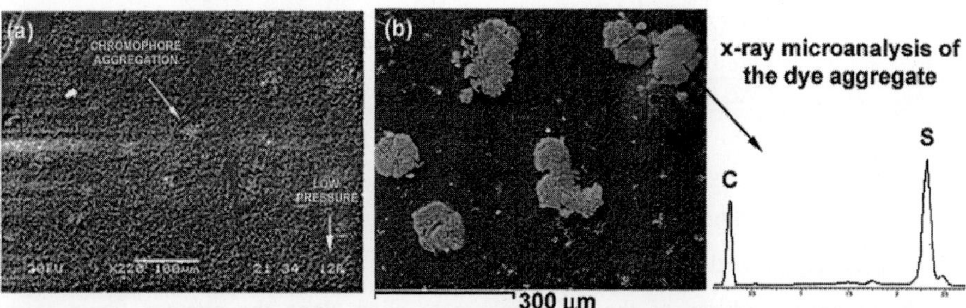

Fig. 3. SEM of UH3-VL (a) and UH3-3 (b) films.

In compatibilized UH3-VL film, the dispersed chromophore particles, smaller than those evidenced in UH3-3, were detectable just increasing the resolution of the instrument using the low pressure technique at 12 Pa. The increased homogeneity of UH3-VL film was also confirmed by differential scanning calorimetry (Figure 4). The melting peak of the chromophore dispersed in UH3-VL recorded by the first heating trace at about 100°C appeared much less pronounced than that showed by the endotherm of UH3-3. In fact the ΔH_f value evaluated for $C_{18}S$-TT-SC_{18} in UH3-VL was 0.15 J/g respect to 1.17 J/g determined for the UH3-3 film.

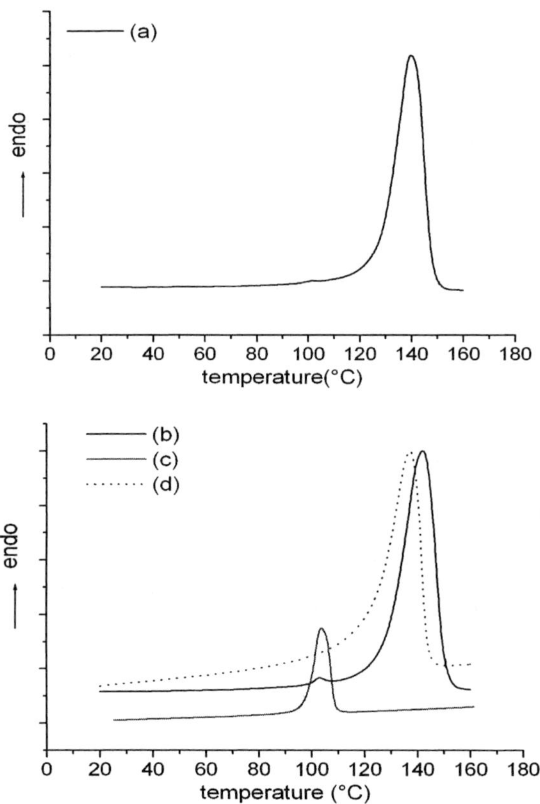

Fig. 4. First heating DSC traces of UH3-VL (a), UH3-3 (b), the neat dye (c) and second heating trace of UH3-3 (d).

UV-Vis spectroscopy in polarized light was performed on oriented UH3-VL and UH3-3 films in order to quantify the influence of the increased chromophore dispersion into UHMWPE on the optical performances of the prepared devices. The dichroic ratio R, related to the chromophore orientation along the stretched macromolecular fibres, is reported in Table 4.

Table 4. Dichroic ratios of oriented UH3-VL and UH3-3 films as a function of λ.

Sample	λ	R
UH3-VL	20	10±0.6
UH3-VL	30	15±0.6
UH3-VL	40	12.2±1.1
UH3-3	20	7.5±1.2
UH3-3	30	10±0.7
UH3-3	40	5±1.3

The optical properties showed by oriented films of UH3-3 and UH3-VL evidenced how the increased chromophore dispersion, due to the presence of polymeric compatibilizer, positively affects the performances of the host-guest devices. Hence, VLDPE-g-DES resulted as a proper compatibilizer between the guest phase of the terthiophene derivative and UHMWPE as evidenced by microscopy images and calorimetry measurements.

Conclusions

The three different strategies performed in obtaining good phase dispersion of terthiophene based guest dyes and the polyethylene host matrix allowed the preparation of multicomponent materials characterized by high degree of homogeneity. Heat of solution measurements of terthiophene derivatives in *n*-heptane, as *analogue* of linear PE, confirmed that functionalization with branched alkyl chains represent the best chromophore modification to increase the dye dispersion in polyethylene matrices. Microscopy and optical analyses evidenced that EVAc as polymer matrix allowed the preparation by melt-processing of binary blends based on $C_{18}S$-TT-SC_{18} as guest with very high dispersion degree. In addition, EVAc showed increased homogeneity respect to blends based on processable polyethylenes (LLDPE and HDPE), already characterized by good homogeneity and optical properties. Moreover, the use of VLDPE-g-DES as polyolefinic compatibilizer represented another important new useful strategy for the preparation of host-guest PE blends by solution-casting characterized by good homogeneity and phase behaviour of the guest chromophoric component. Finally, optical measurements in linearly polarized light performed on oriented blend films is confirmed as an effective investigation to analyse the dispersion degree of these binary host-guest systems. On the other side, the obtained results can provide indications for the preparation of high performance materials for applications as linear absorbing polarizers.

List of Abbreviations

PE: polyethylene; UHMWPE: ultra high molecular weight PE; LLDPE: linear low density PE; HDPE: high density PE; EVAc: ethylene vinyl acetate copolymer; VLDPE: very low density PE; VLDPE-g-DES: VLDPE functionalized with diethylmaleate; 5,5"-bis-thiooctadecyl-2,2': 5',2"-terthiophene: $C_{18}S$-TT-SC_{18}.

Acknowledgments

The authors gratefully acknowledge Dr. Nicola Tirelli and Prof. Francesco Ciardelli for the very helpful discussions.

[1] T. Miteva, L. Minkova, P. Magagnini, *Macromol. Chem. Phys.* **1998**, *199*, 1519.
[2] a) K.P. Raji and C.K.S. Pillai, *Synth. Met.* **2000**, *114*, 27. b) J. Carinhana, R. Faez, A.F. Nogueira, M.A. De Paoli, *Synth. Met.* **2001**, *121*, 1569.
[3] a) P. Uznansky, M. Kryszewski and E.W. Thulstrup, *Eur. Polym. J.* **1991**, *27*, 41. b) R. C. Advincula, E. Fells, M.K. Park, *Chem. Mater.* **2001**, *13*, 2870.
[4] F. P. La Mantia, R. Scaffaro, P.L. Magagnini, M. Paci, *J. Appl. Polym. Sci.* **2000**, *77*, 3027.
[5] E. Passaglia, *PhD thesis*, University of Pisa, **1996**.
[6] N. Tirelli , S. Amabile, C. Cellai, A. Pucci, L. Regoli, G. Ruggeri, F. Ciardelli, *Macromolecules* **2001**, *34*, 2129.
[7] Y. Dirix, T. Trevoort, C. Bastiaansen, *Macromolecules* **1997**, *30*, 2175.
[8] A. Pucci, G. Ruggeri, L. Moretto, S. Bronco, *Polym. Adv. Technol.* **2002**, *13*, 737.
[9] M. Aglietto, R. Bertani, G. Ruggeri, F. Ciardelli, *Makromol. Chem.* **1992**, *193*, 179.
[10] Z.S. Fodor, M. Iring , F. Tudos, T. Kelen, J. Polym. Sci. Polym. Chem. Ed. **1984**, *22*, 2539.
[11] T. Otsu, O. Ito, N. Toyoda, S. Mori, *Makromol. Rapid Commun.* **1981**, *2*, 725.
[12] A. Pucci, L. Moretto, G. Ruggeri, F. Ciardelli, *e-Polymers* **2002**, *015*, http://www.e-polymers.org.
[13] F. Ciardelli, C. Cellai, A. Pucci, L. Regoli, G. Ruggeri, N. Tirelli, C. Cardelli, *Polym. Adv. Technol.* **2001**, *12*, 223.
[14] D.S. Abrams, J.M. Prausnitz, *AIChE J.* **1975**, *21*, 116.
[15] C. Cardelli, G. Conti, P. Gianni, R. Porta, *J. Therm. Anal. Cal.* **2000**, *62*, 1.
[16] L. Sharma, T. Rimura, H. Matsuda, *Polym. Adv. Technol.* **2002**, *13*, 450.
[17] A. Montali, A.R.A. Palmans, M. Eglin, C. Weder, P. Smith, W. Trabesinger, A. Renn, B. Hecht, U.P. Wild, *Macromol. Symp.* **2000**, *154*, 105.

Macromol. Symp. **2003**, *202*, 97–115

Quantitative Analysis of Functional Groups in HDPE Powder by DRIFT Spectroscopy

Enikő Földes,[1]* *Zoltán Szabó,*[2] *Ákos Janecska,*[2] *Gábor Nagy,*[2] *Béla Pukánszky*[1,3]

[1]Research Laboratory of Materials and Environmental Chemistry, Chemical Research Center, Hungarian Academy of Sciences, 1525 Budapest, P.O. Box 17, Hungary
Email: efoldes@chemres.hu
[2]Tisza Chemical Works (TVK), H-3581 Tiszaújváros, P.O. Box 20, Hungary
[3]Department of Plastics and Rubber Technology, Budapest University of Technology and Economics, H-1521 Budapest, P.O. Box 91, Hungary

Summary: FT-IR spectra of ethylene homopolymers and ethylene/1-hexene copolymers polymerized under different conditions were studied by transmission and diffuse reflection (DRIFT) spectroscopy. The absorbance spectra of film samples were compared with the DRIFT spectra of powders ground from the films. For determining the concentration of the methyl and unsaturated (vinyl, vinylidene and *trans*-vinylene) groups of polyethylene powders the DRIFT spectra were calibrated by comparing the IR intensities of the corresponding bands measured by the two methods. The results proved that the effect of differences in scattering of the polymer powder originating from the irregularity of the top surface, as well as the size and shape of the particles can be eliminated by the use of a proper internal standard. Linear correlation was established between the logarithms of the normalized intensities measured in absorbance and Kubelka-Munk units. In the case of polyethylene the selection of the internal reference band affects significantly the accuracy of the calibration due to the difference in the refractive indices of the crystalline and amorphous phases.

Keywords: diffuse reflection; DRIFT; FT-IR; polyethylene; transmission

Introduction

During the processing of polyethylene (PE) chemical reactions take place as an effect of heat, shear and oxygen resulting in some changes of the structure and the properties of the polymer. It was established that all the changes in the various characteristics of the polymer are related to each other.[1,2] A systematic study of the effects of multiple extrusion revealed that the mechanism of the chemical reactions taking place in the first extrusion of PE differs from that in the subsequent processing steps.[3] The chemical structure of the polymer was analyzed by

 DOI: 10.1002/masy.200351210

Fourier Transform Infrared (FT-IR) Spectroscopy. For the determination of the number of the functional groups (methyl and unsaturated) in the nascent polymer powder Diffuse Reflection FT-IR (DRIFT) Spectroscopy was introduced. Initially the DRIFT spectra were evaluated by the method used for the absorbance spectra. The results were handled with care, as the necessity of calibration of the DRIFT spectra was obvious, because of the essential differences between the Kubelka-Munk theory applied to describe diffuse reflection and the Beer's law used in transmission spectroscopy.

According to the Kubelka-Munk theory the reflection characteristics are related to the ratio of the absorption to scattering coefficients:[4,5]

$$F(R_\infty) = \frac{(1-R_\infty)^2}{2R_\infty} = \frac{K}{S} = \frac{2.303\varepsilon C}{S} \qquad (1)$$

where R_∞ is the reflectance from a specimen of "infinite" thickness, i.e. a specimen sufficiently thick so that a further increase of the thickness does not change the reflectance; K is the absorption coefficient (twice the Beer's law absorption coefficient); S is the scattering coefficient of the sample; ε is the absorptivity; and C is the analyte concentration. Experiments showed that for finely ground powders 2 mm deep sample cups fulfil practically the criteria of infinite sample thickness.[6]

The scattering coefficient depends on particle size, shape, refractive index, and wavelength. If S is constant, then $F(R_\infty)$ (Kubelka-Munk units) should change linearly with concentration. Such situation is rarely achieved, but with a careful choice of the experimental conditions and calibration, DRIFT can give quantitative information over a limited range.[4]

The effects of sampling technique and measuring conditions on the reliability of quantitative analysis by DRIFT spectroscopy were studied on inorganic materials. It was established that the packing of the powder and the morphology of the top surface of the sample are two important parameters affecting the reproducibility of the diffuse reflectance measurement.[7] The optical or geometrical irregularities at the surface affect the overall reflectance and transmittance more than an identical perturbation in the bulk or at the bottom of the sample. For the quantitative analysis Boroumand et al.[7] suggested the normalization of the DRIFT spectra by using a reference, and emphasized the importance of the choice of this reference. Studying the IR spectra of quartz and calcite in spectroscopic grade KBr Krivácsy and Hlavay[6,8] showed that the relative standard deviation (RSD) in the DRIFT measurements is high, and depends on the intensities. The smaller the Kubelka-Munk values are the greater the RSD is. The reliability of the quantitative analysis can be improved by using the same reference for the calibration standards and the investigated samples. They found that the same level of reliability as in conventional pressed pellet transmission technique can be achieved by multiple calibration followed by reference reflectance correction.[9]

In macromolecular science the DRIFT technique was applied successfully to study the chemical modification of fibers cut into small pieces. Gulyás et al.[10] investigated carbon fibers subjected to electrochemical oxidation under a wide variety of conditions. Correlation was established between the concentration of certain functional groups on the surface identified by DRIFT and the strength of adhesion between the fiber and an epoxy matrix. Takács et al.[11-13] studied the changes in the chemical structure caused by alkali treatment and irradiation of cotton-cellulose fibers by DRIFT spectroscopy. They used internal standards for normalization of the characteristic bands. The results obtained from the DRIFT analysis correlate well with those of X-ray measurements.

The literature on the quantitative analysis of polyethylene (PE) powder by IR spectroscopy is extremely limited. Earlier Baker at al.[14] studied the chain branching of PE powder by using potassium bromide disc technique. They had to overcome several difficulties - like particles larger than 100 μm led to poor quality discs and inferior results - before obtaining results with a precision of 5-10 % and an accuracy of about 15 %. The DRIFT technique provides a relatively simple sampling method, even if several considerations have to be taken into account in the evaluation of the spectra. Hrebičík et al.[15] investigated the time dependent chemical changes of an ultra high molecular weight polyethylene (UHMWPE) besides rice and oats irradiated by gamma radiation. They used the DRIFT technique and discussed the changes observed in the carbonyl stretching region. The repeatability of the spectra caused the critical problem for them. The change of the carbonyl absorption of UHMWPE was followed by relating the log(1/R) value at 1739 to that at 1717 cm^{-1}, but no attempt was made for the complete quantitative evaluation of the spectra.

The aim of the present work was the development of a method for the quantitative analysis of the methyl and the unsaturated groups of polyethylene powder by DRIFT technique. An assortment of polyethylenes prepared by different catalysts under varying conditions was investigated. The samples were compression molded to films followed by grinding to powder. The DRIFT spectra were calibrated by comparing the intensities of the corresponding bands of the functional groups of the polymer determined from transmission and diffuse reflection spectra of the same material. In the quantitative evaluation of the DRIFT spectra the following considerations were taken into account:

1. The size of the primary powder particles (300 μm-1 mm) and the samples ground for the experiments is much larger than the wavelength of the analytic IR light (2.5-25 μm), therefore little or no wavelength dependence of scattering can be assumed.[16]
2. The light scattering coefficient is affected by the difference in refractive indices of the crystalline and the amorphous phases of the polymer.
3. As the size and shape of the powder particles investigated vary, the regularity and reproducibility of the macroscopic surface layer cannot be ensured.
4. The selection of an adequate reference band for normalization must be crucial in the calibration due to the conditions described.

Experimental

Materials

30 ethylene homopolymers and ethylene/1-hexene copolymers were investigated with different concentrations of methyl and unsaturated groups. Phillips, Ziegler-Natta and metallocene catalysts were used for the polymerization. The molecular weight and the concentration of the functional groups varied in wide ranges. The number of the methyl groups changed from 0.8 to 14 per 1000 carbon atoms, the vinyl content from 0.03 to 1.9, the vinylidene from 0.01 to 0.23, and the *trans*-vinylene from 0 to 0.1. For checking the reliability of the measuring technique and calculations one of the samples was investigated twice at different times (two independent sets of experiments).

Preparation of Samples

Films of 200-300 μm were compression molded from the different polymer powders and granules at 180 °C. Samples with a diameter of 22 mm were cut for transmission measurements. For the diffuse reflection measurements the films were cooled by liquid nitrogen then ground to powder. The powder was not sieved, and no further sample preparation was applied.

In separate sets of experiments the effect of particle size on the DRIFT spectra was investigated by using nascent PE powder samples taken from the polymerization reactor. Before the measurement the polymer powder was sieved into 4 fractions: >1 mm, 1 mm-800 μm, 800-500 μm and <500 μm. It was concluded that the effect of differences in the intensities originating from variations in particle size and packing can be eliminated by normalization with an appropriate internal reference band. It has to be noted that some spectral artifacts can be observed in the case of relatively large particles due to Fresnel reflection.[5] Therefore smaller particles are favorable. However, the quality of the spectrum of larger particles can also be improved by careful rearrangement of the sample in the holder.

Methods

Mattson Galaxy 3020 (*Unicam*) type FT-IR spectrophotometer equipped with a DTGS detector and controlled by the Enhanced FIRST™ FTIR software was used for the experiments. The spectra were measured in the range of 4000-400 cm^{-1} at a resolution of 2 cm^{-1}. In transmission mode 16 scans were accumulated. The DRIFT measurements were carried out with a Baseline Diffuse Reflectance Accessory (*Spectra-Tech*) using 128 scans. Some model experiments revealed that higher number of scans does not improve essentially the signal/noise ratio of the spectra.

For the diffuse reflection measurements the powder was placed into the sample holder then packed loosely by pressing with a flat object. Five parallel measurements were carried out in the case of both techniques. The averages of the peak and integrated intensity values were determined from the transmission and DRIFT spectra.

Results and Discussion

The intensity of the bands in the infrared spectra of polyethylene is strongly affected by the method of investigation, as it can be seen from Figure 1 where the transmission and reflection spectra of the same Phillips type PE are compared. The ratio of the crystalline to amorphous band intensities is much higher in the DRIFT spectrum than in the absorbance spectrum (e.g. CH_2 rocking bands at 730 and 720 cm^{-1}). This indicates that the selection of the reference band for normalization plays a very important role in the quantitative evaluation of the spectra.

Selection of the Reference Band

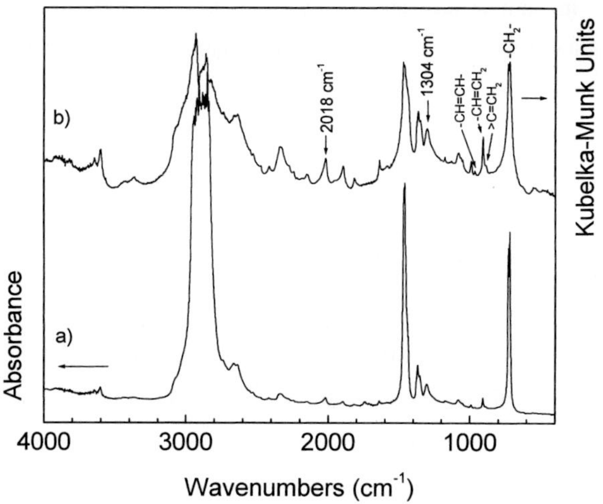

Fig. 1. FT-IR spectra of Phillips type ethylene/1-hexene copolymer measured in transmission (a) and diffuse reflection mode (b)

Two vibrational bands were selected for reference. The band at 2018 cm-1 is a combined overtone band, which originates from both the crystalline and the amorphous phases.[17] The band at 1304 cm^{-1} corresponds to the CH_2 wagging mode vibration in the disordered phase of

PE.[17-19] The ratio of the intensity at 2018 cm^{-1} (A_{2018}) to that at 1304 cm^{-1} (A_{1304}) is essentially larger in the DRIFT spectrum than in the transmission spectrum (Figure 1).

Correlation was established between the thickness of the film samples and the intensities of the selected reference bands measured in transmittance mode. Table 1 contains the calibration constants calculated by Equation (2) for both reference bands, and Figure 2 shows the relationship between A_{2018} and the film thickness.

$$A_{ref} = B \cdot l \qquad (2)$$

where A_{ref} is the absorbance of the reference band, l is the sample thickness measured (cm), and B is the proportionality factor.

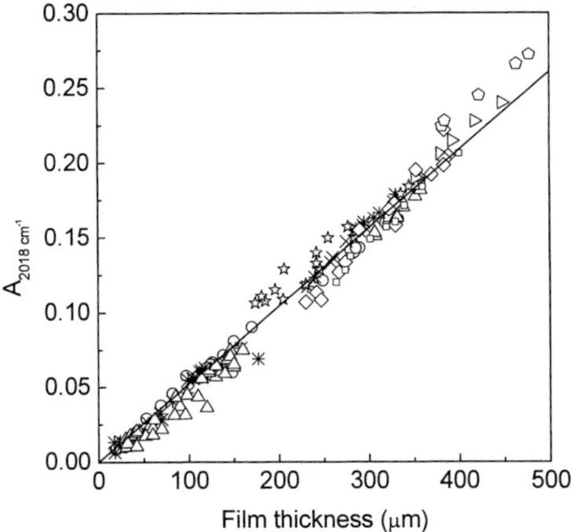

Fig. 2. Relationship between sample thickness and absorbance at 2018 cm^{-1} of different polyethylene types.

Table 1 reveals that the correlation of the film thickness with A_{2018} is closer than with A_{1304}. The reason is that the density of the polymer is affected by both the crystalline and disordered phases. However, the correlation coefficient for the amorphous band at 1304 cm^{-1} is also acceptable (>0.967).

Table 1. Calibration constants calculated for the correlation between the thickness of PE film and the FT-IR absorption intensities of two bands according to Equation (2).

Reference Band	B		Correlation Coefficient	Standard Deviation of Fitting	Number of Points
	Value	Standard Deviation			
A_{2018} [a)]	5.21	0.03	0.9916	0.0092	175
A_{1304} [b)]	6.24	0.06	0.9673	0.0192	174

a) Baseline between 2100-1980 cm^{-1}
b) Baseline between 1329-1275 cm^{-1}

Calibration for the Methyl Band

Willbourn[20] introduced a compensation method for the quantitative determination of the total number of chain branches in polyethylene from the methyl symmetrical deformation band around 1378 cm^{-1} in the absorbance spectra, which was later developed into a standard test method.[21] The position and the molar absorptivity of the methyl band at 1378 cm^{-1} varies with the length of the chain attached,[22-26] moreover the absorptivity depends also on the degree of crystallinity.[25] Nevertheless, the chemical changes of polyethylene during processing are well reflected in the modification of the intensity of this band.[1]

In the present work the intensity of the methyl symmetric deformation band around 1378 cm^{-1} (A_{1378}) was determined by the compensation method, i.e. subtracting the spectrum of a very high molecular weight linear polyethylene (M_n=130,000 g/mole) from that of the investigated sample (Figure 3). The absorbance spectra of the high molecular weight HDPE film were applied for compensation even in the case of DRIFT measurements, as it yielded difference spectra of better quality than the subtraction of DRIFT spectra. It is interesting to note that in their study of chain branching of polyethylene in KBr disc Baker at al.[14] also recommended a polymethylene film for the compensation instead of a disc.

The measured A_{1378} values were related to the intensities of the two reference bands (A_{2018} and A_{1304}), respectively, then the normalized intensities determined from the transmission and diffuse reflection spectra were plotted against each other. The correlation is non-linear, as it can be seen in Figure 4a where the A_{1378}/A_{1304} values are plotted. Linear correlation can be obtained in double logarithmic plots for both reference bands. Figure 4b illustrates the relationship in the case of normalization to A_{1304}.

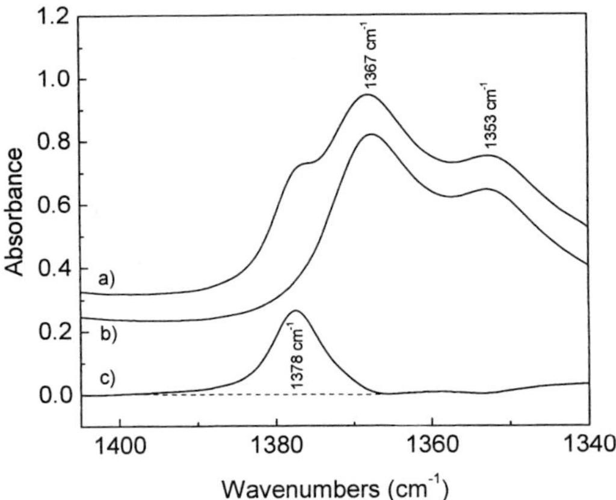

Fig. 3. Determination of the methyl absorption at 1378 cm^{-1} by compensation. Spectra of a) ethylene/1-hexene copolymer, b) reference HDPE, c) difference.

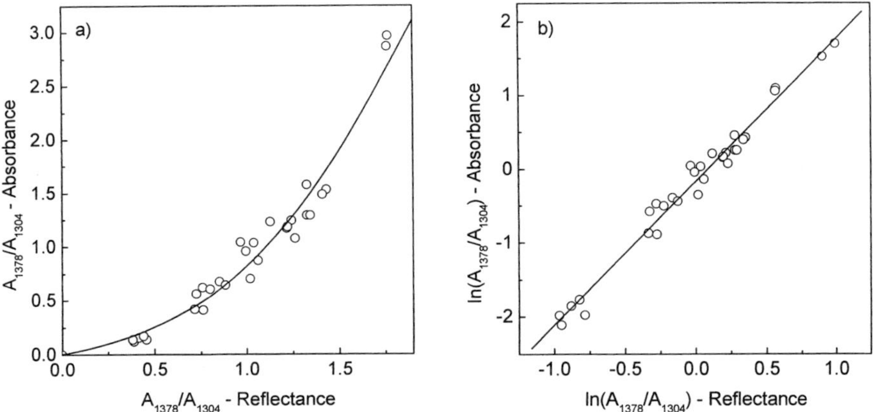

Fig. 4. Relationship between the normalized peak intensities of methyl group determined by DRIFT and transmission spectroscopy; a) linear plot, b) double logarithmic plot.

The relation between absorbance and Kubelka-Munk units can be expressed quantitatively by the following equation:

$$\ln\left(\frac{A_x}{A_{ref}}\right)_{abs} = C_1 + C_2 \cdot \ln\left(\frac{A_x}{A_{ref}}\right)_{refl} \qquad (3)$$

where A_x and A_{ref} are the peak intensities of the investigated and the reference bands, respectively. Subscript *abs* and *refl* refer to the absorbance and reflectance values, respectively. C_1 and C_2 are the calibration constants.

The normalization and correlation can be carried out also by the integrated intensities, which yield a relationship expressed by Equation (4):

$$\ln\left(\frac{I_x}{I_{ref}}\right)_{abs} = C_1' + C_2' \cdot \ln\left(\frac{I_x}{I_{ref}}\right)_{refl} \qquad (4)$$

where I_x and I_{ref} are the integrated intensities of the investigated and the reference bands, respectively. C_1' and C_2' are the calibration constants for the integrated intensities.

Table 2. Calibration constants calculated for the correlation between absorbance and Kubelka-Munk units according to Equations (3) and (4).

Investigated band	Reference band	C_1 and C_1'	C_2 and C_2'	Correlation Coefficient	Equation Nr.
A_{1378}	A_{1304}	-0.17 ± 0.03	1.94 ± 0.05	0.9894	3
A_{1378}	A_{2018}	1.50 ± 0.14	2.21 ± 0.16	0.9324	3
I_{1378}	I_{2018}	3.02 ± 0.35	2.12 ± 0.16	0.9243	4
A_{908}	A_{1304}	-0.85 ± 0.05	1.22 ± 0.05	0.9758	3
A_{908}	A_{2018}	0.15 ± 0.03	1.50 ± 0.03	0.9939	3
I_{908}	I_{2018}	0.36 ± 0.04	1.30 ± 0.03	0.9939	4
A_{890}	A_{2018}	-0.13 ± 0.08	1.13 ± 0.04	0.9799	3
I_{890}	I_{2018}	0.39 ± 0.16	1.18 ± 0.06	0.9704	4
A_{888}	A_{1304}	-1.09 ± 0.09	1.14 ± 0.06	0.9594	3
A_{888}	A_{2018}	-0.28 ± 0.08	1.12 ± 0.04	0.9816	3
A_{965}	A_{1304}	-1.18 ± 0.14	1.17 ± 0.06	0.9627	3
A_{965}	A_{2018}	0.06 ± 0.16	1.26 ± 0.06	0.9768	3
I_{965}	I_{2018}	0.31 ± 0.25	1.16 ± 0.06	0.9758	4

The calibration constants determined by Equations (3) and (4) for the methyl group intensities are given in Table 2. By comparing the correlation coefficients of the calculation we can conclude that the normalization to the band at 1304 cm^{-1} gives a better fit between the absorbance values and the Kubelka-Munk units than the use of the 2018 cm^{-1} reference band.

This result confirms that the deformation band around 1378 cm^{-1} originates intrinsically from the methyl groups in the amorphous phase of the polymer.[17]

For checking the reliability of Equation (3) the methyl concentrations of the samples were calculated from both the transmission and the reflection (transformed Kubelka-Munk units) spectra. For calibration ^{13}C NMR method was used. The concentrations are compared in Figure 5. The deviation of the data determined by the two methods is less than 25 %, which can be considered satisfactory. The scatter of the values originates from different sources. It is affected by the precision of spectrum compensation, the chemical heterogeneity of the polymers (RSD ~5 %), the sensitivity of the deformation band at 1378 cm^{-1} to the structure of the polymer chain, as well as the optical and geometrical irregularities of the polymer powders investigated (RSD ~10%).

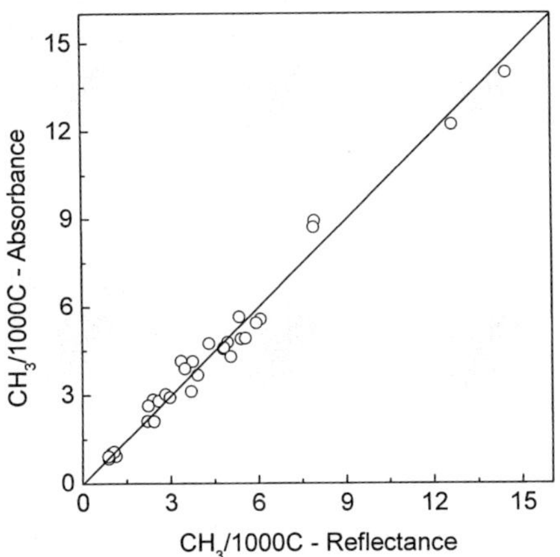

Fig. 5. Correlation between the methyl group concentrations derived from DRIFT and transmission spectra (normalization to the peak intensity of the reference band at 1304 cm^{-1}).

Calibration for the Vinyl Band

Except for some PE samples investigated, the bands of the vinyl and vinylidene groups overlap. For the determination of the concentrations of these unsaturated groups first the bands at 908 cm^{-1} [vibration of vinyl group[27]] and 890 cm^{-1} [vibrations of vinylidene group and butyl branches[28]] were separated by curve fitting (Figure 6).

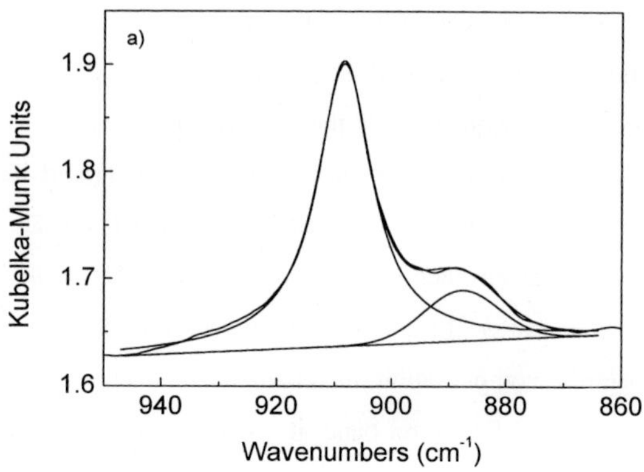

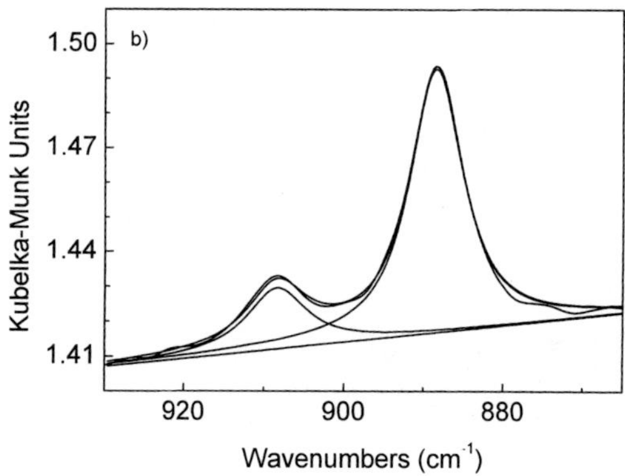

Fig. 6. Separation of the bands of vinyl and vinylidene groups by curve fitting. Spectra of a) Phillips type PE, b) metallocene PE.

In the next step the intensities at 908 cm^{-1} measured in transmission and diffuse reflection modes were normalized to the reference bands and plotted against each other. Similarly to the methyl vibration, non-linear correlation was obtained which could be linearized by double logarithmic representation. The calibration constants calculated by Equations (3) and (4) are summarized in Table 2. Comparing the correlation coefficients of the calculation we can conclude that the peak intensities and the integrated intensities normalized to the band at

2018 cm^{-1} yield similar accuracy and closer correlation than the use of the A_{1304} reference. The relationship between the A_{908}/A_{2018} relative intensities determined from the transmission and the DRIFT spectra is shown in Figure 7.

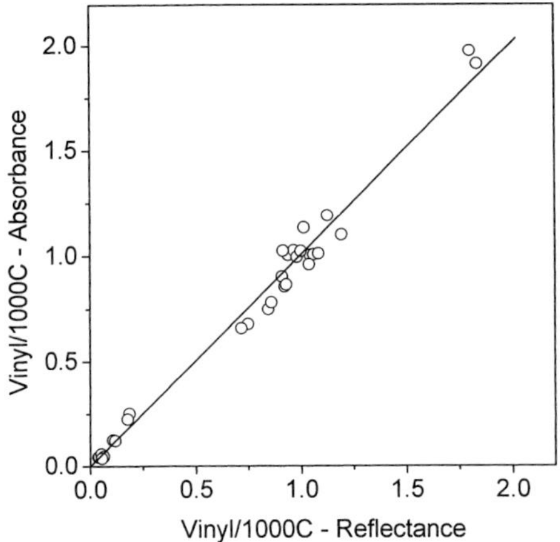

Fig. 7. Double logarithmic plot of the normalized peak intensities of vinyl group determined from DRIFT and transmission spectra.

The concentration of the vinyl group was calculated from both the absorbance values and the transformed Kubelka-Munk units normalized to the 2018 cm^{-1} reference band by Equations (5) and (6) using the absorptivity of 123 cm^2/mmol given by de Kock and Hol.[29] The sample thickness was substituted by Equation (2).

$$n_{vinyl} = \frac{0.593}{\rho}\left(\frac{A_{908}}{A_{2018}}\right)_{abs} \quad (5)$$

$$n_{vinyl} = \frac{1.5}{\rho}\left(\frac{I_{908}}{I_{2018}}\right)_{abs} \quad (6)$$

where n_{vinyl} is the number of vinyl groups per 1000 carbon atoms; and ρ is the density of the sample.

The deviation between the vinyl concentrations derived from the peak intensities and the integrated intensities, respectively, in the transmission and reflection spectra is less than 20 %. The most reliable calibration of the Kubelka-Munk units is provided by the I_{908}/I_{2018} integrated intensity ratios (Figure 8).

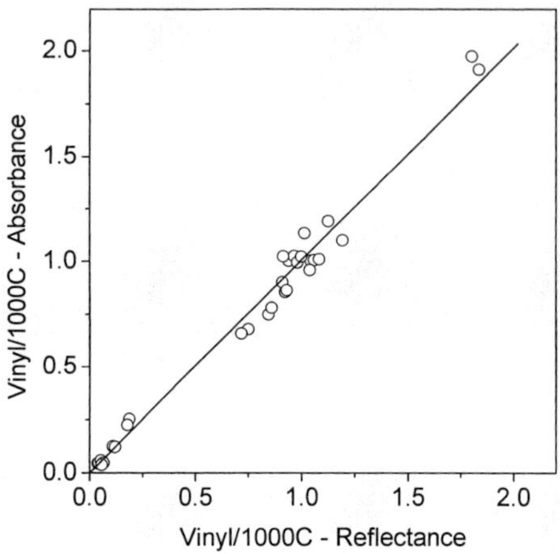

Fig. 8. Correlation between the vinyl group concentrations derived from DRIFT and transmission spectra (normalization to the integrated intensity of the reference band at 2018 cm^{-1}).

Calibration for the Vinylidene Band

The concentration of the vinylidene group can be obtained from the intensity of the band at 890 cm^{-1} (A_{890}), which is an overlap of the vinylidene band at 888 cm^{-1} and the 895 cm^{-1} band originating predominantly from butyl branches.[28,30] For the determination of the vinylidene concentration from the transmission spectrum of PE Lomonte[28] developed a calculation method. According to that the intensity of the butyl branch can be derived from the absorbance of the CH_3 group at 1378 cm^{-1} (A_{1378}). Considering the overlap of the two bands at 888 cm^{-1} the peak intensity of the vinylidene absorption (A_{888}) can be calculated by the following equation:

$$A_{888} = A_{890} - 0.85 \cdot \frac{A_{1378}}{28.9} \qquad (7)$$

After separation of the bands at 908 and 890 cm^{-1} according to Figure 6 the intensity of the vinylidene group was calculated by Equation (7), as the overlapping bands of the vinylidene group and the butyl branch cannot be separated reliably by curve fitting. In the evaluation of the DRIFT spectra the question arose whether the ratio of the methyl intensities at 1378 and 895 cm^{-1} are similar in the transmission and reflection spectra. For the resolution two calculation methods were applied:

1. The intensity of the band at 890 cm^{-1} measured in Kubelka-Munk units was calibrated according to the absorbance values, then A_{888} was calculated by Equation (7).
2. First the intensity of the 888 cm^{-1} band was derived from the reflection spectra by Equation (7), then the calibration was carried out for the Kubelka-Munk units.

Both reference bands (2018 cm^{-1} and 1304 cm^{-1}) were tried for the calibration. Linear relationship was obtained between the logarithms of the normalized values determined from the transmission and reflection spectra in all cases. Figure 9a illustrates the double logarithmic correlation of the A_{888}/A_{2018} normalized peak intensities (method 2). From the results given in Table 2 the following conclusions can be drawn:

1. The calibration with the reference band at 2018 cm^{-1} results in better correlation than the normalization to the 1304 cm^{-1} band.
2. Calculation method 2 gives closer correlation between the absorbance values and the Kubelka-Munk units than method 1.
3. The integrated intensities do not provide higher correlation coefficient than the peak intensities.

Figure 9b reveals that the A_{888}/A_{2018} normalized intensities determined from the transmission and reflection spectra by method 2 relate close to linearly in the concentration range investigated. The correlation can be described by Equation (8):

$$\left(\frac{A_{888}}{A_{2018}}\right)_{abs} = C'' \cdot \left(\frac{A_{888}}{A_{2018}}\right)_{refl} \quad (8)$$

where the value of the proportionality factor C'' is 0.61 ± 0.02, and the correlation coefficient of the fitting $R = 0.9754$.

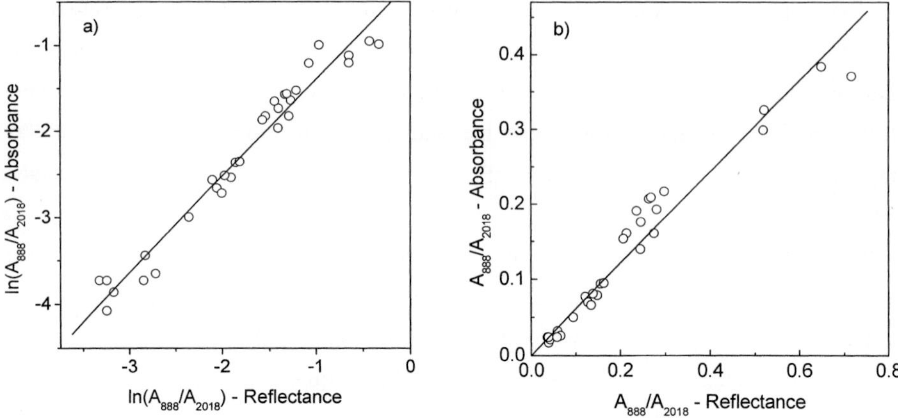

Fig. 9. Correlation between the normalized peak intensities of vinylidene group determined by DRIFT and trasmission spectroscopy; a) double logarithmic plot, b) linear plot.

The linear relationship between the absorbance and reflectance values shown in Figure 9b means, that the intensity of the vinylidene group is determined primarily by the absorption coefficient and affected less by the scattering coefficient of the sample.

The concentration of the vinylidene group was calculated from the transmission and the calibrated reflection spectra using the absorptivity given by de Kock and Hol.[29] Except for some samples the comparison reveled less than 20 % deviation between the concentrations determined by the two methods. The larger relative errors obtained for some samples originate mainly from the complex procedure used for the determination of the concentration of the vinylidene group.

Calibration for the Trans-vinylene Band

The *trans*-vinylene group of PE gives a vibration band at 965 cm^{-1}, which is well separated from the out-of-plane vibration bands of the vinyl group at 908 and 990 cm^{-1}.[27,29] The correlation between the normalized band intensities (reference bands at 2018 and 1304 cm^{-1}) determined from the transmission and reflection spectra is non-linear. Linear relationship can be obtained in double logarithmic representation also for this vibration. The calibration constants calculated are listed in Table 2, while Figure 10 shows the comparison of the A_{965}/A_{2018} normalized intensities derived from the DRIFT and transmission spectra. The correlation coefficient of the fit is larger in the case of the 2018 cm^{-1} reference band than for the normalization to the band at 1304 cm^{-1}. The peak intensities and integrated intensities yield correspondences of similar reliability.

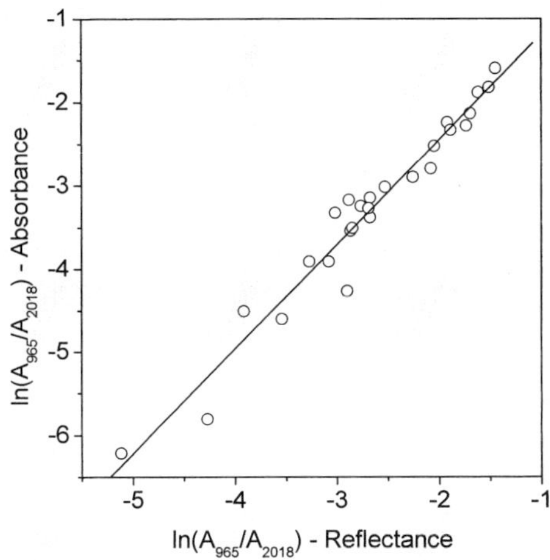

Fig. 10. Double logarithmic plot of the normalized peak intensities of *trans*-vinylene group determined from DRIFT and transmission spectra.

The concentration of the *trans*-vinylene group was calculated from the absorbance and calibrated reflectance values according to Reference 29. Except for the very low concentrations (<0.005 *trans*-vinylene/1000C) the deviation of the data determined by the two methods is less than 20 %.

Conclusions

Comparison of the transmission and reflection spectra of ethylene homopolymers and ethylene/1-hexene copolymers of different chemical structure revealed that the DRIFT technique can be applied successfully for the quantitative measurement of the functional (methyl and unsaturated) groups in polyethylene powder samples. The normalized intensities determined from the transmission and reflection spectra correlate linearly in double logarithmic plots. The relationship between the normalized intensities measured in Kubelka-Munk units by DRIFT technique (R_n) and in absorbance values by transmission method (A_n) can be described by the following empirical equation:

$$A_n = a \cdot R_n^b \tag{9}$$

where a and b are constants characteristic for the investigated band.

As the amorphous and the crystalline phases of the polymer have different refractive indices, the selection of the internal reference band used for normalization affects strongly the accuracy of the calibration of the reflection spectra. Two internal reference bands were selected for the calibration, one of them (at 1304 cm^{-1}) originating from the amorphous phase of the polymer and the other (at 2018 cm^{-1}) affected by both phases. For the methyl groups the normalization to the amorphous CH_2 vibration at 1304 cm^{-1} yields more reliable calibration. On the other hand, for the calculation of the concentrations of the unsaturated groups the combined overtone band at 2018 cm^{-1} can be used with higher accuracy as an internal standard.

The precision of the concentration values of the functional groups determined from the reflection spectra is around 20-25 % or better, which can be considered satisfactory when the difficulties in sampling and calculations are considered.

One of the most important conclusions of the present work is that the use of an appropriate internal standard for normalization of the reflection spectra can eliminate the effects of differences in scattering caused by the irregularity of the polymer powder and the variations in the investigated surface layer.

Acknowledgement

The authors are grateful for the financial support of TVK and the National Scientific Research Fund of Hungary (Grant No. OTKA T 037687). They also thank Gábor Keresztury for advicing and correcting the manuscript.

[1] E. Epacher, E. Fekete, M. Gahleiter, B. Pukánszky, *Polym. Degrad. Stab.* **1999**, *63*, 489.
[2] E. Epacher, E. Fekete, M. Gahleiter, B. Pukánszky, *Polym. Degrad. Stab.* **1999**, *63*, 499.
[3] E. Epacher, J. Tolvéth, K. Stoll, B. Pukánszky, *J. Appl. Polym. Sci.* **1999**, *74*, 1596.
[4] A. Garton, *"Infrared Spectroscopy of Polymer Blends, Composites and Surfaces"*, Hanser, Munich 1992, p. 107.
[5] J. P. Blitz, in *"Modern Techniques in Applied Molecular Spectroscopy"*, F. M. Mirabella, Ed., John Wiley & Sons, New York 1998, p. 185.
[6] Z. Krivácsy, J. Hlavay, *Talanta* **1994,** *41*, 1143.
[7] F. Boroumand, H. van den Bergh, J. E. Moser, **1994**, *Anal. Chem. 66*, 2260.
[8] Z. Krivácsy, J. Hlavay, *Magy. Kém. Folyórat* **1995**, *101*, 47.
[9] Z. Krivácsy, J. Hlavay, *Talanta 42*, **1995**, 613.
[10] J. Gulyás, E. Földes, A. Lázár, B. Pukánszky, *Composites: Part A* **2001**, *32*, 353.
[11] E. Takács, L. Wojnárovits, Cs. Földváry, P. Hargittai, J. Borsa, I. Sajó, *Res. Chem. Intermed.* **2001,** *27*, 837.
[12] J. Borsa, T. Tóth, E. Takács, P. Hargittai, *Radiat. Phys. Chem.* **2003**, in press.
[13] T. Tóth, J. Borsa, E. Takács, *Radiat. Phys. Chem.* **2003**, in press.
[14] C. Baker, P. David, W. F. Maddams, *Makromol. Chem.* **1979**, *180*, 975.
[15] M. Hrebičík, M. Suchánek, K. Volka, P. Novák, C. N. G. Scotter, *J. Molecular Structure*, **1995**, *347*, 485.
[16] R. W. Duerst, M. D. Duerst, W. L. Stebbings, in *"Modern Techniques in Applied Molecular Spectroscopy"*, F. M. Mirabella, Ed., John Wiley & Sons, New York 1998, p. 11.
[17] S. Krimm, **1960**, *Fortschr. Hochpolym.-Forsch*, *2*, 51.
[18] S. Krimm, *Phys. Rev.* **1954**, *94*, 1426.
[19] S. Krimm, C. Y. Liang, G. B. B. M. Sutherland, *J. Chem. Phys.* **1956**, *25*, 549.
[20] A. H. Willbourn, *J. Polym. Sci.* **1959**, *34*, 569.
[21] ASTM D 2238-68 (Reapproved 1979). *"Absorbance of polyethylene due to methyl groups at 1378 cm^{-1}"*.
[22] D. L. Wood, J. P. Luongo, *Modern Plastics*, **1961**, (*March*), 132.
[23] C. Baker, W. F. Maddams, G. S. Park, B. Robertson, *Makromol. Chem.* **1973**, *165*, 321.
[24] T. Usami, S. Takayama, *Polym. J.* **1984**, *16*, 731.
[25] A. Solti, D. O. Hummel, P. Simak, *Makromol. Chem., Macromol. Symp.* **1986**, *5*, 105.
[26] G. C. Pandey, *Process Control Quality* **1995**, *7*, 173.
[27] G. Socrates, *"Infrared Characteristic Group Frequencies"*, John Wiley & Sons, Chichester 1980.
[28] J. N. Lomonte, *Anal. Chem.* **1962**, *34*, 129.
[29] R. J. de Kock, P. A. H. M. Hol, *J. Polym. Sci. B, Polym. Lett.* **1964**, *2*, 339.
[30] J. P. Blitz, D. C. McFaddin, *J. Appl. Polym. Sci.* **1994**, *51*, 13.

Reactive Compatibilization of Recycled Low Density Polyethylene/Butadiene Rubber Blends During Dynamic Vulcanization

Alexander Fainleib, Olga Grigoryeva, Olga Starostenko, Inna Danilenko, Lubov Bardash*

Institute of Macromolecular Chemistry of the National Academy of Sciences of Ukraine, Kharkivske shose 48, 02160 Kyiv, Ukraine
E-mail: fainleib@i.kiev.ua

Summary: For reactive compatibilization of the recycled LDPE with butadiene rubber (BR) an equal quantity of few couples of reactive polyethylene copolymer/reactive polybutadiene (1/1) were introduced into the corresponding phases before the dynamic vulcanization. The LDPE/BR thermoplastic dynamic vulcanizates (TDVs) produced using the poly(ethylene-co-acrylic acid), PE-AA/polybutadiene terminated with isocyanate groups, PB-NCO compatibilizing couple with different ratio of functional groups have demonstrated the best mechanical properties and have been characterized by X-Ray analysis and DMTA measurements. For all of systems studied the increasing components compatibility due to the formation of the essential interface layer have been observed. The PB-NCO modifier participates in two processes: it is co-vulcanised with BR in rubber phase and reacts in the interface with the PE-AA dissolved in LDPE. The amorphous phase of LDPE is dissolved by rubber phase, i.e. the morphology with dual phase continuity is formed that provides an improvement of mechanical characteristics of material obtained. The best combination of mechanical characteristics was obtained for LDPE(PE-AA)/BR(PB-NCO), PB-NCO=7.5 wt.% per PB, COOH/NCO=1/1. The tensile strength and an elongation at break for these blends were 3.9 MPa and 353 % and for the basic non-compatibilized blend 3.2 MPa and 217 %, relatively.

Keywords: butadiene rubber; dynamic vulcanization; low density polyethylene; reactive compatibilization; recycling

Introduction

It is known that thermoplastic elastomers can be produced from polymer blends consisting of non-vulcanized fresh rubber and thermoplastic polymers such as polyolefins.[1-9] Their properties can highly be improved by exploiting the technology of dynamic or in-situ curing.[2-5,7,9] During dynamic vulcanization, carried out by intense mixing above the melt

 DOI: 10.1002/masy.200351211

temperature of the thermoplastic polymer, the rubber phase will be crosslinked (vulcanized) and finely dispersed (mean particle size in few microns) in the thermoplastic, which overtakes the role of the matrix.[4] The resulting thermoplastic dynamic vulcanizate (TDV) exhibits rubbery characteristics by maintaining the thermoplasticity of the matrix. As a consequence the TDV is melt (re)processable. A further benefit of this TDV compound is that it represents a high value-added product if its components are derived from waste sources ("upcycling"). Preliminary results showed that the TDV concept could be adopted for well-selected post-consumer goods. As far as a thermoplastic and rubber are usually incompatible the further improvement of TDV properties can be achieved by increase of components compatibility via an enhancement of interfacial adhesion.[10-13]

The reactive compatibilization can be realized by few ways.[10, 12] The method used in this work consisted of introduction of reactive polyethylene copolymer into thermoplastic phase and reactive polybutadiene rubber into rubber phase. Herewith, the functional groups of polymer additives used should be reactive to each other. During the intensive mixing of components at high temperature the chemical reaction occurs in the interface leading to increase of adhesion between thermoplastic and rubber phases. The fine morphology of such compatibilized TDVs provides a reinforcement of their mechanical properties.

Experimental

Materials

A fresh butadiene rubber (BR) (M_w = 21,000) was a trademark SKD-2 (Voronezhsintez-kauchuk, Russia), its Mooney viscosity, ML(1+4)130°C, was 46. As a polyolefin the post-consumer recycled low density polyethylene (LDPE) was selected. LDPE was produced from used greenhouse films of composition LDPE – 65-70%, linear low density polyethylene (LLDPE) – 12-17%, ethylene/vinyl acetate copolymer (EVA) – 12-15%, additives (kaolin, talc, silica) ≈ 500 ppm, UV stabilisers (amine, benzophenone) ≈ 2500 ppm, E-modulus – 180 MPa, tensile strength – 16 MPa, ultimate tensile elongation – 500%, melt flow index – 0.29 and 0.95 g/10 min at 190 and 230°C, respectively, under the load of 2.16 kg.

The reactive thermoplastics used were poly(ethylene-co-acrylic acid) (PE-AA), poly(ethylene-co-glycidyl methacrylate) (PE-GMA), poly(ethylene-g-maleic anhydride) (PE-g-MAN), poly(ethylene-co-vinyl acetate-co-acrylic acid) (PE-VA-AA), and reactive rubbers were polybutadiene, terminated with epoxy (PB-E), hydrazide, (PB-Hz), carboxyl, (PB-COOH), amine, (PB-NH_2) and isocyanate groups (PB-NCO). Thus, the reactive couples of functional groups can be given as follows: carboxyl – epoxy (1), carboxyl – isocyanate (2), epoxy – hydrazide (3), epoxy – amine (4), epoxy – isocyanate (5), anhydride – amine (6), anhydride – isocyanate (7) (Figure 1).

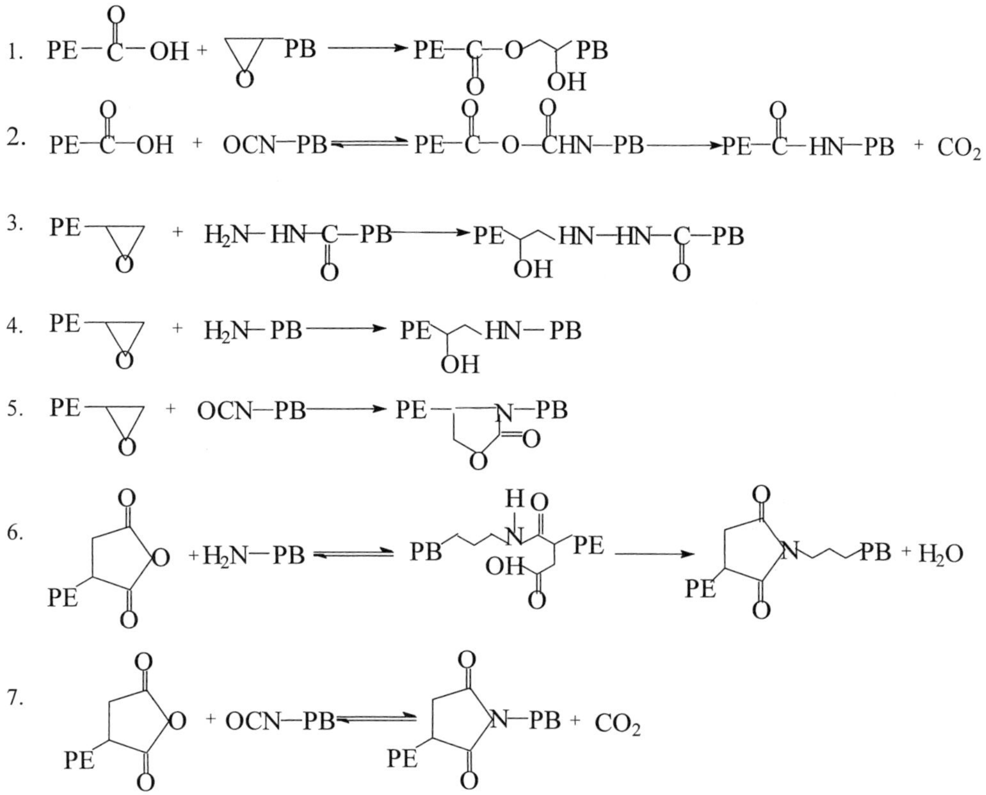

1. PE–C(=O)–OH + (epoxide)–PB ⟶ PE–C(=O)–O–CH₂–CH(OH)–PB
2. PE–C(=O)–OH + OCN–PB ⇌ PE–C(=O)–O–C(=O)HN–PB ⟶ PE–C(=O)–HN–PB + CO_2
3. PE–(epoxide) + H_2N–HN–C(=O)–PB ⟶ PE–CH(OH)–CH₂–HN–HN–C(=O)–PB
4. PE–(epoxide) + H_2N–PB ⟶ PE–CH(OH)–CH₂–HN–PB
5. PE–(epoxide) + OCN–PB ⟶ PE–(oxazolidinone)N–PB
6. PE–(succinic anhydride) + H_2N–PB ⇌ PB–NH–C(=O)–CH(PE)–CH₂–C(=O)OH ⟶ PE–(succinimide)N–PB + H_2O
7. PE–(succinic anhydride) + OCN–PB ⇌ PE–(succinimide)N–PB + CO_2

Fig. 1. Reactions used for compatibilization of polyethylene / rubber phases during TDVs preparing.

All reactive thermoplastics and rubbers have been used as received. The curatives used were (wt% per BR): Sulfur (3.0), bis-(2-benzothiazolyl)disulphide (1.0), zinc oxide (5.0), stearic acid (1.0).

Reactive Compatibilization and TDVs Preparation

The components of the blends LDPE/BR = 60/40 wt % were mixed in a twin-rotor mixer of the Brabender type at 180 °C, 100 rpm for 10 min. In all cases, the LDPE was melted first for 2 min before the addition of BR. For dynamic vulcanization, curatives as well as reactive couple chosen were introduced after 2 min of mixing BR with molten LDPE and mixed for further 6 min. Timing terms (10 min) were determined from the curve "torsion torque-time" as the time of the maximum value of torsion torque.

The resulting blends were pressed in the form of 1 mm-thick plates by molding at 150 °C under the pressure of 10 MPa followed by cooling under pressure at a rate of about 15 °C/min.

Dynamic Mechanical Thermal Analysis (DMTA)

DMTA measurements in tensile test mode were obtained on a viscoelastometer (Rheovibron type) with temperature scans from –100 to 170 °C at a frequency of 100 Hz and the heating rate of 2.0 °C/min. Dimensions of samples were 5.0 x 0.5 x 0.1 cm^3. The temperature corresponding to the maximum of the E" peak position was considered as the dynamic glass transition temperature (T_g).

Wide-angle X-ray Scattering (WAXS)

Wide angle X-ray scattering curves were recorded by X-ray diffractometers DRON-4-07. Nickel-filtered Cu-K_α radiation (monochromatized by Ni filter, radiation wavelength, $\alpha = 0.154$ nm) was produced by an IRIS-M7 generator at an operating voltage of 30 kV and a current of 30 mA. The scattering intensities were measured by a scintillation detector scanning in 0.2° steps over the range of angles of 5-40°. The degree of crystallinity <X> was calculated from the WAXS data by Matthews method.[14]

Mechanical Testing

Mechanical measurements were performed with an Instron 1122 testing machine at the ambient temperature at the elongation rate (the speed of upper cross-arm) 50 mm/min. The average data for 6-7 specimens were taken for consideration. The parameters such as tensile strength (σ_b) and elongation at break (ε_b) were determined.

Results and Discussion

Effect of Reactive Compatibilization on Properties of Reclaimed LDPE/BR TDVs

As mentioned above the TDVs are the blends of heterogeneous structure because of thermodynamic incompatibility of components, therefore an approach to development of TDVs by dynamic vulcanization is a multi-level problem. The selection of curatives in case of using post-consumer blend components is complicate problem, because it is necessary to avoid, as far as possible, the degradation of a pretreated ingredient in the utilization processes. In this work, the most effective agents, in terms of reaction time and final properties of vulcanizates, were used. The properties of TDVs and the blends with an uncured rubber phase were analyzed and given in Figure 2.

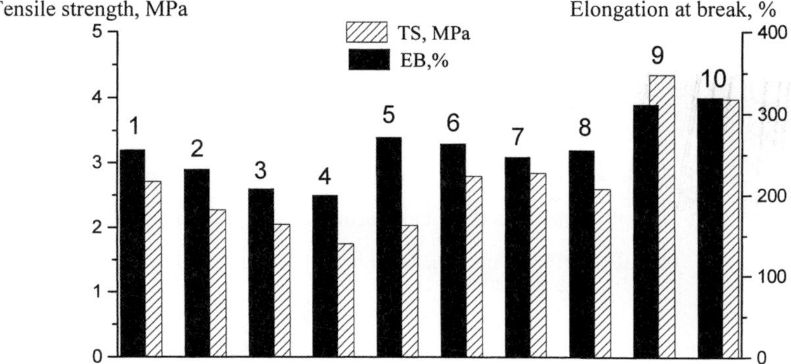

Fig. 2. Effect of reactive compatibilization on tensile strength and elongation at break of different LDPE/BR=60/40 wt.% TDVs (1) compatibilized by: PB-E/PE-AA (2); PB-Hz/PE-GMA (3); PB-COOH/PE-GMA (4); PB-NH_2/PE-GMA (5); PB-NH_2/PE-GMA (6); PB-NH_2/PE-g-MAN (7); PB-NCO/PE-g-MAN (8); PB-NCO/PE-AA (9); PB-NCO/PE-VA-AA (10). All PB-modifiers were used in amount of 7.5 wt.% per BR. The ratio of functional groups of PB and PE was 1/1.

One can see that the best mechanical characteristics were obtained for LDPE/BR TDVs compatibilized by couple PE-AA/PB-NCO (PB-NCO=7.5 wt.% per PB) with equivalent functionalities. The tensile strength and an elongation at break for these blends were 3.9 MPa and 353 % and for the basic non-compatibilized blend 3.2 MPa and 217 %, relatively, indicating a better compatibility between the rubber and polyolefin phases in system. This fact is explained by the effective reaction between PE-AA and PB-NCO in the interface leading to increase of interfacial adhesion and therefore compatibilization effect of PE-AA/PB-NCO. It was found that the value of tensile strength reaches a plateau with value of 4.2 MPa at 10 wt. % PB-NCO content and further increase of PB-NCO content up to 15 wt. % provides a reduction of elongation at break to the value approximately equal to the one of non-compatibilized LDPE/BR TDV. One can suppose that an excess of PB-NCO in the blend acts as a curing agent increasing a degree of vulcanization of the rubber phase through the reaction of NCO-groups excess with unsaturated bonds of the BR. It was found that the excess of PE-AA (PE-AA=1.5 e.e.w.) does not influence significantly on values of tensile strength and elongation at break. It can be explained by low reactivity of COOH-group towards BR chains.

Dynamic Mechanical Thermal Analysis

Two main transition regions (Figure 3a and b) evidence about two–phase morphology of the LDPE/BR TDVs studied: the first one from 50 to 110°C corresponds to α-relaxation of LDPE (T_g = 79°C); the second one is observed in the temperature interval –50°C to 25°C and it is a result of superposition of both the T_g of BR (T_g = –20°C) and the β-relaxation of LDPE (region from -75 to 50°C). A partial compatibility has been fixed between amorphous phases of LDPE and BR as a result of their chemical and structure affinities. So, it is supposed that one phase of TDV is crystalline one of LDPE and the other one is that formed by BR and LDPE amorphous parts.

A shift of transitions toward one another is observed in $E'' = f(T)$ dependencies that evidences of increase of component compatibility in the LDPE (PE-AA) / BR (PB-NCO) TDVs due to the formation of the essential interface layers. The PB-NCO modifier participates in two processes: it is co-vulcanized with BR in rubber phase and reacts in the interface with the PE-AA dissolved in LDPE. The amorphous phase of LDPE is dissolved by rubber phase. This leads to the transformation of the morphology of the basic blend (continuous LDPE phase and dispersed rubber phase) to the morphology with dual phase continuity that provides an improvement of mechanical characteristics of TDVs obtained. The glass transition temperatures for different TDVs produced are summarized in Table 1.

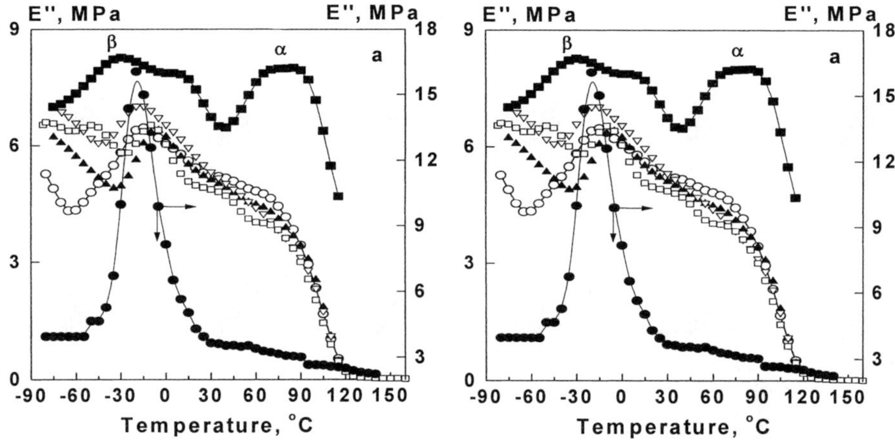

Fig. 3. Temperature dependence of the loss modulus, E´´, (a) and storage modulus, E´, (b) from DMTA measurements in tensile mode for: ■ - LDPE; ○ - LDPE/BR; ▲ - LDPE (PE-AA)/BR (PB-NCO = 1.5%); ▽ - LDPE (PE-AA)/BR (PB-NCO = 7.5%); □ - LDPE (PE-AA)/BR (PB-NCO = 10 %); ● - BR.

Table 1.Glass transition temperatures for different TDVs studied.

Composition, wt.%	T_g [a)], °C
LDPE	79
BR	-20
LDPE/BR[b)] = 60 / 40	-15, 35, 60
LDPE (PE-AA) / BR (PB-NCO = 1.5 %)	-6, 40, 57
LDPE (PE-AA) / BR (PB-NCO = 7.5 %)	-18, 56, 73
LDPE (PE-AA) / BR (PB-NCO = 10.0 %)	-47, -9, 37, 70

[a)] T_g values taken from the E'' peaks.
[b)] LDPE/BR = 60 / 40 is constant for all samples studied.

We consider that the LDPE (PE-AA) / BR (PB-NCO) TDVs studied can be considered as Interpenetrating Polymer Networks (IPNs).[15]

Wide-angle X-ray Scattering

All WAXS diffractograms of the modified LDPE (PE-AA) / BR (PB-NCO) TDVs (Figure 4) show two sharp peaks located at scattering angles of 21.5 and 23.8° (characteristic for orthorombic crystal cell of LDPE) and the diffuse maximum located at 19.5° which corresponds to LDPE amorphous phase scattering maximum. Note that three sharp peaks in the region from 31 to 36° (observed in diffractogram of the cured BR) are not characteristic for amorphous BR and can be attributed to low molecular weight additives used in the process of curing.

The value of degree of crystallinity represents the overall crystallinity of blend material and can be compared with the theoretical ones calculated in an assumption of retaining by LDPE component its original value of crystallinity ($x = 30.7\%$). The results of the calculations are 30.7 multipled by 0.6 (60 wt.% of LDPE) is equal to 18.4 %, Δx values in Table 2 show the difference between the experimental crystallinity value and the corresponding theoretical ones. Comparison of the figures allows concluding that the only consequence of BR introducing is the changing of the relative intensities of crystalline LDPE peaks. Blending LDPE with BR leads to increase of crystallinity degree by 5.1%. The modifier (PE-AA/PB-NCO) introduction leads to lowering the $\Delta\, \boldsymbol{x}$ difference to 0.5% (1.5% of modifier), -0.8% (7.5% of modifier). Further increasing the modifier content to 10% leads to restoring the crystallinity values to the level characteristic of unmodified LDPE/BR TDV.

Table 2. Degree of crystallinity for TDVs produced.

Compositions	x [a], %	Δ x [b], %
BR	0	-
LDPE	27.5	-
LDPE / BR (unmodified)	19.3	5.1
LDPE (PE-AA)/BR (PB-NCO = 1.5%)	18.5	0.5
LDPE (PE-AA)/BR (PB-NCO = 7.5%)	18.3	-0.8
LDPE (PE-AA)/BR (PB-NCO = 10 %)	19.3	4.7

[a] Experimental degree of crystallinity.
[b] Difference between experimental and theoretical degree of crystallinity.

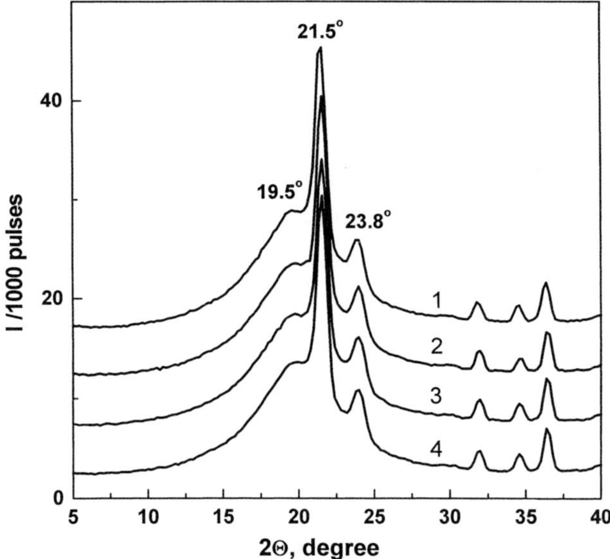

Fig. 4. WAXS curves for LDPE / BR (1); LDPE (PE-AA)/BR (PB-NCO= 1.5%) (2); LDPE (PE-AA)/BR (PB-NCO= 7.5%) (3); LDPE (PE-AA)/BR (PB-NCO= 10 %) (4). Beginning from the second curve from the bottom, each next curve was shifted upwards by 5 digits.

The reason for the crystallinity values increasing for unmodified LDPE / BR TDVs can be the influence of the rubber component on the LDPE component crystallization conditions, i.e. on phase separation between amorphous and crystalline phases. In such a case a lower value of difference between the experimental and the corresponding theoretical crystallinity values can be attributed to the increase of the blend components compatibility. Thus, the TDVs modified by PE-AA/PB-NCO are characterized by higher compatibility of LDPE and BR. Optimal content of the PE-AA/PB-NCO modifier corresponds to 7.5% PB-NCO per BR.

Conclusions

A partial compatibility has been fixed between amorphous phases of LDPE and BR as a result of their chemical and structure affinities. It is supposed that one phase of TDV is crystalline one of LDPE and the other one is that formed by BR and LDPE amorphous parts.

A shift of T_g of the components toward one another evidences of an increase of component compatibility in the LDPE (PE-AA) / BR (PB-NCO) TDVs due to the formation of the essential interface layers. The PB-NCO modifier participates in two processes: it is co-vulcanized with BR in rubber phase and reacts in the interface with the PE-AA dissolved in LDPE.

The amorphous phase of LDPE is dissolved by rubber phase. This leads to the transformation of the morphology of the basic blend (continuous LDPE phase and dispersed rubber phase) to the morphology with a dual phase continuity that provides an improvement of mechanical characteristics of material obtained. LDPE (PE-AA) / BR (PB-NCO) TDVs studied can be considered as IPNs.

The best combination of mechanical characteristics was obtained for LDPE/BR TDVs compatibilized by couple PE-AA/PB-NCO (PB-NCO=7.5 wt.% per PB, COOH/NCO=1/1). The tensile strength and an elongation at break for these blends were 3.9 MPa and 353 % and for the basic non-compatibilized blend 3.2 MPa and 217 %, relatively.

Acknowledgements

The authors are thankful to the European Union (Inco-Copernicus project - Contract No.: ICA2-CT-2001-10003) for the financial support of this project.

[1] B. Adhikari, D. De, S. Maiti, *Prog. Polym. Sci.* **2000**, *25*, 909.
[2] J. George, K. T. Varughese, S. Thomas, *Polymer* **2000**, *41*, 1507.
[3] P. Nevatia, T. S. Banerjee, B. Dutta, A. Jha, A. K. Naskar, A. K. Bhowmick, *J. Appl. Polym. Sci.* **2002,** *83*, 2035.
[4] J. Karger-Kocsis, in: *"Polymer Blends and Alloys"*, G. O. Shonaike, G. P. Simon, Eds., Marcel Dekker, New York 1999, p.125.
[5] J. George, K. Ramamurthy, K. T. Varughese, S. Thomas, *J. Appl. Polym. Sci.* **2000,** *38*, 1104.
[6] S. N. Bhattacharya, I. Sbarski, *Plast., Rubb. & Comp. Proc. Appl.* **1998**, *27*, 317.
[7] Y. Yang, T. Chiba, H. Saito, T. Inoue, *Polymer* **1998**, *39*, 3365.
[8] R. J. Spontak, N. P. Patel, *Current Opin. in Coll. & Interf. Sci.* **2000**, *5*, 334.
[9] S. Abdou-Sabet, R. C. Ruydak, C.P. Rader, *Rubber Chem. Technol.* **1996**, *69*, 476.
[10] H.-J. Radusch, B. Corley, L. H. Hai, *Proceedings of 14th Bratislava Intern. Conf. on Modified Polymers* **2000**, p. 27.
[11] B. Corley, H.-J. Radusch, *J. Macromol. Sci.-Phys.* **1998**, *B37*, 265.
[12] C. A. Orr, J. J. Cernohous, P. Guegan, A. Hirao, H. K. Jeon, C. W. Macosko, *Polymer* **2001**, *42*, 8171.
[13] T. Bray, S. Damiris, A. Grace, G. Moad, M. O'Shea, E. Rizzardo, G.V. Diepen, *Macromol. Symp.* **1998**, *129*, 109.
[14] J. L. Matthews, H. S. Peiser, R. B. Richards, *Acta Crystallogr.* **1949**, *2* , 85.
[15] Y. S. Lipatov, L. M. Sergeeva, *Interpenetrating Polymer Networks*, Naukova Dumka, Kiev, **1979**.

Creep Relaxation and Stress Relaxation of PS-HI/SEBS Blends

Vesna Rek,[1] *Tamara Holjevac Grgurić,**[1] *Želimir Jelčić*[2]

[1]Faculty of Chemical Engineering and Technology, University of Zagreb, Marulićev trg19, Zagreb, Croatia
[2]DIOKI d.o.o., Žitnjak bb., Zagreb, Croatia

Summary: For the selection of the polymer materials and polymer blends for various fields of applications the stability of material under constant deformation and constant load are very important. In this paper, the copolymers high-impact polystyrene, PS-HI, styrene-ethylene/buthylene-styrene block copolymer, SEBS, and their blends PS-HI/SEBS were investigated. The investigations were done by DMA analysis. The secondary viscoelastic functions, creep, creep modulus, stress and flexural relaxation modulus were investigated in creep and stress relaxation experiment at temperatures 25, 35, 45, 55 and 65 °C during 1 h. The master curves were created by time-temperature correspondence principle, TTC. The correlation of the secondary viscoelastic functions with time, temperature and content of the hard, PS, phase was discussed.

Keywords: creep; dynamic-mechanical behavior; high-impact polystyrene; master curve; PS-HI/SEBS blends; stress relaxation; styrene-ethylene/butylene-styrene block copolymer; time-temperature superposition

Introduction

Creep experiments have a great importance in the design of plastic products, because they reflect the load-bearing capacity of end-products. Chemical engineers have to care about creep failure in polymer structures, and they should be able to estimate the polymer material durability in order to prevent premature failures and to avoid large deformation.

Stress-relaxation experiments are also important for consideration of time-dependent performance of polymer materials. Today, metals are more and more replaced with plastics in structural components, so the study of the creep and stress relaxation behaviour in polymers and metals is important for further development of advanced long-life materials.

Time-temperature correspondence principle, TTC, gives the possibility to determine the rheological behaviour of polymer material from creep and stress relaxation data, over a much

 DOI: 10.1002/masy.200351212

longer experiment range of time on the base of measurements at a higher temperatures.[1-4] It was found that the distribution of relaxation times of a polymer is shifted by a change in temperature through the induced change in free volume.[5] Time-temperature correspondence principle holds for thermorheologicaly simple polymeric materials. There are many reasons why the TTC principle doesn´t valid for some polymer material, like the occurrence of more than one relaxation mechanisms with distinct temperature dependence.[6-10] For example, in inhomogeneous polymer blends each component has a different temperature-dependence rheology and TTC principle, usually fails. Although, many authors reports that the time-temperature correspondence principle holds for miscible[11-16] and also immiscible blends.[17-21]

Experimental

Materials

The investigations were done with high impact polystyrene, PS-HI 417, DIOKI, Zagreb, Croatia, with the content of polybutadiene, PB, 8 % weight, styrene-ethylene/buthylene-styrene block copolymer, SEBS, Kraton 1650, Kraton Polymers, Germany, with the content of polystyrene, PS, 29 % weight (with ethylene/buthylene, EB, as rubbery mid block) and with their blends. The compositions of the blends studied are shown in Table 1.

Table 1. The compositions of investigated samples.

SAMPLE	PS-HI/SEBS Weight %	SOFT PHASE PB weight%	SOFT PHASE EB weight%	HARD PHASE PS weight%
1	100/0	8.0	/	92.0
2	75/25	6.0	17.7	76.3
3	30/70	2.4	49.7	47.9
4	0/100	/	71.0	29.0

Specimens Preparation

The blends were prepared by using Haake Record 90 twin extruder with the intensive mixing profile, Haake TW 100, with the following temperatures in zone 150/200/200/150°C and the frequency of rotation 60 min^{-1}. The specimens were obtained by compression molding at 220°C; the mold temperature was 40°C.

Measurements

Creep relaxation and stress relaxation experiments were done by Dynamic Mechanical Analyser 983, TA Instruments in creep fatigue regime. In creep experiments, the specimens were stressed for 1 h at constant stress, and then allowed to relax for 1 h. The temperature was then increased in increments of 5°C followed by an equilibration period of 15 min before of the next displacement/recovery cycle. The creep experiments were done at 25, 35, 45, 55 and 65°C. Using the time-temperature superposition correspondence principle the master curves for reference temperature, $T_{ref.}$ = 25 °C were created, from short-therm experiments performed at 25,35,45,55 and 65 °C.

Stress relaxation experiments were performed under a constant strain in the temperature range from 25 °C – 65 °C, in 10 °C intervals during 1 h. The master curves for reference temperature, $T_{ref.}$ = 25 °C were constructed also from short-therm experiments performed at 25, 35, 45, 55 and 65 °C using the time-temperature correspondence principle, TTC.

Results and Discussion

The creep curves for all examined samples were determined under a constant load during 1 h at each temperature. Then, the load was removed and the recovery curves were determined. The example of obtained isothermal creep and recovery curves is shown in Fig.1. Viscoelastic data, creep and recovery are given in Table 2.

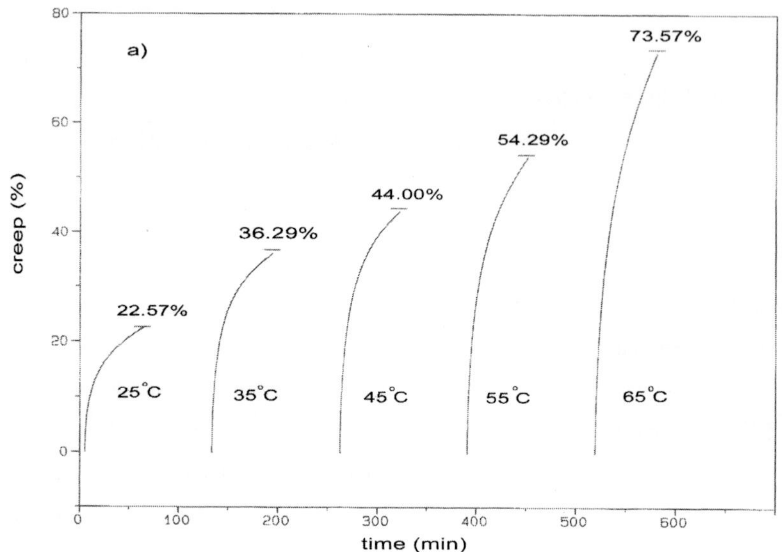
a)
creep (%)
time (min)
22.57%
36.29%
44.00%
54.29%
73.57%
25°C
35°C
45°C
55°C
65°C
80
60
40
20
0
0
100
200
300
400
500
600

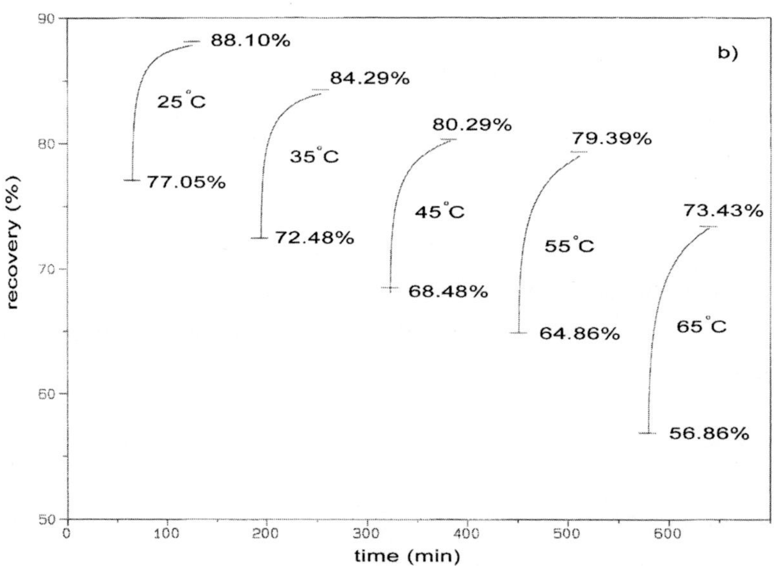
b)
recovery (%)
time (min)
88.10%
77.05%
25°C
84.29%
72.48%
35°C
80.29%
68.48%
45°C
79.39%
64.86%
55°C
73.43%
56.86%
65°C
90
80
70
60
50
0
100
200
300
400
500
600

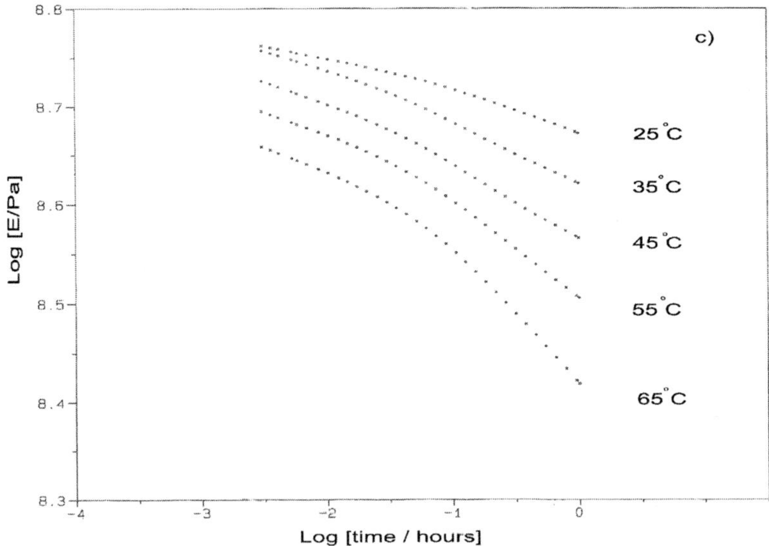

Fig. 1. The example of creep (a), recovery (b) and creep modulus vs. time (c) at different temperatures for PS-HI/SEBS 75:25 blend.

Table 2. The creep and recovery data for PS-HI, SEBS and PS-HI/SEBS blends.

SAMPLE	PS-HI/ SEBS weight%	HARD PHASE PS weight%	CREEP (%)					RECOVERY (%)				
			25°C	35°C	45°C	55°C	65°C	25°C	35°C	45°C	55°C	65°C
1	100	92.0	6.33	8.67	14.67	24.67	57.00	8.10	9.48	11.19	14.27	20.00
2	75:25	76.3	22.57	36.29	44.00	54.29	73.57	11.04	11.62	12.19	14.28	16.57
3	30:70	47.9	38.57	44.29	51.57	71.71	96.86	12.95	14.86	17.72	24.19	26.85
4	100	29.0	17.24	30.33	45.00	80.62	93.19	11.66	14.52	20.72	29.05	36.43

In all examined samples the creep and recovery values depend upon the time. On the basis of creep data the isothermal curves of the creep modulus E vs. time were obtained at 25, 35, 45, 55 and 65 °C an example is given in Fig.1. The creep values increase, while the creep modulus decrease with time under the constant temperature. Under the constant load polymers undergo molecular rearrangements in attempt to minimize local stress.[22-24] Those processes are faster at higher temperature. As a consequence of this the creep and recovery values increase as the temperature increases (Table 2., Fig. 1a-1b.), while creep modulus decreases with temperature (Fig. 1c.). This behaviour indicates the dependence of viscoelastic properties of copolymers and their blends on time and temperature.

High-impact polystyrene, PS-HI, has the lowest creep and recovery values at all examined temperatures (Table 2.). It was found that the styrene-ethylene/buthylene-styrene block copolymer, SEBS, shows higher creep and recovery than PS-HI. It is due to the higher content of the soft phase. Soft, ethylene/buthylene, EB, phase has negative glass transition temperature, so it is more flexible and liable to deformation than hard, polystyrene phase at examined creep temperatures.

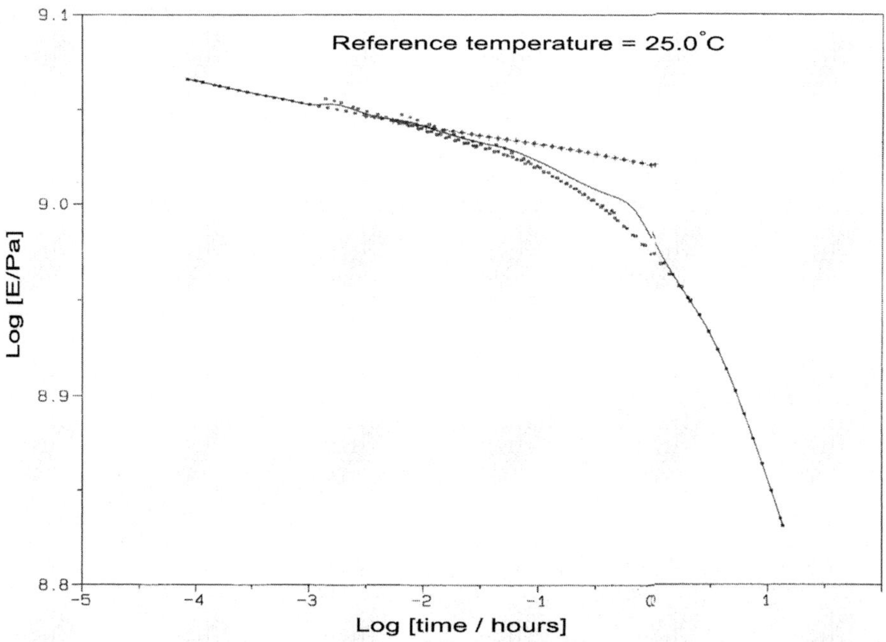

Fig. 2. Master curve at 25°C for PS-HI; creep relaxation measurement.

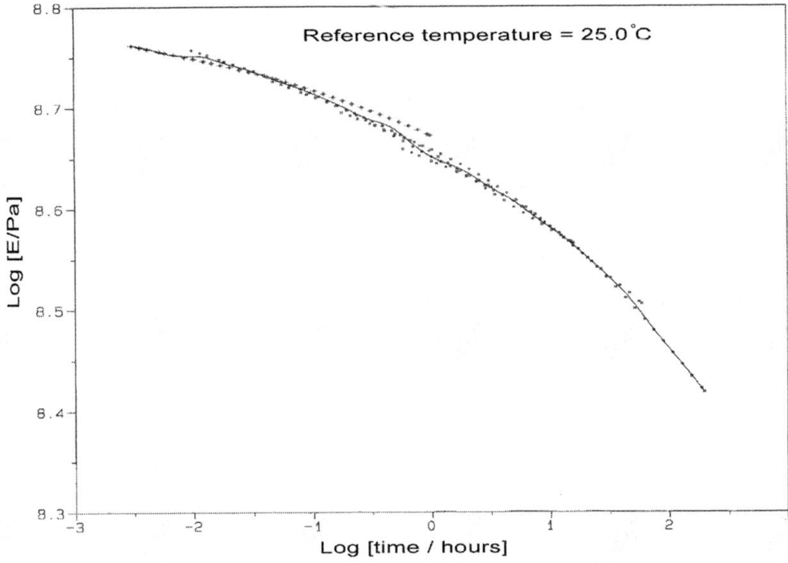

Fig. 3. Master curve at 25°C for PS-HI/SEBS 75:25 blend; creep relaxation measurement.

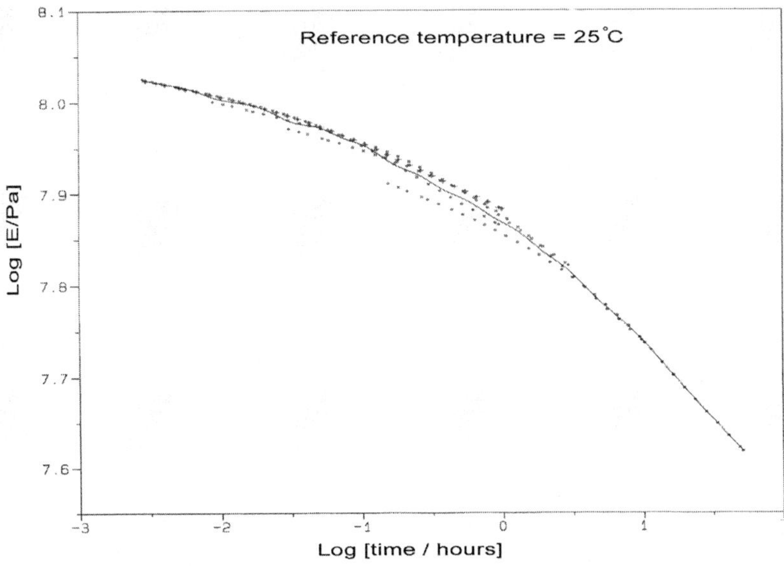

Fig. 4. Master curve at 25°C for PS-HI/SEBS 30:70 blend; creep relaxation measurement.

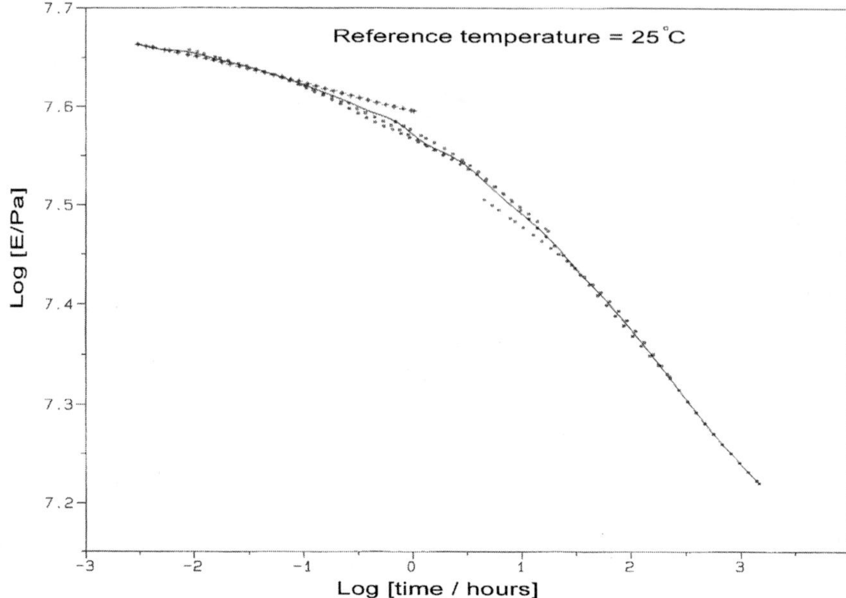

Fig. 5. Master curve at 25°C for SEBS; creep relaxation measurement.

As the content of the hard, polystyrene, PS, phase in PS-HI/SEBS blends decrease, the creep and recovery increase at the constant temperature. At all investigated temperatures the creep and recovery are more pronounced in the samples with lower content of the hard, PS, phase. The creep modulus decreases at all isothermal measurements as the content of hard, PS, phase decrease. The curves E` vs. time become more sheer with decreasing polystyrene content in samples. Such dependence of the creep modulus on the hard phase content is due to faster molecular rearrangements, changes of the conformations, in soft phase what is connected with negative Tg of soft phases, EB and PB.

The lower deformation at 25°C and 35°C of SEBS block copolymer in comparison with the blends may be assumed to the more influence of soft, polybutadiene phase in PS-HI/SEBS blends, which is more liable to deformation than EB phase due to its lower T_g.

By selecting as the reference the curve for 25 °C, and than shifting all other isothermal curves of the creep modulus vs. time obtained at 25,35,45,55 and 65 °C with respect to time, the master curves of creep modulus vs. time at references temperature are generated (Figs. 2-5).

The stress relaxation was determined under a constant strain at the same temperatures as the creep experiment during 1 h. The example of the stress relaxation results are shown on Fig. 6. The stress and recovery values are given in Table 3.

It was found that at the constant deformation the stress decreases with time in all investigated samples. The stress needs for holding the constant deformation decreases with increasing the temperatures for all samples (Table 3, Fig 6.). Such behaviours are connected with relaxation process in our samples which occur under the constant deformation and are more pronounced with time and at higher temperature.[25]

Table 3. The stress and recovery data for PS-HI, SEBS and PS-HI/SEBS blends.

SAMPLE	PS-HI/ SEBS weight%	HARD PHASE PS weight%	STRESS (%)					RECOVERY (%)				
			25°C	35°C	45°C	55°C	65°C	25°C	35°C	45°C	55°C	65°C
1	100	92.0	93.81	92.62	89.52	80.48	60.48	6.66	5.72	5.71	7.62	10.47
2	75:25	76.3	85.29	80.14	73.29	66.13	50.43	13.33	13.72	11.81	14.86	16.76
3	30:70	47.9	83.00	76.00	60.00	50.33	44.33	14.00	17.14	15.43	21.72	27.43
4	100	29.0	87.21	84.64	78.00	69.43	62.57	11.14	17.44	30.04	38.59	46.60

The stress of styrene-ethylene/buthylene-styrene block copolymer, SEBS is lower than the stress of high-impact polystyrene, PS-HI, but at the same time higher than stress of PS-HI/SEBS blends. It can be assumed that the flexibility of the soft PB phase (T_g –67°C) is higher than that of the soft, EB, phase (T_g –40°C). At the same time soft PB and EB phases have greater mobility, faster changes of the conformations, than hard PS phase.

Recovery of PS-HI and SEBS copolymers, as well as their blends, mainly increase with temperature at the same content of the hard, PS phase and also increases with decreasing of hard PS segments at the same temperature (Table 3).

The stiffness of the investigated systems decrease with time and temperature under a constant deformation, which is evident from isothermal curves flexural modulus, E\` vs. time (Fig. 6d). With decreasing the hard, PS, phase content those curves are shifted to lower values. From the isothermal curves of the flexural modulus vs. time at temperatures at 25,35,45,55 and 65 °C the master curves were created by selecting as the reference the curve for 25 °C and than shifting all other isothermal curves of the creep modulus vs. time with respect to time.

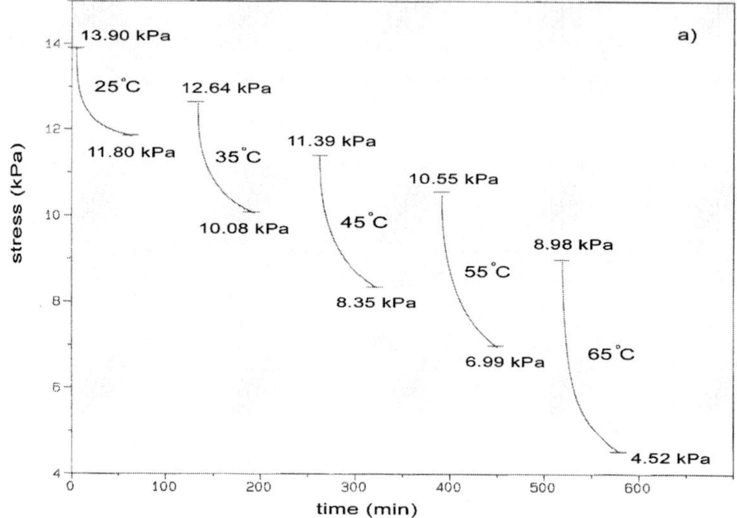

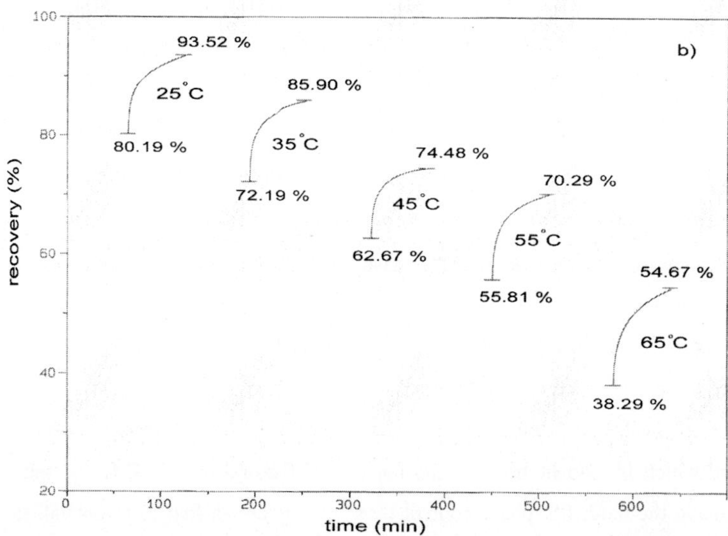

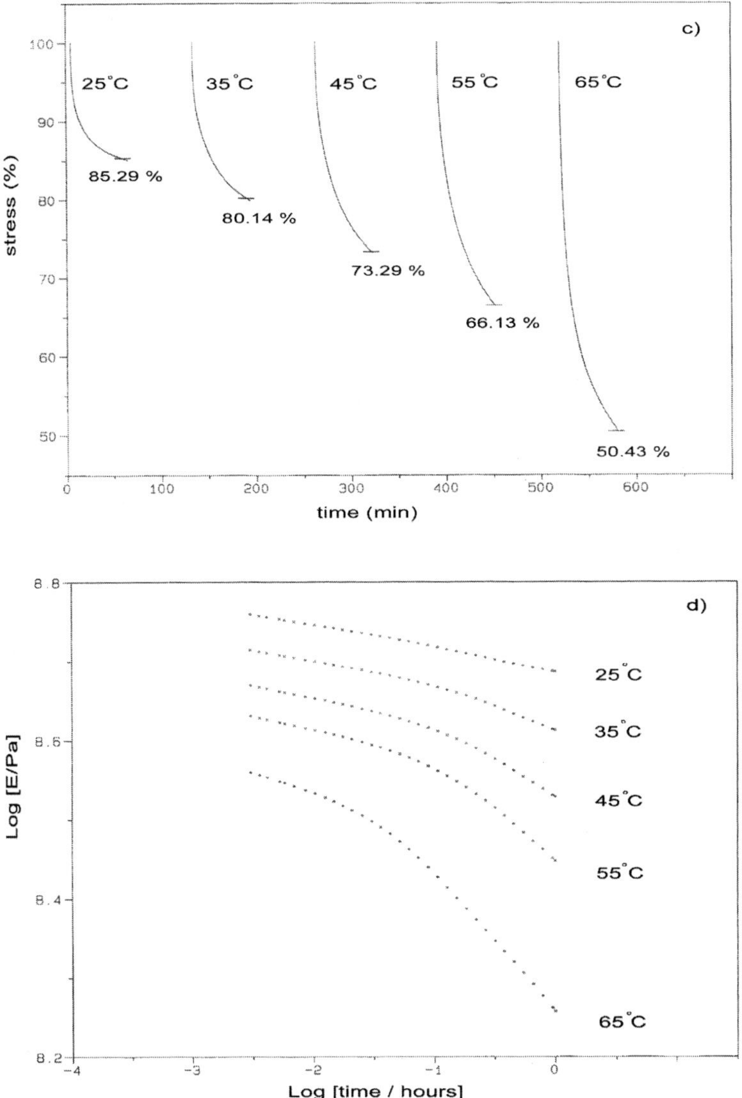

Fig. 6. The isothermal curves of stress (a), recovery (b), relative stress (c) and modulus (d) for sample PS-HI/SEBS 75:25.

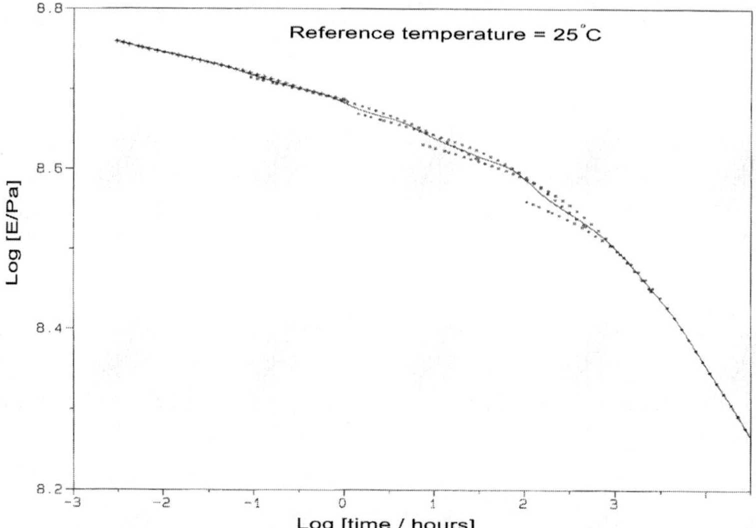

Fig. 7. Master curve at 25 °C for sample PS-HI; stress relaxation measurement.

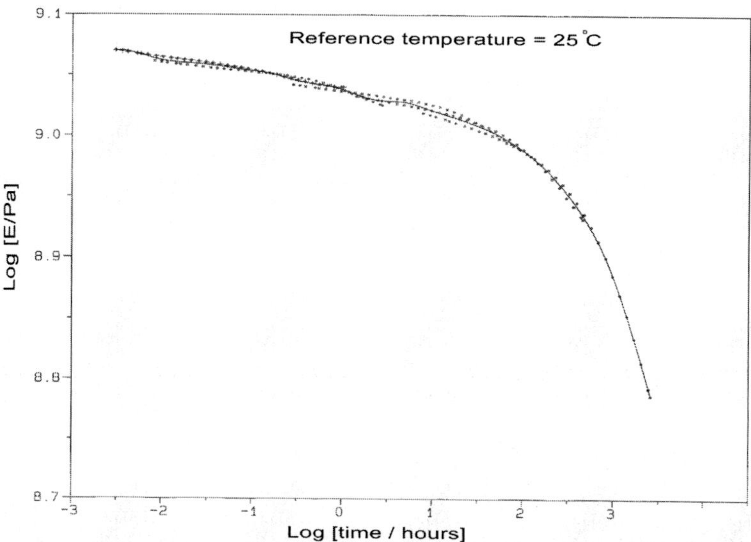

Fig. 8. Master curve at 25 °C for PS-HI/SEBS 75:25 blend; stress relaxation measurement.

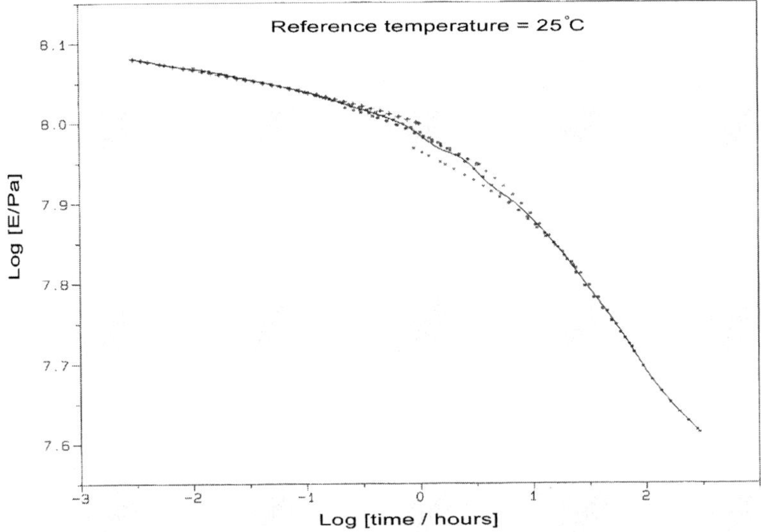

Fig. 9. Master curve at 25 °C for PS-HI/SEBS 30:70 blend; stress relaxation measurement.

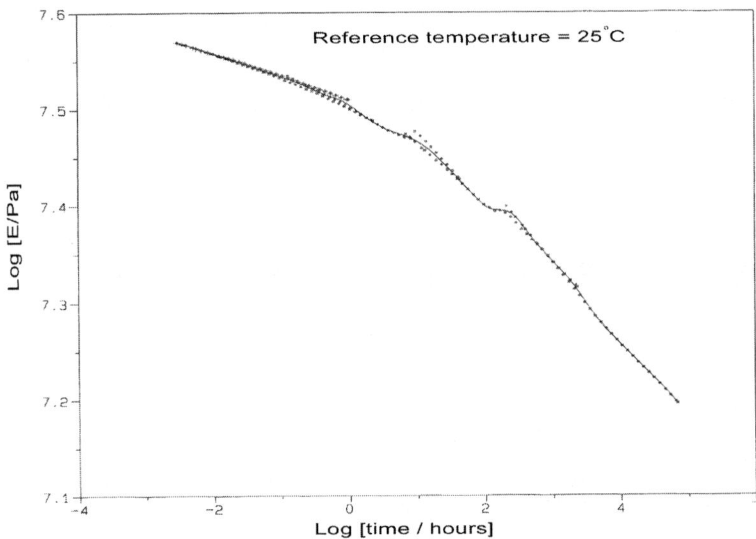

Fig. 10. Master curve at 25 °C for sample SEBS; stress relaxation measurement.

The master curves are shown on Fig. 7– 10. The master curves give us the possibility to examine the creep and flexural modulus of our samples and the possibility to predict useful life of examined samples in a wide range of time and temperature. At the short time interval the materials examined exhibit a relatively high creep modulus (Fig. 2-5.) as well as flexural modulus (Fig. 7-10.). At longer time period the viscous flow occurs, the modulus become lower what is in agreement with faster relaxation process which are more pronounced with time and are faster at higher temperatures.

With the lower content of hard, PS, phase in examined samples the master curves are in the lower region of creep and flexural modulus values, what is due to the faster relaxation process in those samples.

Conclusion

The correlation of the creep and stress relaxation and time, temperature as well as the content of the hard and soft phases in copolymers PS-HI, SEBS and their blends was obtained.

The secondary viscoelastic functions obtained from creep relaxation experiments (creep, recovery and creep modulus) and stress relaxation experiments (stress, recovery and flexural relaxation modulus) show dependence on time and temperature.

The creep increase with time and temperature, while the creep modulus decreases. The stress and flexural relaxation modulus decrease with time and temperature. Such behaviour is due to the relaxation process, which are more pronounced with time and are faster at higher temperature.

The creep and stress relaxation are more pronounced in samples with lower content of the hard, PS, phase. As a consequence of this creep and recovery values increase and stress value decrease in samples with lower content of hard phase, PS, while creep modulus and flexural relaxation modulus decrease. The lower creep and higher stress of SEBS in comparison with the PS-HI/SEBS blends may be assumed to the higher flexibility of soft, PB, phase.

The master curves were created using a TTC principle from creep and stress relaxation measurements.

[1] W. Brostow, N. A.D`Souza, J. Kubat, R. Maksimov, *J. Chem. Phys.*, **1999**, 110, 9706.
[2] Y. M. Boiko, W. Brostow, A. Y. Goldman, A.C. Ramamurthy, *Polymer*, **1995**, 36, 7, 1383.
[3] J. Lai, A. Bakker, *Polymer*, **1995**, 36, 1, 93.
[4] V. Rek, T.H. Grgurić, Ž. Jelčić, *Macromol. Symp.*, **1999**, 148, 425.
[5] J. D. Ferry, *Viscoelastic properties of Polymers*, Wiley, New York, **1960**
[6] Roovers, Toporowski, *Macromolecules*, **1992**, 25, 3454.
[7] Han, Rim, *Polymer*, **1993**, 34, 2533.
[8] Colby, *Polymer*, **1989**, 30, 1275.
[9] Arendt, *Rheol. Acta*, **1994**, 33, 322.
[10] Kannan, Komfield, *J. Rheol.*, **1994**, 38, 1127.
[11] R. Stadler, L. L. Freitas, V. Krieger, S. Klotz, Polymer, **1988**, 29, 1643.
[12] Han, Chuang, *JAPS*, **1985**, 30, 4431.
[13] Mani, *J. Rheol.*, **1992**, 36, 1625.
[14] S. Wu, *JPS:B:PP*, **1987**, 25, 557.
[15] C. Friedrich, C. Schwarzwälder, R. E. Riemann, *Polymer,* **1996**, 37, 2499.
[16] R. Stadler, D. Araujo, *Makromol. Chem. Macromol. Symp.*, **1990**, 38, 243.
[17] H. Watanbe, T. Kotake, *Macromolecules*, **1983**, 16, 769.
[18] Lipatov, *JAPS*, **1981**, 26, 499.
[19] K. J. Wang, L. J. Lee, *JAPS,* **1987**, 33, 431.
[20] Alle, L. Jorgensen, *Rheol. Acta*, **1980**, 19, 94.
[21] Cassagneau, *JAPS*, **1995**, 58, 1393.
[22] T. Murayama, *Dynamic Mechanical Analysisi of Polymeric Materials*, Elsevier, New York, 1978.
[23] A. A. Collyer, L.A. Utracki, *Polymer Rheology and Processing*, Elsevier, New York, 1990.
[24] A. V. Tobolsky, *Properties and Structures of Polymers*, Wiley, New York, 1960.
[25] S. Blonski, W. Brostow, J. Kubat, *APS, Phys. Rew. B*, **1994**, 49, 10, 6494.

Macromol. Symp. **2003**, *202*, 143—150

Processing and Dynamic Mechanical Properties of PS-HI/SEBS Blends

Vesna Rek,[1] *Tamara Holjevac Grgurić,**[1] *Želimir Jelčić*[2]

[1]Faculty of Chemical Engineering and Technology, University of Zagreb, Marulićev trg19, Zagreb, Croatia
[2]DIOKI d.o.o., Žitnjak bb., Zagreb, Croatia

Summary: Thermoplastic elastomers have been successfully used as impact modifiers for thermoplastics. Dynamic mechanical properties and phase transitions are very important for their application. In this paper, the copolymers high-impact polystyrene, PS-HI, styrene-ethylene/butylene-styrene block copolymer, SEBS, and their blends PS-HI/SEBS were investigated. Rheological behaviour of those samples during processing was followed by measuring the torque vs. time in the extruder Haake Record 90. The investigations were done by DMA analysis. The primary viscoelastic functions, the storage modulus, E`, the loss modulus, Eˇ, and loss tangent, tgδ, were determined in the temperature range –150°C to 160°C. The correlation of the processing parameters and primary viscoelastic functions with content of the hard, PS, phase were discussed.

Keywords: dynamic mechanical properties; high-impact polystyrene; processing; PS-HI/SEBS blends; styrene-ethylene/butylene-styrene block copolymer

Introduction

Blending two or more polymers offer a great possibility to modify thermoplastic materials with aim to improve their properties. Thermoplastic elastomers, especially styrene-ethylene/butylene-styrene block copolymer, SEBS, styrene-butadiene-styrene block copolymer, SBS, and styrene-isoprene-styrene block copolymer, SIS, have been succesfully used as a impact modifiers for thermoplastics.[1-4] Those triblock copolymers are convenient for thermoplastic modification because of their unique combination of mechanical properties and processability. Thermodynamic incompatibility of the styrene and elastomeric midblock cause the microphase separation of the material. The polystyrene microdomains, PS act as physical crosslinks between the elastomer sequencies, which results with high tensile strenght of material, while rubber midblock gives its elasticity.

Styrene-ethylene/buthylene-styrene block copolymer also has a great application as a compatibilizer, what is more and more important in recycling of polymer materials.[5-9]

 DOI: 10.1002/masy.200351213

Polymer scrap ussually contain two or more thermoplastics with poor impact strenght due to their immiscibility. Compatibilisation results with better morphology and properties of blends. In this paper, the processing behaviour and dynamic mechanical properties of copolymers PS-HI, SEBS and their blends are investigated.

Experimental

Materials

The investigations were done with high impact polystyrene, PS-HI 417, DIOKI, Zagreb, Croatia, with the content of polybutadiene, PB, 8 % weight, styrene-ethylene/buthylene-styrene block copolymer, SEBS, Kraton 1650, Kraton Polymers, Germany, with the content of polystyrene, PS, 29 % weight (with ethylene/buthylene, EB, as rubbery midblock) and with their blends. The compositions of the blends studied are shown in Table 1.

Table 1. Compositions of the investigated samples.

SAMPLE	PS-HI/ SEBS weight%	SOFT PHASE PB weight%	SOFT PHASE EB weight%	HARD PHASE PS weight%
1	100/0	8.0	/	92.0
2	75/25	6.0	17.7	76.3
3	30/70	2.4	49.7	47.9
4	0/100	/	71.0	29.0

Specimens Preparation

The blends were prepared by using Haake Record 90 twin extruder with the intensive mixing profile, Haake TW 100, with the following temperatures in zone 150/200/200/150°C and the frequency of rotation 60 min^{-1}. The specimens were obtained by compression molding at 220°C; the mold temperature was 40°C.

Measurements

Torque vs. time curves were recorded during the processing of the blends PS-HI/SEBS in whole range of compositions, in the extruder Haake Record 90. Output, Q, as well as pressures, in front die pressure, p_3, die pressure, p_4, and back pressure, p_5, were followed by computer.

The storage modulus, E`, loss modulus, E”, and loss tangent, tanδ, of PS-HI, SEBS and PS-HI/SEBS blends were measured using a Dynamic Mechanical Analyser 983, TA Instruments, in resonant frequencies mode in temperature range from –150°C to 160°C, with amplitude 0.2 mm and heating rate 5 °C/min.

Results and Discussion

Rheological behaviour of the blends PS-HI/SEBS during processing was followed by torque, TQ, output, Q, and viscosity, TQ/Q values. The change of the torque values with the SEBS content is presented in the curve on the Fig. 1.

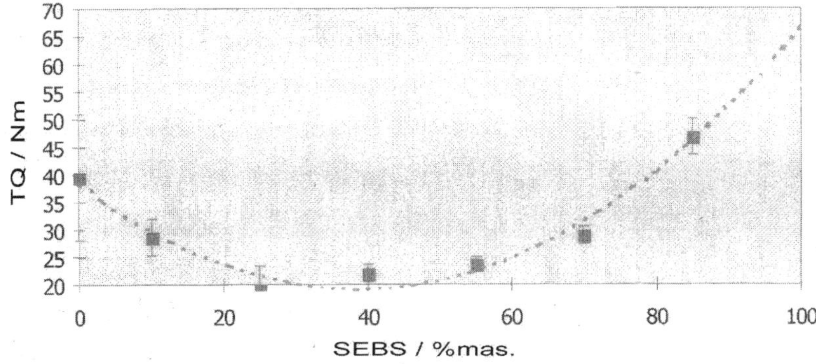

Fig. 1. The influence of the composition of blends on torque value.

It was found that the torque value decrease with increasing content of styrene-ethylene/buthylene-styrene block copolymer until 40 % mas. in PS-HI/SEBS blends. After that, the torque increasing is due to the higher viscosity of the rubbery component. Such behaviour indicate better processability of blends PS-HI/SEBS with lower content of styrene-ethylene/buthylene-styrene copolymer, what is also visible from curve output vs. SEBS content

(Fig. 2).

The viscosity obtained from the TQ/Q ratio increase with increasing content of thermoplastic elastomer, as well as the ratio of pressure differences in the extruder and Q (Fig. 2).

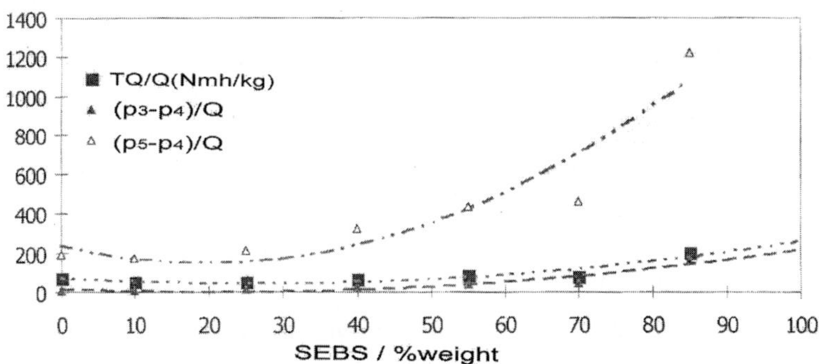

Fig. 2. The process parameters for PS-HI, SEBS and PS-HI/SEBS blends; torque, TQ, output, Q, pressures in the extruder, in front die pressure, p_3, die pressure, p_4, and back pressure, p_5.

The results of dynamic mechanical behaviour of copolymers PS-HI and SEBS and their blends PS-HI/SEBS are reported in Table 2. and Fig 3.

PS-HI and SEBS copolymer consist of two phases what is evident from DMA results. (Table 2) PS-HI has two glass transitions; Tg of the soft, polybutadiene phase at –67°C, and Tg of the hard, polystyrene phase at 111°C.

SEBS block copolymer has the glass transitions of the soft, ethylene/buthylene phase at –40°C, and Tg of the hard, polystyrene phase at 110°C. The same values for those glass transitions was found in previous reports.[10]

PS-HI/SEBS blends consist of hard, polystyrene phase and two soft phases, ethylene/buthylene and polybutadine phase (Table 2). It can be seen that PS-HI/SEBS blends have a single glass transitions of the hard, polystyrene phase at positive temperatures, which are above the glass transition of the PS block in the neat SEBS (110°C) and PS-HI copolymer (111°C). The glass transition, Tg of the polystyrene component is higher in the blend with lower content of PS-HI. Tg of the soft, ethylene/buthylene, EB phase decrease with increasing hard segments in the PS-HI/SEBS blends. Similar observatins for the glass transition of the soft phase in blends was reported in many papers.[11-13]

Table 2. Glass transitios, T_g, storage modulus, E`, and intensities of loss tangent, tgδ, for PS-HI, SEBS and PS-HI/SEBS blends.

SAMPLE	PS-HI/ SEBS weight%	HARD PHASE PS weight%	T_g PB °C	T_g EB °C	T_g PS °C	E' / GPa		$I\alpha_s$ PB	$I\alpha_s$ EB	$I\alpha_s$ PS
			E"	E"	E"	-150°C	25°C	tg δ	tg δ	tg δ
1	100	92.0	-67.6	/	111.0	3.287	1.395	0.109	/	1.794
2	75:25	76.3	-121.3	-59.8	114.0	3.108	0.889	0.034	0.098	1.211
3	30:70	47.9	-122.1	-34.4	120.0	5.288	0.076	0.059	0.359	0.373
4	100	29.0	/	-40.1	110.2	1.688	0.052	/	0.416	0.201

Kim et al. found that in SEBS/PPE blends glass transition of ethylene/buthylene phase decrease as the amount of PPE was increased.[10] Paul and co-workers also found in SBS/PPE blends that Tg of the soft segment is shifted to lower temperature in blends with higher content of hard segments.[14] Some authors explained such behaviour by increased difference in the thermal expansion coefficient between soft and hard phase in the blends.[10,12,13] As a consequence of this increased thermal stress around the soft phase results, which cause the dilation of the soft domains. Therefore, the glass transition of the soft segments decrease with decreasing the amount of the soft phase in the blends.

It wasn`t observed significant difference between the glass transitions of polybutadiene phase in the investigated PS-HI/SEBS blends.

It can be seen that the storage modulus, E´, in platou region increase with increasing content of the hard, polystyrene phase (Fig. 3).

The intensity of the loss tangent in α_s maximum of polybutadiene phase decreases as the amount of styrene-ethylene/buthylene-styrene block copolymer increase, as well as the intensity of hard, polystyrene phase (Fig. 3.). At the same time, the intensity of α_s on tgδ/T curve of soft, ethylene/buthylene phase increase. It is obvious the influence of styrene-ethylene/buthylene-styrene block copolymer on the toughness and impact properties of high-impact polystyrene.

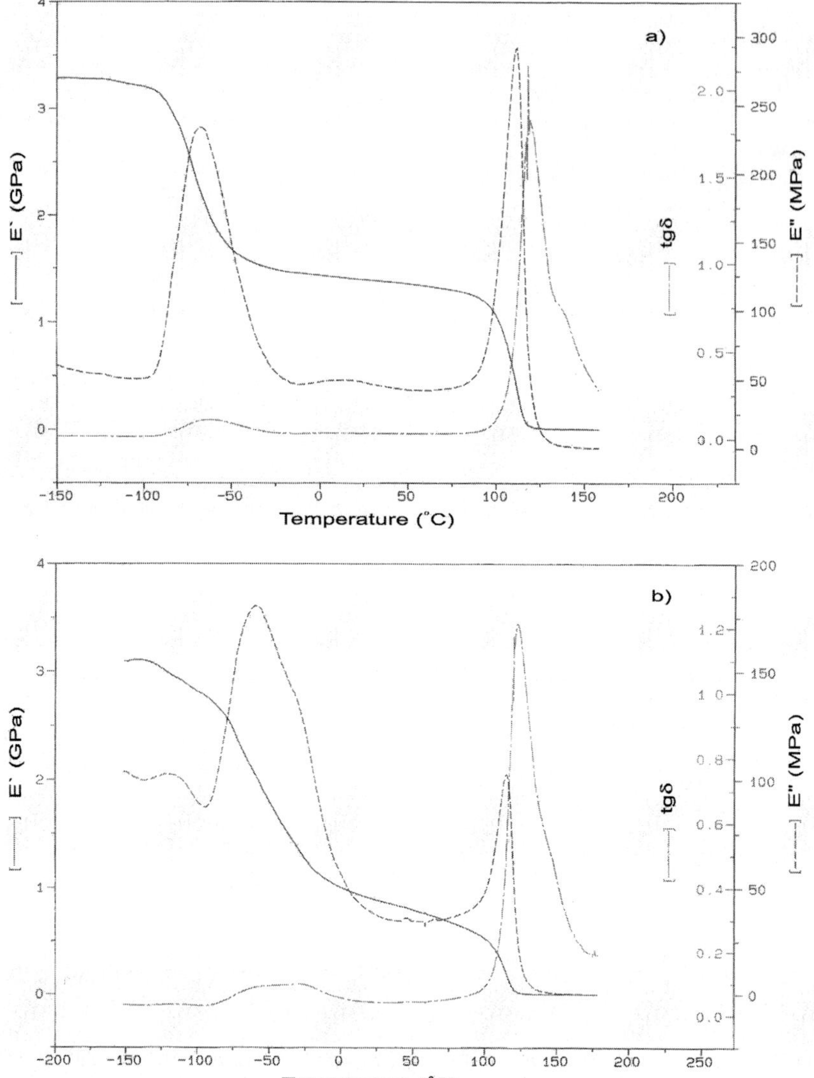
a)
E' (GPa)
tgδ
E" (MPa)
Temperature (°C)
b)
E' (GPa)
tgδ
E" (MPa)
Temperature (°C)

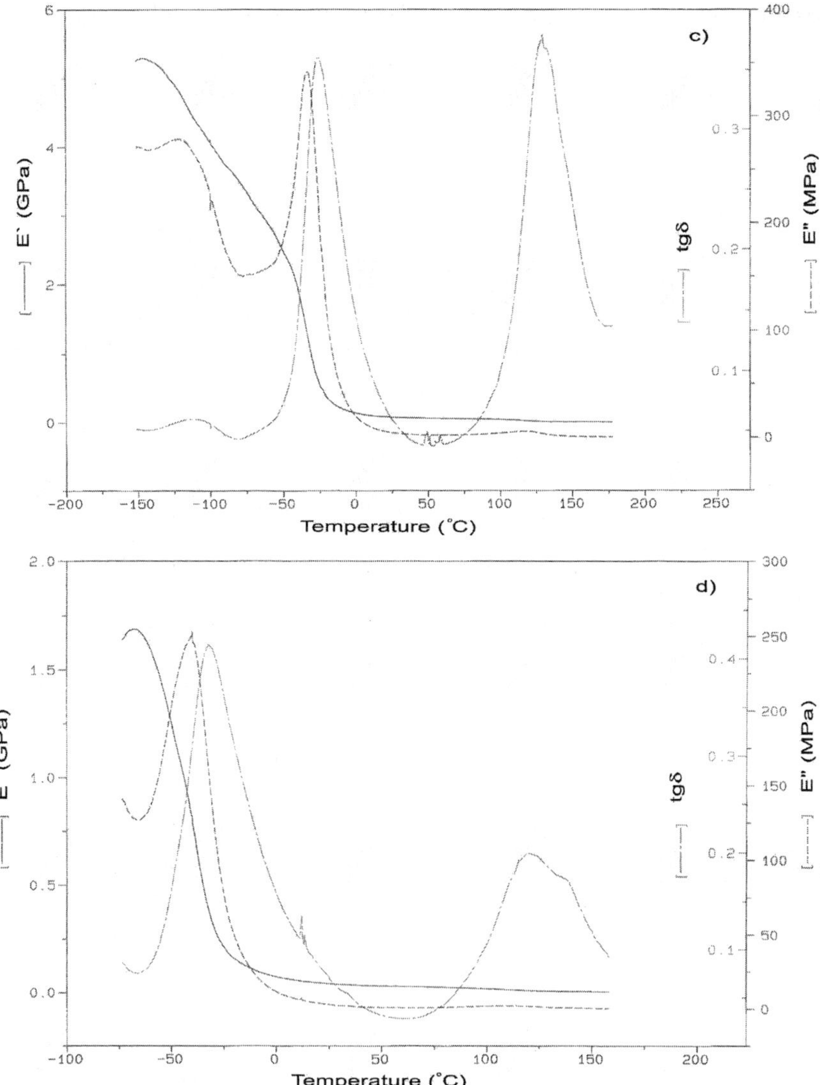

Fig. 3. DMA spectras for PS-HI (a), PS-HI/SEBS 75:25 blend (b), PS-HI/SEBS 30/70 (c) and SEBS (d).

Conclusion

The process parameters torque, TQ, output, Q, viscosity, TQ/Q and pressures in the extruder, in front die pressure, die pressure and back pressure, of copolymers PS-HI, SEBS and blends PS-HI/SEBS were obtained.

The process parameters, particularly torque vs. SEBS content changes against the curve with minimum at 40 % weight SEBS content.

PS-HI and SEBS are two phases systems with T_g of the hard, PS, phase at ~110°C and T_g of the soft phase, PB, at –67°C and T_g of EB at –40°C.

Blends PS-HI/SEBS have a single T_g for hard, PS, phase and for soft phases, EB and PB. The intensity of the soft, PB, phase and hard, PS, phase in α_s maximum on tgδ curve decrease as the SEBS content in the investigated systems increases.

The storage modulus, E′, in platou region increase with increasing content of the hard, polystyrene phase.

[1] S.Ghosh, D. Khastgir, A. K. Bhowmick, *J. Appl. Polym. Sci.,* **1998**, 67, 2015.
[2] K. R. Srinivasan, A. K. Gupta, *J. Appl. Polym. Sci.*, **1994**, 53, 1.
[3] V. Rek, T. H. Grgurić, Ž. Jelčić, *Macromol. Symp.,* **1999**, 148, 425.
[4] C. D. Han, D. M. Baek, J. Kim, K.Kimishima, T. Hashimoto, *Macromolecules*, **1992**, 25, 3052.
[5] M. Heino, J. Kirjva, P. Hietaoja, J. Seppälä, *J. Appl. Polym. Sci.*, **1997**, 65, 241.
[6] B. Ohlsson, H. Hassander, B. Törnell, *Polymer,* **1998**, 26, 6705.
[7] K. K. Shull, E. J. Kramer, *Macromolecules,* **1990**, 23, 4769.
[8] T. A. Vilgis, J. Noolandi, *Macromolecules,* **1990**, 23, 2341.
[9] L. Leibler, *Macromol. Chem. Macromol. Symp.*, **1988**, 16, 1.
[10] J. K. Kim, D. S. Jung, J. Kim, *Polymer,* **1993**, 34, 22, 4613.
[11] L. Moribitzer, G. H. Kranz, K. H. Ott, *J. Appl. Polym. Sci.*, **1976**, 20, 2691.
[12] L. Bohn, Angew. *Makromol. Chem.,* **1971**., 20, 129.
[13] L. Bohn, *Adv. Chem. Ser.*, **1975**, 142, 66.
[14] P. S. Tucker, J. W. Barlow, D. R. Paul, *Macromolecules*, **1988**, 21, 1678.

Macromol. Symp. **2003**, *202,* 151—165

Study of the Properties of Rigid and Plasticized PVC/PMMA Blends

*N. Belhaneche-Bensemra,**[1] *A. Bedda,*[1] *B. Belaabed*[2]

[1] Laboratoire de recherche en sciences et techniques de l'Environnement, Ecole Nationale Polytechnique, BP 182, El-Harrach, Alger, Algérie
E-mail: nbelhaneche@yahoo.fr
[2] Ecole Supérieure des Techniciens de l'Aéronautique, Dar- El- Beida, Alger, Algérie

Summary: The aim of this work is to study the structure-properties relationship of rigid and plasticized PVC/PMMA blends. For that purpose, blends of variable compositions were prepared in the absence and in the presence of a plasticizer di (ethyl-2 hexyl) phtalate or DEHP. The miscibility of the two polymers was investigated by differential scanning calorimetric analysis (DSC) and Fourier transform infrared spectroscopy. The weight loss from 30 to 600 °C was investigated by thermogravimetric analysis (TGA). The thermal degradation under nitrogen at 185 °C was studied and the amount of HCl released from PVC was measured by the pH method. Furthermore, the variation of mechanical properties such as tensile behavior, hardness and impact resistance was investigated for all blend compositions.

Keywords: DEHP; DSC; PVC/PMMA; structure-property relations; TGA

Introduction

Poly(vinylchloride) (PVC) is one of the most important and widely used thermoplastics. Its principal drawback, however, is low thermal stability at processing temperatures. Several workers[1-6] have reported on the degradation and stabilization of PVC. Several polymers are mixed with PVC as polymer plasticizers or processing aids. Polymethacrylates, specially polymethylmethacrylate (PMMA), are used as processing aids for PVC. The first study on PVC and PMMA blends was done by Shurer et al.,[7] who concluded that PVC was partially miscible with atactic and syndiotactic PMMA but almost completely immiscible with isotactic PMMA. The miscibility is due to a specific interaction of hydrogen bonding type between carbonyl groups (C=O) of PMMA and hydrogen from (CHCl) groups of PVC.[8,9] Polymers and their blends are often processed in the melt, which makes the thermal stability of these materials of primary importance. Mc Neill et al.[10,11] have found that the PVC is slightly less stable and

 DOI: 10.1002/masy.200351214

gives monomer at temperatures corresponding to PVC dehydrochlorination (DHC). Braun et al.,[12,13] studying the thermal degradation of PVC with various polymethacrylates, have shown that longer n-alkylester groups and higher concentrations of the respective polymethylacrylate exhibit some stabilization of PVC, whereas smaller ester chains and low concentrations lead to destabilization. More recently, Braun et al.[5] synthetized copolymers based on methacrylates and investigated them as costabilisers for PVC. They found that these polymers are capable of improving the induction time of PVC dehydrochlorination.

In a previous work,[9] we have found that PMMA exerted a stabilizing effect on the thermal degradation of PVC by reducing the zip dehydrochlorination and by leading to the formation of short polyenes.

In this paper, further investigations are done on the miscibility, the thermal degradation and the mechanical properties such as tensile behavior, hardness and impact resistance of rigid and plasticized PVC/PMMA blends.

Experimental

Materials

Commercial grades of resins and additives listed in Table 1 were used as received. The K wert value of PVC is 67 according to the DIN 53-726; ρ(PVC) = 0.54 g/cm^3; ρ(PMMA) = 1.18 g/cm^3.

Table 1. Compounds used in this study.

Compound	Source
PVC 4000M	ENIP – Skikda (ALGERIA)
PMMA	BASF (GERMANY)
Lead bibasic phosphite (heat stabilizer)	HENKEL (GERMANY)
Stearic acid (lubricant)	HENKEL (GERMANY)
Di (ethyl-2 hexyl) phtalate (plasticizer)	BASF (GERMANY)

Polymer Blends

Rigid blends of PVC and PMMA of variable compositions from 0 to 100 wt % were realized. Melt mixing was performed at 185 °C on a two-roll mill, using 1 part per 100 parts PVC of lubricant and 4 parts per 100 parts PVC of stabilizer. The blends were then pressed in a hydraulic press at 185 °C for 3 min under a pressure of 150 MPa. They were water cooled under the same pressure for 2 min. The same blends of PVC and PMMA were prepared in the same conditions in the presence of 30 parts per 100 parts PVC of plasticizer (plasticized blends).

Glass Transitions

The glass transitions temperatures were measured with a Perkin Elmer DSC-7 apparatus at a heating rate of 20 °C/min.

Thermogravimetric and Differential Thermogravimetric Analysis

The thermogravimetric (TG) and differential thermogravimetric (DTG) curves were recorded in a static nitrogen atmosphere from 30 to 600 °C at a heating rate of 10 °C/ min using a Perkin-Elmer DSC-7 apparatus.

Dehydrochlorination of the Blends

The amount of liberated HCl in nitrogen (72 ml/min) was determined by using the continuous pH method at 185 ± 1 °C according to the ISO 182-2 (1990).

Purification of the Samples

The samples were purified before and after dehydrochlorination by dissolution in tetrahydrofuran (THF), precipitation with distilled water and drying under vacuum at 40 °C during 72 h. The purified samples were used for FTIR analysis.

FTIR Analysis

The infrared spectra were recorded with a Philips type PU 9800 FTIR spectrophotometer at a resolution of 2 cm^{-1} using KBr pellets (1 wt %). For the deconvolution of the carbonyl bands a method using the software Grams 386 was used.

Mechanical Characterization

Measurements of tensile properties were undertaken using an ADAMEL LHOMARGY DY 29 testing machine according to the ISO 37 (1994).

A shore D type durometer was used for the determination of the hardness of the blends according to the NFT 51-109 (1981). The Charpy impact resistance's were determined using a WOLPERT pendulum according to the NFT 51-035 (1976).

Results and Discussion

The glass transitions temperatures (Tg) of rigid and plasticized PVC/PMMA blends are reported in Table 2. The glass transition is an important feature of a blend. For miscible systems a single Tg is observed. From the obtained results and according to the criterion of miscibility, this polymer pair appears to be miscible up to 50 wt % PMMA. It is known that the presence of a plasticizer decreases the Tg value. Table 2 shows that for a same composition all the plasticized blends have lower values of Tg in comparison with the rigid ones.

Table 2. Glass transition temperatures (Tg) of rigid and plasticized PVC/PMMA blends.

wt % of PMMA	Tg of rigid blends (°C)		Tg of plasticized blends (°C)	
0	83.00		51.80	
10	84.00		57.00	
20	84.50		69.50	
30	85.00		75.00	
40	85.80		84.80	
50	96.00		97.00	
60	88.70	112.40	85.50	111.30
70	96.00	113.00	86.00	112.00
80	100.50	113.60	98.50	113.80
90	103.50	114.50	104.50	114.50
100	115.00			

FTIR analysis of the blends after purification showed a shift of the carbonyl band of PMMA to lower wavenumbers. The shift of the peak is about 3.5-3.6 cm^{-1} within the domain of miscibility of the two polymers. This feature indicates that the miscibility of PVC/PMMA blends is due to a specific interaction of hydrogen bonding type between carbonyl groups (C=O) of PMMA and hydrogen from (CHCl) groups of PVC. In the case of carbony stretching bands, the high mass of the oxygen atom and the great rigidity of the bond lead to a new band which permits not only the identification of associated and non-associated groups, but also their quantification.

The deconvolution of the carbonyl band within the miscibility domain showed two contributions one corresponding to non bonded carbonyl groups in the blend (1733-1737 cm^{-1}) and one at lower wavenumbers corresponding to hydrogen bonded carbonyl groups (1720-1723 cm^{-1}) (Figs. 1 and 2).

The deconvolution of the carbonyl band outside the miscibility domain showed only contribution corresponding to non bonded carbonyl groups in the blend (Figs. 3 and 4).

Thermogravimetric analysis (TGA) was used to study the thermal decomposition of the rigid and plasticized PVC/PMMA blends from 30 to 600 °C. The thermal degradation of PVC in a broad range of temperatures (up to and above 727 °C) is a two-step process. The first step (up to 350 °C) mainly involves dehydrochlorination of the polymer and formation of macro-molecules with conjugated double bonds that suffer cracking during the second step (up to 550 °C).[14-16] In the first step HCl is the main volatile product (96 to 99,5 %), the amount of other products being very low including quantities of benzene and some other hydrocarbons (1 to 3 %).[14,15] In the second step the degradation of the polymer which has already become the dehydrochlorinated product continues with cracking and pyrolysis to low hydrocarbons of linear or cyclic structure (more than 170 products C_1-C_7 have been identified).[14]

Concerning PMMA, it is generally assumed that its thermal decomposition at temperatures higher than 300 °C gives statistical chains scissions followed by depolymerisation and liberation of the pure monomer.[17]

Data obtained from the thermogravimetric curves of Fig. 5 and 6 are given in Tables 3 and 4. They show clearly that for the two types of blends considered the weight losses (%) corres-ponding to the first step of the thermal decomposition are lower than those of PVC and PMMA. Hence it seems that PMMA exert a stabilising effect on the thermal degradation of PVC by reducing the dehydrochlorination. On the other hand, the weight losses corresponding to the second step of the thermal decomposition are higher than those of PVC. It seems that chains scissions, cracking of double bonds and depolymerisation of PMMA are more important during this second step. However, the first process is the most interesting because temperatures used in industrial processing never exceed 230 °C.

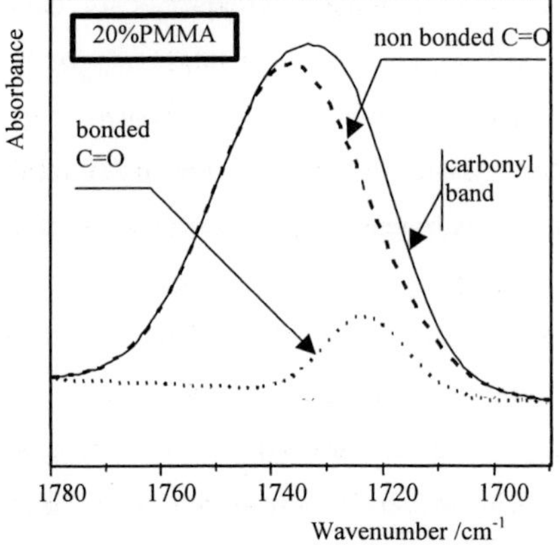

Fig. 1: Deconvolution of the carbonyl band within the miscibility domain of rigid PVC/PMMA blends.

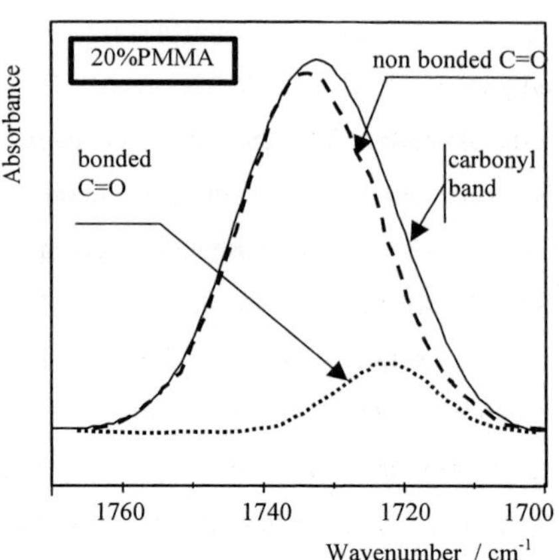

Fig. 2: Deconvolution of the carbonyl band within the miscibility domain of plasticized PVC/PMMA blends.

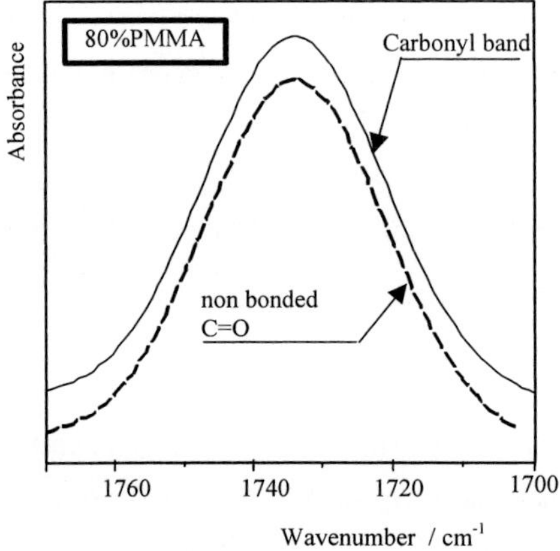

Fig. 3: Deconvolution of the carbonyl band outside the miscibility domain of rigid PVC/PMMA blends.

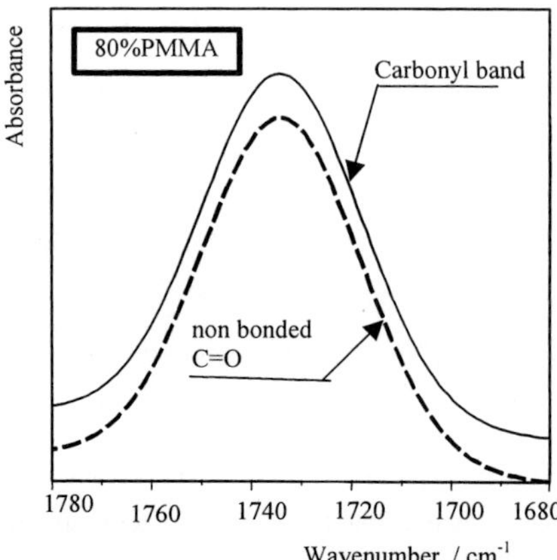

Fig. 4: Deconvolution of the carbonyl band outside the miscibility domain of plasticized PVC/PMMA blends.

Fig. 5 and 6 show the results of thermogravimetric analysis of the rigid and plasticized PVC/PMMA blends, respectively.

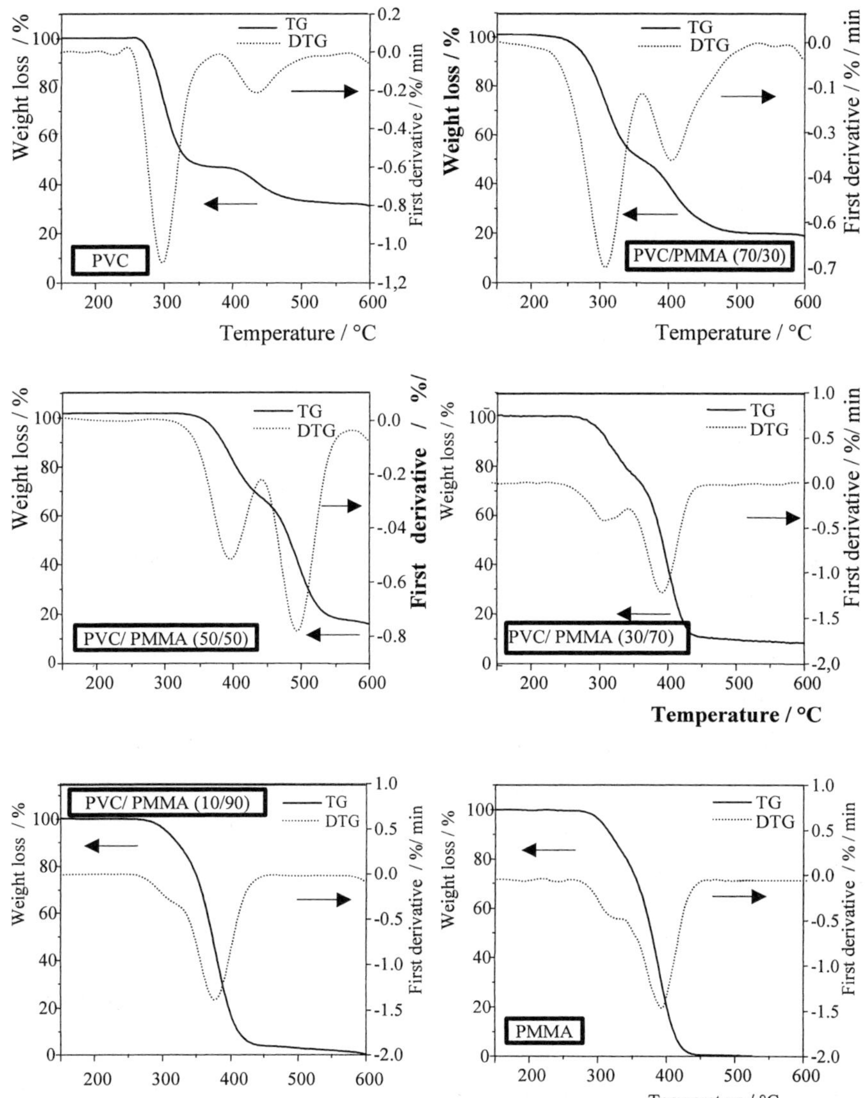

Fig. 5: TG and DTG curves for rigid PVC/PMMA blends.

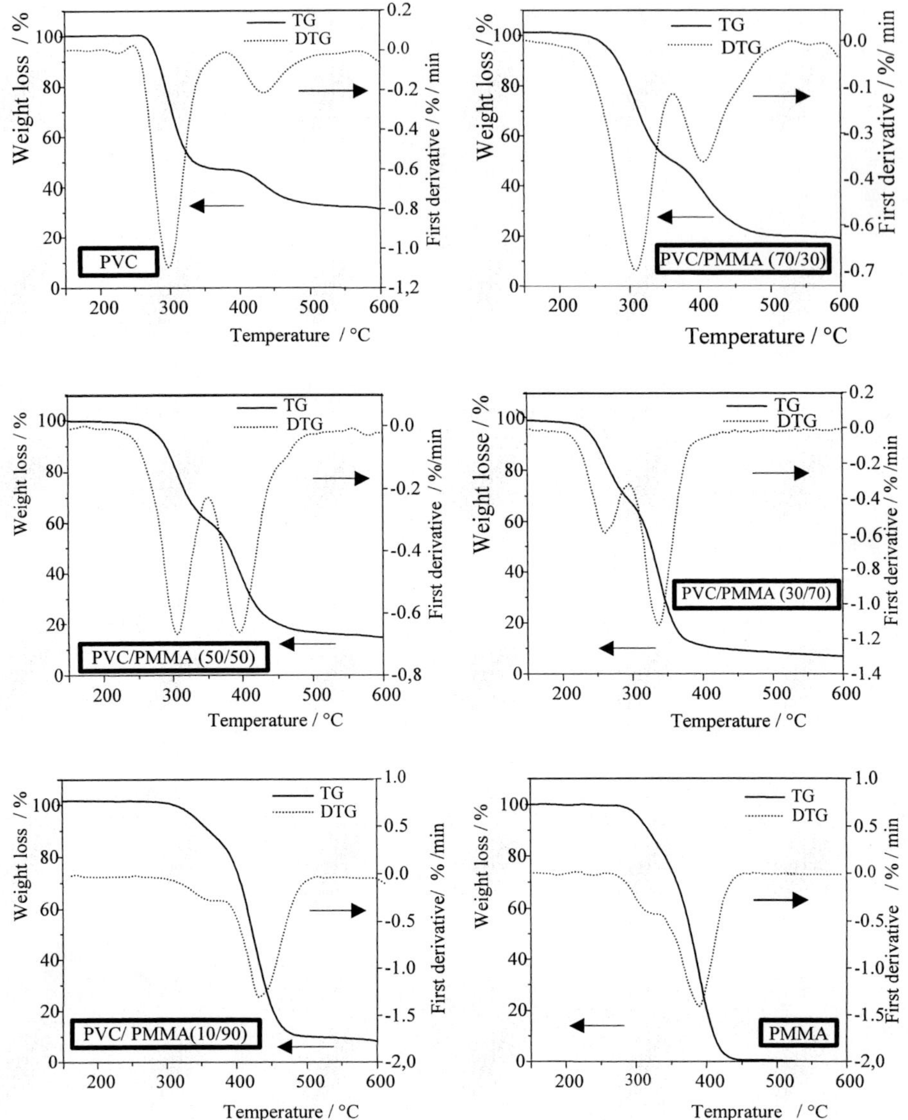

Fig. 6: TG and DTG curves for plasticized PVC/PMMA blends.

Table 3. Results of thermogravimetric analysis of rigid PVC/PMMA blends.

Sample		PVC	70 / 30	50 / 50	30 / 70	10 / 90	PMMA
Weight $loss_{(1)}$	(%)	58	40	30	15	12	100
$Tmax_{(1)}$	(°C)	340	400	395	325	305	395
Weight loss $_{(2)}$	(%)	75	88	78	90	95	-
$Tmax_{(2)}$	(°C)	490	510	490	400	375	-

Table 4. Results of thermogravimetric analysis of plasticized PVC/PMMA blends.

Sample		PVC	70 / 30	50 / 50	30 / 70	10 / 90	PMMA
Weight $loss_{(1)}$	(%)	52	40	30	28	18	100
$Tmax_{(1)}$	(°C)	295	315	305	265	330	385
Weight loss $_{(2)}$	(%)	68	80	78	90	90	-
$Tmax_{(2)}$	(°C)	425	405	395	335	420	-

Weight $loss_{(1)}$ = weight loss at the end of the first step.
Weight $loss_{(2)}$ = total weight loss at the end of the second step.
Tmax $_{(1)}$ and Tmax $_{(2)}$: temperatures of maximum weight loss in the steps 1 and 2, respectively.

The thermal dehydrochlorination (DHC) of the rigid and plasticized PVC/ PMMA blends was measured at a constant temperature (185 °C) in an inert atmosphere to exclude disturbing reactions of oxygen.

Fig. 7 and 8 show the corresponding degradation curves. All the blends are more stable than PVC and the thermal stability increases with increasing the amount of PMMA in the blend. Furthermore, for a same blend composition, plasticized mixtures showed longer induction times than rigid mixtures (Fig. 9). The induction time in defined as the time before the accelerated DHC occurs. MC NEILL et al.[10, 11] explained the better stability of PVC in presence of PMMA in terms of two processes occurring simultaneously during the degradation of the blend. The first is attack on PMMA by chlorine radicals produced during the DHC of the PVC; the second is the reaction between methacrylate ester groups and the hydrogen chloride. In our case, the presence of the heat stabilizer, which reacts with evolved HCl, gives a supplementary explanation of the observed features.

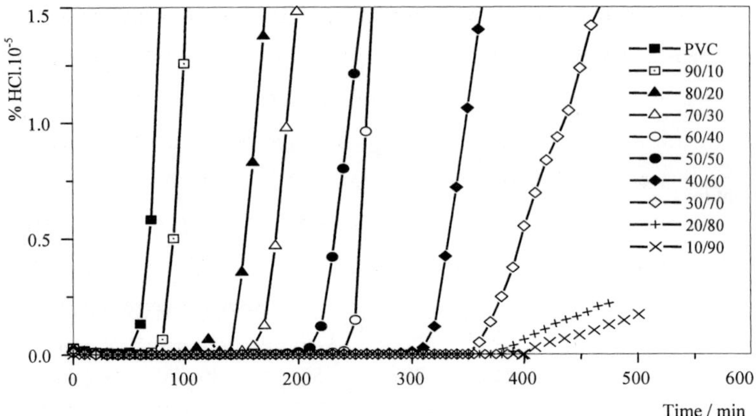

Fig. 7. Dehydrochlorination curves of rigid PVC/PMMA blends at 185 °C in nitrogen.

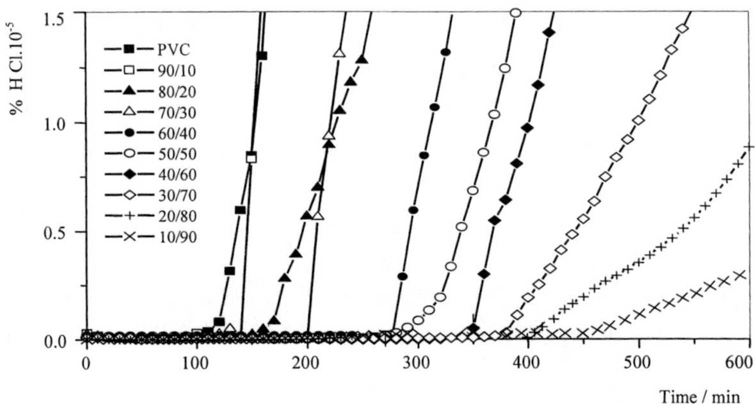

Fig. 8. Dehydrochlorination curves of plasticized PVC/PMMA blends at 185 °C in nitrogen.

Furthermore, the linear variation of the induction time as a function of the amount of PMMA for both rigid and plasticized mixtures indicates the absence of a direct effect of miscibility on the thermal degradation of the blends.

Examples of FTIR spectra of degraded samples after purification are shown in Fig. 10 and 11. The appearance or disappearance of characteristic bands were not observed, but a noticeable decrease of the intensity of all the regions of the spectra was observed showing that DHC of

PVC and depolymerisation of PMMA have occurred. However, the relative decrease of the carbonyl band of PMMA at 1730 cm^{-1} which can be related to the depolymerisation became apparent from 100 minutes (Fig. 10) and 180 minutes (Fig.11) of thermal degradation for rigid and plasticized PVC/PMMA (70/30) blends, respectively. These features indicate that chlorine radicals evolved from PVC reacted with the thermal stabilizer present in the blend. When the thermal stabilizer is consumed chlorine radicals attack PMMA which begin to depolymerise.

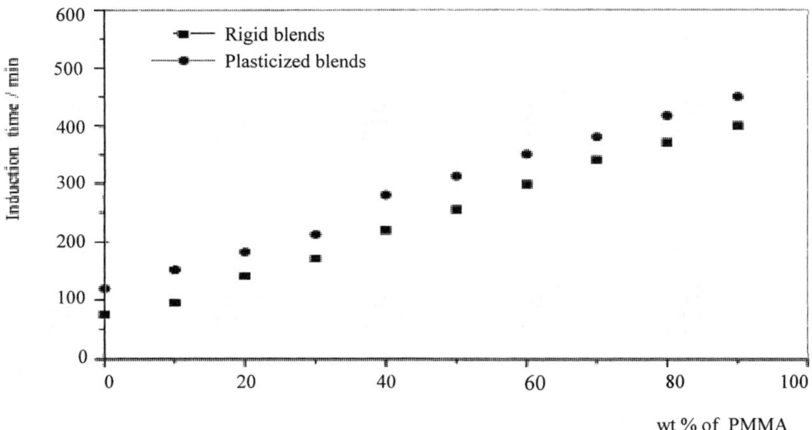

Fig. 9. Variation of the induction time (ti) as a function of the amount of PMMA.

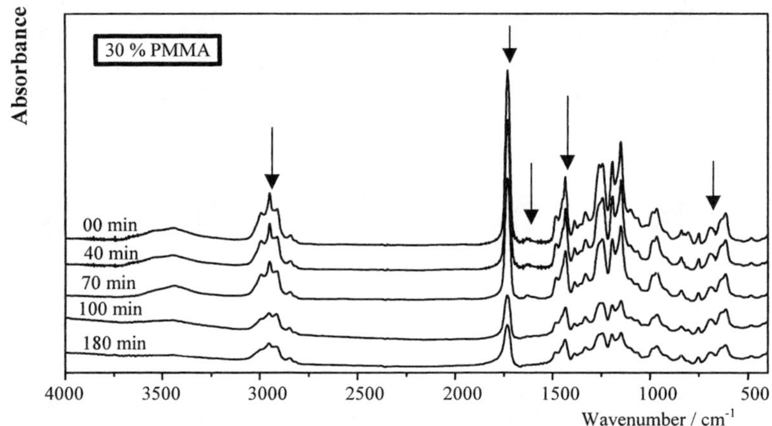

Fig. 10. FTIR spectra of the rigid blend PVC/PMMA (70/30) at various times of degradation at 185 °C.

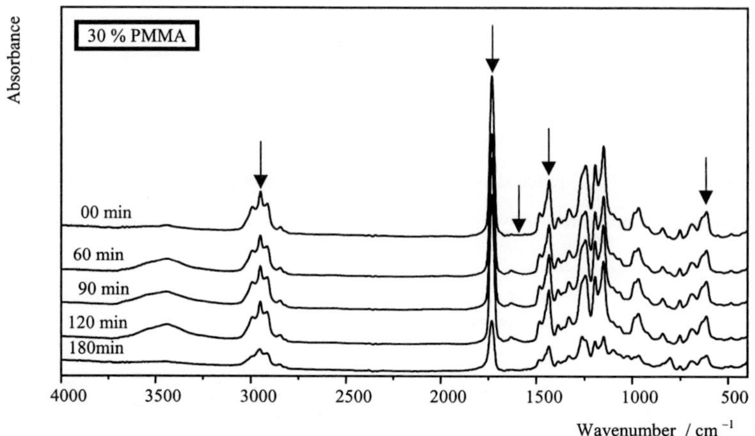

Fig. 11. FTIR spectra of the plasticized blend PVC/PMMA (70/30) at various times of degradation at 185 °C.

All the previous results indicated that the presence of the plasticizer has an important influence on the behavior of the PVC/PMMA blends.

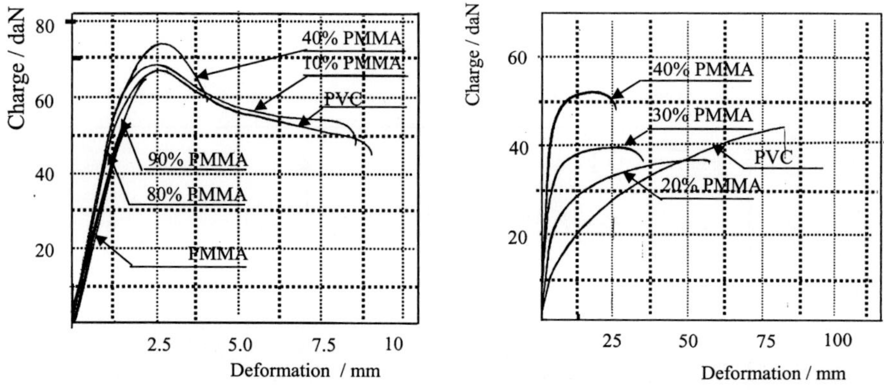

Fig. 12. Tensile curves of rigid PVC/PMMA blends.

Fig. 13. Tensile curves of plasticized PVC/PMMA blends.

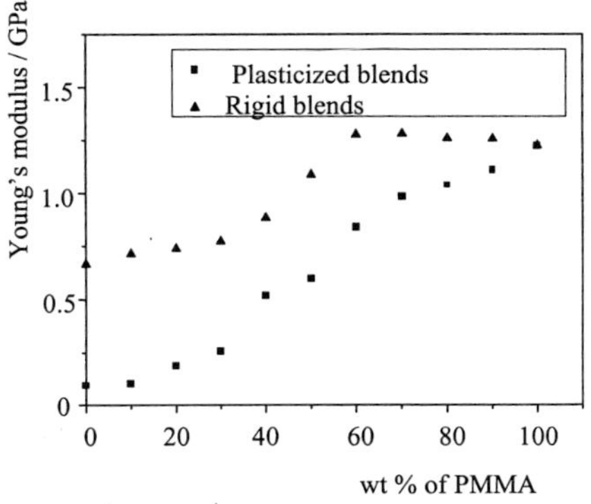

Fig. 14. Variation of Young's modulus with blends composition.

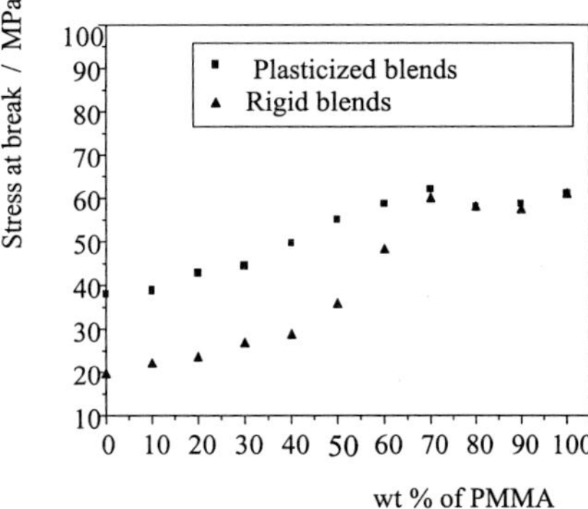

Fig. 15. Variation of stress at break with blends composition.

The mechanical characterization confirmed this feature. Tensile behavior of the various blends which were studied is illustrated in Fig. 12 (without plasticizer) and Fig. 13 (with plasticizer). The comparison between the charge- elongation curves shows that the presence of the plasticizer has modified tensile behavior of PVC/PMMA blends. From the stress-strain curves, Young's modulus (E), which was computed from the initial slope of the curves, stress at break and strain at break were determined and evaluated as a function of the blend composition.

E values obtained for rigid blends are higher than those of plasticized blends (Fig. 14). This feature indicates the diminution of the chains interactions due to the plasticizer.

Fig. 15 shows that the values of stress at break obtained for plasticized blends are higher than those of rigid blends. Furthermore this property presents a maximum corresponding to about 70 wt % of PMMA, decreases slowly after and the values obtained become identical for both rigid and plasticized blends.

Fig. 16 illustrates the variation of strain at break with blends composition. The higher values of strain at break are observed for plasticized blends while very weak values practically constants are observed for rigid blends. From about 70 wt % of PMMA the values obtained become identical for both considered blends.

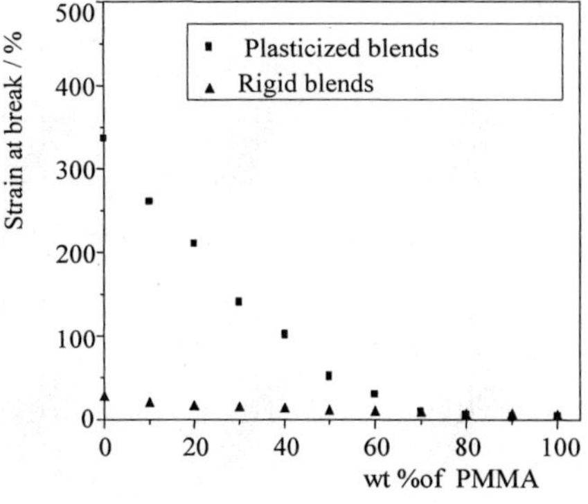

Fig. 16: Variation of strain at break with blends composition.

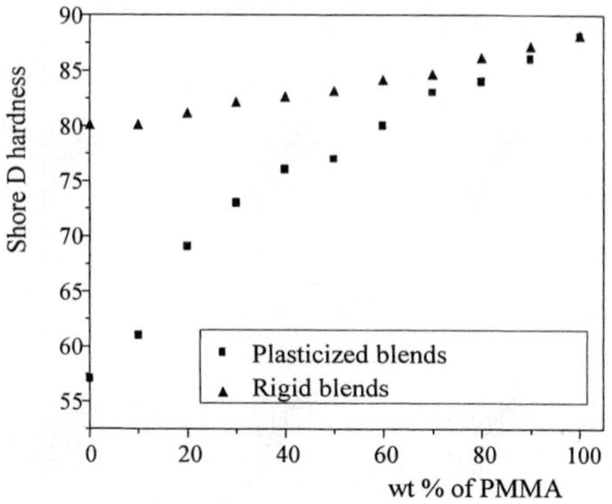

Fig. 17: Variation of hardness with blends composition.

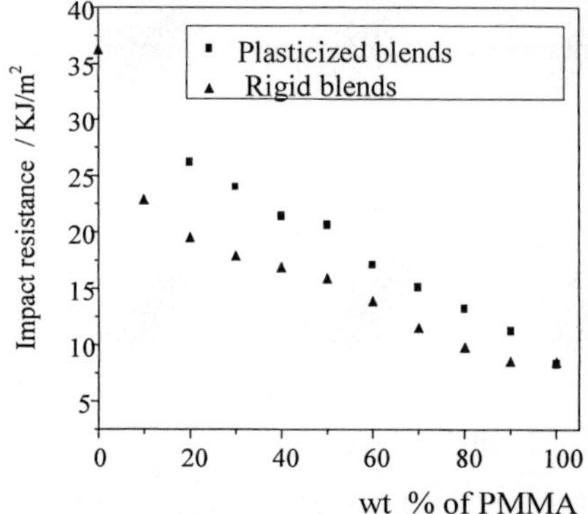

Fig.18: variation of Charpy impact resistance with blends composition.

Shore D hardness is influenced by the rigidity of PMMA (Fig.17). However the presence of the plasticizer decreases it strongly because of the flexibility of the polymeric chains. On the contrary the plasticizer increases the impact resistance's (Fig.18), but globally this property decreases with increasing the amount of PMMA in both considered blends.

Conclusions

The analysis of thermograms showed miscibility up to 50 wt % PMMA. This miscibility is due to a specific interaction of hydrogen bonding type between carbonyl groups (C=O) of PMMA and hydrogen from (CHCl) groups of PVC.

Thermogravimetric analysis showed that the weights losses of the blends are lower than those of PVC and PMMA in the first step of degradation indicating that PMMA exerted a stabilizing effect on the degradation of PVC. Dehydrochlorination curves showed that the blends are more stable than PVC and the thermal stability increases with increasing the amount of PMMA in the blend. Furthermore, for a same blend composition, plasticized mixtures showed longer induction times than rigid mixtures.

Young's modulus, stress and strain at break, hardness and impact resistance are dependant on the blend composition and on the presence or the absence of plasticizer. Miscibility of the two polymers seems to have an effect on these properties also. All the obtained results show that a set of mechanical properties can be generated by blending rigid or plasticized PVC with PMMA.

[1] Z. Mayer, *J. Macromol. Sci.- Revs. Macromol. Chem. C* **1974**, 10 (2), 263.
[2] M. K. Naqvi, *JMS- REV. Macromol. Chem. Phys.* **1985**, *C* 25 (1), 119.
[3] M. Beltran, A. *Marcilla, Eur. Polym. J.* **1997**, 33, (7), 1135.
[4] N. Bensemra, T.V. Hoang, A. Michel, M. Bartholin, A. Guyot, *Polym. Degr. Stab.* **1989**, 23, 33.
[5] D. Braun, P. Belik, E. Richter, Die Angew. *Mackromol. Chem.* **1999**, 268 (4672), 81.
[6] M.T. Benaniba, N. Belhaneche- Bensemra, G. Gelbard, *Polym. Degr. Stab.*, **2001**, 501.
[7] J. W. Shurer, A. de Boer, G. Challa, *Polymer* **1975**, 16, 201.
[8] H. Jager, E. J. Vorenkamp, G. Challa, *Polymer Commun.* **1983**, 24, 290.
[9] N. Belhaneche- Bensemra, B. Belaabed, A. Bedda, *Macromol. Symp.* **2002**, 180, 203.
[10] I. C. Mc Neill, D. Neill, *Eur. Polym. J.* **1970**, 6, 143.
[11] I. C. Mc Neill, D. Neill, *Eur. Polym. J.* **1970**, 6, 569.
[12] D. Braun, B. Böhringer, W. Knoll, N. Eidam, W. Mao, *Die Angew. Makromol. Chem.* **1990**, 181, 23.
[13] D. Braun, B. Böhringer, N. Eidam, M. Fisher, S. Kömmerling, *Die Angew. Makromol. Chem.* **1994**, 216, 1.
[14] A. Jimenez, V. Berenguer, J. Lopez, A. Sanchez, *J. Appl. Polym. Sci.* **1993**, 50, 1565.
[15] H. Bockhorn, A. Hornung, V. Hornung, P. Jakobströer, *Combust. Sci. And Tech.* **1998**, 134, 7.
[16] J. P. Metayer, B. Beccard, J. M. Brocas, *Caoutchoucs et Plastiques*, **1998**, 675, 84 .
[17] O. Chiantore, M. P. Luda Di Cortemiglia, M. Guaita, *Polym. Degr. Stab.* **1989**, 24 (2), 113.

Cation-Exchange Fabric Prepared by Electron Beam-Induced Graft Copolymerization of Binary Monomer Mixture

Yuliya Bondar,[1,2*] *Hong Je Kim,*[2] *Yong Jin Lim,*[3] *Lyubov' Kravets*[4]

[1]Institute of Environmental Geochemistry, 34A Palladin ave., 03142 Kiev, Ukraine, Email: juliavad@yahoo.com
[2]Korea Dyeing Technology Center, 404-7, Pyongri-6Dong, Seo-Gu, Daegu, Korea
[3]Department of Dyeing and Finishing, Kyungpook National University, Daegu, Korea
[4] Joint Institute for Nuclear Research, Dubna, Russia

Summary: Applying the electron-beam preirradiation method in air the cation - exchange fabric (CEF) containing sulfonic acid ($R\text{-}SO_3H$) groups was prepared by graft copolymerization of sodium styrenesulfonate with acrylic acid onto non woven polypropylene fabric. The effect of reaction conditions on the grafting yield and reaction mechanism was examined.
The ion-exchange properties towards Cu(II) and Co(II) ions of the CEF were investigated depending on the form of the CEF and a pH of the solution. It was found that the synthesized CEF contains both strong acid groups ($R\text{-}SO_3H$) and weak acid (R-COOH) groups in almost equal proportion. The utilization of the CEF in Na^+ form allows to make the best use of its ion-exchange capacity.

Keywords: cation-exchange fabric; electron beam irradiation; preirradiation graft copolymerization; polypropylene fabric; sodium styrenesulfonate

Introduction

The production of sorption-active natural and synthetic fibers and textile materials is of great scientific and practical interest. This is determined by the fact that such kinds of materials, owing to their highly developed specific surface area and due to better ion-exchange and complexation parameters exhibit multiple advantages compared with synthetic ion-exchange resins and have gained great importance in solving of a wide range of ecological problems.

For the purpose of production such kind of materials the application of economical and ecologically clean radiation technologies is now under the intense attention of researchers.[1-3] In particular, the utilization of radiation-induced graft polymerization technique allows the modification of an inert polymeric matrix by grafting of a monomer with desirable functional group, or a precursor-monomer with subsequent chemical modification.

 DOI: 10.1002/masy.200351215

For the practical purpose the production of strong acid cation-exchanger with sulfonate functional groups is an object of intense interest.[4,5] This work aims to synthesize the cation-exchange fabric with strong acid functional groups by means of radiation-induced graft copolymerization of acrylic acid and sodium styrenesulfonate onto nonwoven polypropylene fabric, to investigate the mechanism of binary monomer mixture graft copolymerization and to characterize cation-exchange properties of the resultant.

Experimental

Materials

Commercial nonwoven polypropylene fabric – PP - (120g/m^2, degree of crystallinity of 65%, Toray-Saehan Co.Ltd) was used for grafting. Reagent grade acrylic acid -Aa- (CH_2=CHCOOH, Junsei Chemical Co.Ltd), sodium styrenesulfonate -SSS- (CH_2=$CHC_6H_4SO_3Na$, Tokyo Kasei Kogyo Co.Ltd), Mohr's salt ($(NH_4)_2Fe(SO_4)_2x6H_2O$, Merck, Germany) were used as received.

Graft Copolymerization Procedure

Preirradiation grafting technique was employed for graft copolymerization of monomer mixture. Polypropylene fabric was irradiated in air at ambient temperature by an electron beam from an electron beam accelerator ELV-04 with accelerated energy of 1MeV for different time period.

After irradiation PP fabric was stored at room temperature in a dessicator. The irradiated fabric (about of 0.2 g) was weighted and immersed into a 100 ml glass flask with the monomer solution (Aa:SSS = 0.1M (7.2g) : 0.1M (20.6g) in 100 ml of water) and Mohr' salt additive. The flask was purged then by bubbling nitrogen, sealed up and placed in a water bath maintained at a constant temperature for different time periods. After the stipulated time period the grafted PP fabric was washed thoroughly with hot distilled water and methanol to remove homopolymers and unreacted monomer, dried in an oven at 50°C until constant weight was obtained.

The degree of cografting was calculated from the weight gain:

$$DG\% = [(W_1 - W_0)/W_0] \cdot 100$$

where W_0 and W_1 are the weight (g) of original and grafted fabric, respectively.

The preparation scheme is shown in Figure 1.

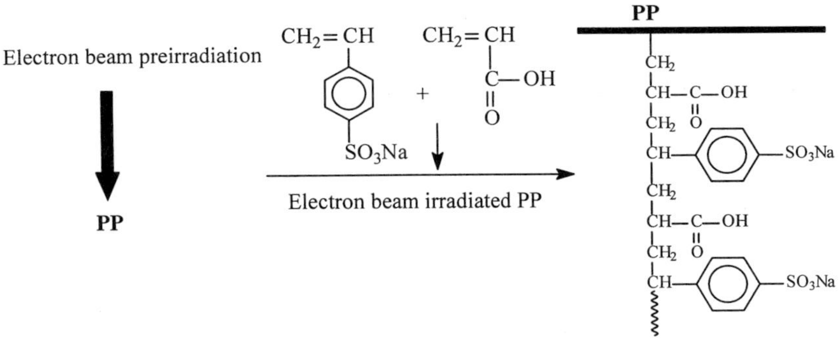

Fig. 1. Scheme of preparation of cation-exchange fabric by radiation-induced graft copolymerization of monomer mixture.

FT-IR Analysis

In order to confirm the introduction of functional groups onto PP fabric the IR spectra of modified fabric were obtained from FT-IR spectrometer (Perkin-Elmer, 1725).

Cation-exchange Capacity

Total cation-exchange capacity of the modified fabric, which reflects general content of acidic functional groups grafted onto PP trunk, was determined by backward titration experiments. For this purpose the grafted PP fabric (about 0.2g) was completely converted to the H^+ form by treatment with excess solution of 0.1M HCl for 12h. The sample was than washed with water to remove sorbed HCl, until the washing solution became neutral and dried at 60° C to a constant weigh. Subsequently the H^+ form sample was weighted and put into a flask with 0.20 ml of standard 0.5M NaOH solution stoppered down and shaken for about 12h. After shaking, 2ml aliquots of the solution was titrated to the phenolphthalein end point with standard 0.1 M HCl solution.

To separate the contribution to the total cation-exchange capacity of both carboxylic and sulfonic acid groups a set of kinetic-exchange experiments depending on the form (H^+- or Na^+) and a pH of the solution was carried out.

Me(II) Ions Exchange Capacity

The rate of exchange of Cu(II) and Co(II) ions from Me(II)-ion solutions with different pH (at 4 and 8) was determined by contacting about of 0.2 g modified PP fabric with 50 ml Me(II)-ion solution at 22°C under stirring (140 rpm). 1 ml solution was collected at appropriate time intervals for metal-ion determination by using of atomic absorption spectroscopy with wavelength set at 324.8 nm for Cu and 240.7 nm for Co (AA-spectrophotometer, Model UNICAM 989, England).

The quantities of metal ions adsorbed per unit of dry weight of the CEF (mg/g) were calculated by the following expression:

Metal ions adsorbed = $(C_o - C)\ V/W$

Where, C_o and C are the Me(II)-ion concentrations in the Me(II)-ion solution before and after the incubation period, respectively (mg/ml); V is the volume of the solution (ml); W is the weight of the adsorbent used (g).

Results and Discussion

Copolymerization Process

Strong acid cation exchanger was prepared by radiation-induced graft copolymerization of acrylic acid and sodium styrenesulfonate monomers onto PP fabric. We, like other investigators,[6] were unable to graft SSS monomer onto PP fabric from water solution, but copolymerization of SSS with Aa occurred successfully. Many factors affect the radiation grafting process of binary monomer mixture. Among them, investigated here, are total exposure dose, reaction time and temperature, concentration of specific additives (Mohr's salt). The effects of these factors has been evaluated from the viewpoint of obtaining cation-exchange fabric with reasonable grafting degree.

When polypropylene fabric is irradiated in the presence of air, mainly hydroperoxides are formed onto the polymeric backbone. These hydroperoxides decompose upon heating to generate hydroxyl radical ($OH^{\cdot}$) and the macroradical ($PPO^{\cdot}$). The macroradical offers a site for grafting, while the hydroxyl radical initiates the polymerization of a monomer (homopolymerization). The various processes occurring during the irradiation and grafting reaction may be outlined as follows:[7]

$$PP\text{-}H \xrightarrow{O_2} PP\text{-}OOH \xrightarrow{HEAT} PPO^{\cdot} + OH^{\cdot}$$

$$PPO^{\cdot} + nM \rightarrow PPO\text{-}M^{\cdot}_n$$

$$M + OH^{\cdot} \rightarrow M^{\cdot}\text{-}OH \xrightarrow{nM} M^{\cdot}\text{-}(M)_n\text{-}OH$$

$$PPO^{\cdot} + M^{\cdot}\text{-}(M)_n\text{-}OH \rightarrow PPO\text{-}M\text{-}(M)_n\text{-}OH$$

It can be seen that the hydroxyl radical can initiate undesirable homopolymerization, that leads to lower grafting yield. To overcome homopolymerization some small amount of a transition metal salt is usually used as an inhibitor:[8,9]

$$OH^{\cdot} + Fe^{2+} \rightarrow OH^{-} + Fe^{3+}$$

In our case, no homopolymer formation was observed in the reaction medium at the reaction temperature 50°C. The degree of monomer mixture grafting onto PP fabric during the reaction time of 12 h reached only 19%, while the addition of small amount of Morh's salt to the reaction medium leaded to a significant increase of grafting. Apparently, it is connected with well known fact, that transition metal ions favour the decomposition of the hydroperoxides:

$$PP\text{-}OOH + Fe^{2+} \rightarrow PPO^{\cdot} + OH^{-} + Fe^{3+},$$

The excessive adding of metallic salt may cause a deactivation process and lead to the decrease of grafting yield:

$$PPO^{\cdot} + Fe^{2+} \rightarrow PPO^{-} + Fe^{3+}$$

The influence of Mohr's salt concentration on the degree of graft copolymerization of Aa and SSS monomer mixture was studied and the results are presented in Table 1. It can be seen that the addition of Mohr's salt to the reaction medium exhibits an increasing trend of grafting yield up to the maximum value at 10^{-3} w% of Mohr's salt content. It should be noted that the trend of the influence of Mohr's salt content onto degree of the monomer mixture grafting is quite similar to that when only acrylic acid has been grafted under the same experimental conditions.

The effect of absorbed dose on the monomer mixture grafting is presented in Figure 2. It is observed that the degree of grafting increases slowly up to 40.3 kGy, and then begins to rise significantly up to 70.5 kGy. In the IR spectra of grafted samples, irradiated with doses less than 40.3 kGy, only a characteristic band at 1708 cm^{-1} assigned to -C=O group of acrylic acid was observed, while the IR spectra of the samples, irradiated with higher doses, exhibited also characteristic absorption bands attributed to SSS group.

Table 1. Effect of Mohr's salt concentration on the degree of graft copolymerization.

Content of Mohr's salt, w%	Degree of grafting	
	SSS+Aa monomer mixture*	7% Aa
0	18.6	57.7
$5x10^{-5}$		61.7
10^{-4}	29.3	113.0
$5x10^{-4}$	86.5	150.4
10^{-3}	110.0	164.6
$5x10^{-3}$	82.2	108.3
10^{-2}	79.7	101.4
10^{-1}	19.9	34

Reaction conditions: Absorbed dose-70.5 kGy; reaction temperature–50°C; reaction time – 12h.
*Monomer mixture consists of 0.1M SSS (20.6g): 0.1M Aa (7.2g) in 100ml H_2O

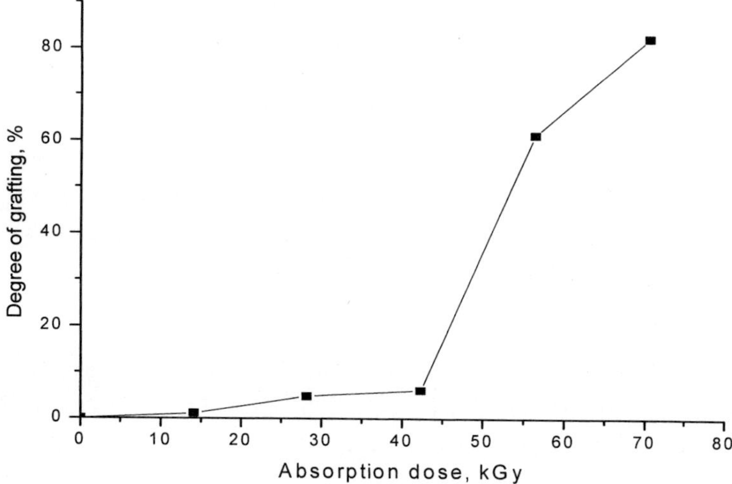

Fig. 2. Effect of absorbed dose on the on the degree of graft copolymerization of Aa-SSS monomer mixture (reaction temperature–50°C; reaction time-12h; 10^{-2} w% Mohr's salt additive).

In order to clarify the mechanism of graft copolymerization of the monomer mixture the kinetic curve of grafting at 50°C as well as corresponding IR spectra were investigated. It can be seen from Figure 3 that the kinetic curve of graft copolymerization consists of different stages (within 0-9 h; 9-12 h, 12-33 h), that probably may reflect the phased development of cografting process. At the first stage (0-9 h) of the kinetic curve, a slow increase of the copolymerization rate occurs, while the second stage (9-12 h) is characterized by an intensive growth of the copolymerization rate, after which there is a slowing down on the following stage (12-30 h), reaching a saturation at 28-32 h.

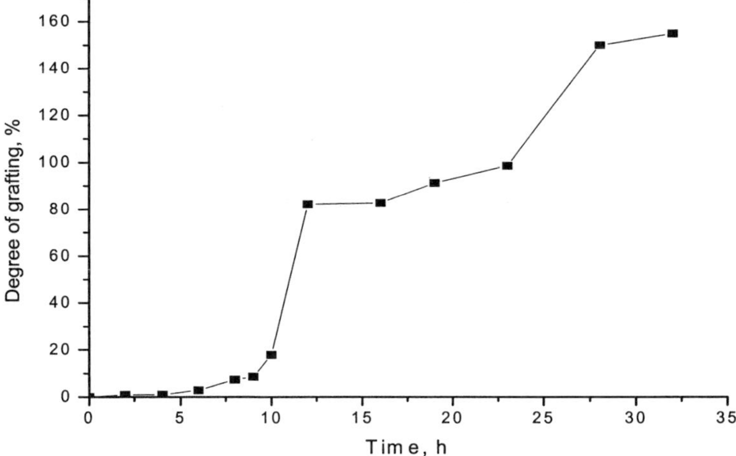

Fig. 3. Kinetic curve of graft copolymerization of Aa-SSS monomer mixture (70.5 kGy absorbed dose; 50°C reaction temperature; 10^{-2} w% Mohr's salt additive).

Figure 4 shows the IR spectra of the samples of appropriate stages. It can be seen that the IR spectra of the first stage (2,9%, 7.3%, 8.6, 17.8) exhibit only the absorption band at 1715 cm^{-1} due to carbonyl group (-C=O) of acrylic acid. As the intensity of this absorption band extends with the increase of grafting degree, so these data testify that the kinetic curve at this stage reflects the peculiarity of grafting of acrylic acid. The characteristic absorption bands of sulfonic acid group {1150 cm^{-1} (asymmetric stretch), 1030 cm^{-1} (symmetric stretch), 670 cm^{-1}} appear in the IR spectra of the subsequent stages. Their intensity extends with the increase of grafting degree, as well as the intensity of carbonyl absorption band.

These data confirmed our assumption about stepwise development of copolymerization process and allowed us to judge about its mechanism. The process begins with the grafting of acrylic acid monomers onto PP backbone (at the first stage) and goes on after that by the inclusion of the SSS monomers into the growing chains.

The maximal degree of grafting (155%) was obtained under the following conditions – absorbed dose 70.5 KGy, reaction temperature 50°C and reaction time 32h.

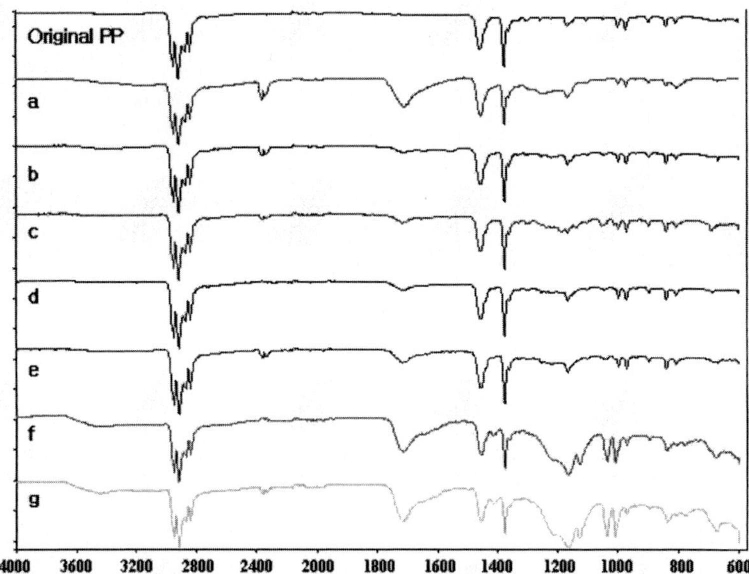

Fig. 4. IR-spectra of original polypropylene fabric (PP); (a) acrylic acid grafted PP fabric (DG=95%); (b-g) Aa-SSS monomer mixture grafted PP fabric with DG=2.9%; 7.7%; 8.8%; 17.8%; 82.6%; 98.6%, respectively. Reaction conditions: absorbed dose: 70.5 kGy; reaction temperature: 50°C.

Cation-exchange Properties

As a result of radiation-induced graft copolymerization of sodium styrene sulfonate and acrylic acid onto PP fabric the cation–exchanger was obtained. The main specificity of the synthesized CEF is that it represents a bifunctional cation-exchanger, carrying two types of functional groups

– both week acid and strong acid ones and strictly speaking can't be classified as strong acid – or week acid cation-exchanger. Strong acid exchanger are so named because their chemical behaviour similar to that of a strong acid: it is highly ionizied in both the acid (R-SO_3H) and salt (R-SO_3Na) form; its hydrogen and sodium forms are highly dissociated and the exchangeable Na^+ and H^+ are readily available for exchange over the entire pH range. So the exchange capacity of strong acid cation-exchanger is independent of a solution pH. In a week acid cation-exchanger the ionizable group is a carboxylic acid group, so that it behaves similarly to weak organic acids in that the degree of dissociation is strongly influenced by pH. As a rule it has limited capacity below a pH of 6.0.

As the cation-exchanger considered here is a bifunctional cation-exchanger it follows as a logical consequence that its behaviour will depend on a solution pH therefore for its practical application the evaluation of the content of both strong (sulfonic) and weak (carboxylic) acid groups would be extremely important.

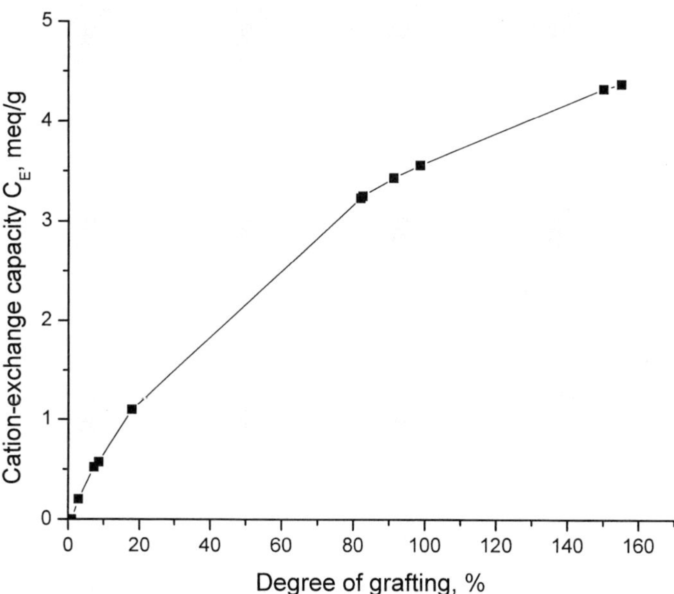

Fig. 5. Dependence of the total cation-exchange capacity of Aa-SSS grafted fabric upon the degree of graft copolymerization.

The total exchange capacity of the synthesized CEF, which reflects the total content of acidic functional groups (both carboxylic and sulfonic), has been determined by backward titration method. A relationship between the total cation exchange capacity of the CEF and the degree of graft copolymerization is shown in Figure 5. Maximum value of the total exchange capacity of 4.30 meq/g was obtained for the CEF with 155% grafting degree.

To determine the separate contribution of each of two acidic functional groups to the total exchange capacity a number of experiments in binding of Cu(II) and Co(II) ions from the acidic solutions (pH=4) were carried out. The equilibrium value of Me(II) ions binding capacity measured in acidic solution allows us to determine the amount of strong acid (sulfonic acid) functional groups, owing to weak dissociating of the carboxylic groups and unavailability of their hydrogen ions for exchange in acidic solution. Subsequently, taking into account a value of total exchange capacity we can easily calculate the amount of carboxylic groups.

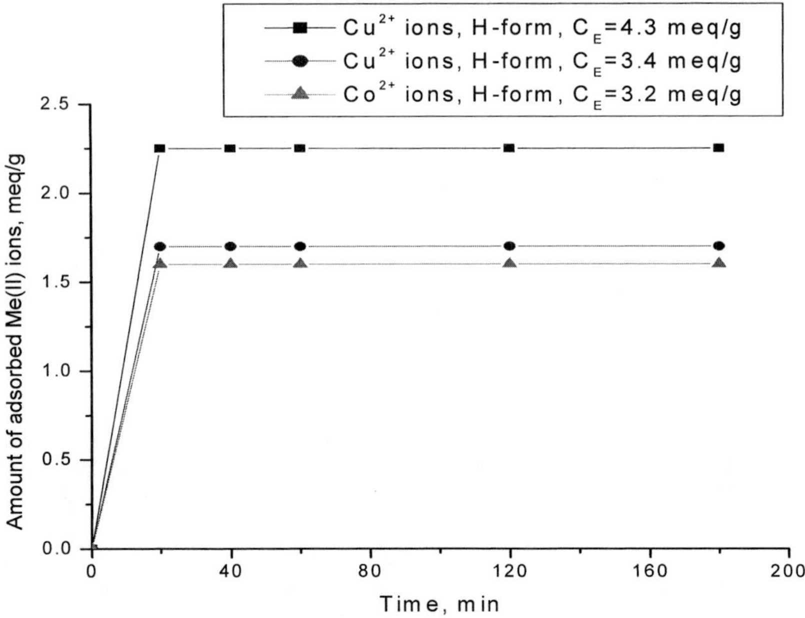

Fig. 6. Sorption rates of Me(II) ions on the Aa-SSS grafted fabric samples in H-form with different values of the total exchange capacity from acidic solution (pH=4).

Three samples of the synthesized CEF in the H^+ form with value of total exchange capacity of 3.2 meq/g, 3.4 and 4.3 meq/g were taken for the experiments of Me(II) ions binding from acidic solution. The kinetic curves of Cu (II) and Co(II) ions sorption from acidic solution on the synthesized CEF in the H^+ form are presented in Figure 6. It can be seen that the adsorption capacity of the investigated samples was two times lower than their total exchange capacity (1.55 meq/g, 1.68 and 2.25 meq/g, correspondingly). It means as a logical consequence that the CEF obtained by radiation-induced graft copolymerization of sodium styrenesulfonate and acrylic acid contains groups of strong acid (R-SO3H) and weak acid (R-COOH) in almost equal proportion.

Figure 7 shows the adsorption rate of Co(II) ions from basic solution (pH=7.8) on the CEF in Na^+ form with total exchange capacity of 3.6 meq/g. As seen here, the cobalt adsorption capacity increases with the time during the first 20 min and then levels off towards the equilibrium adsorption capacity of 3.67 meq/g. Thus, the application of grafted fabric in Na^+ form in basic solution allows to utilize its total exchange capacity completely.

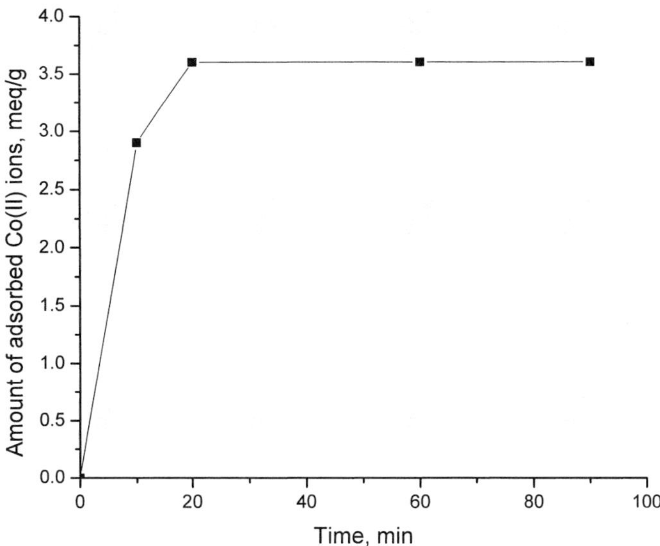

Fig. 7. Sorption rates of Co(II) ions on the AA-SSS grafted fabric in Na-form from basic solution.

Conclusions

Cation–exchange PP fabric with strong acid ($R-SO_3H$) groups has been prepared by radiation-induced graft copolymerization of binary monomer mixture, consisting of sodium styrene sulfonate and acrylic acid, onto nonwoven PP fabric. The main factors affecting the radiation grafting process of binary monomer mixture onto PP fabric have been considered. Comparing the kinetic curve of graft copolymerization of Aa-SSS monomer mixture at 50°C with the IR spectra of the appropriate samples a scheme of graft copolymerization process has been proposed. The process begins with the grafting of acrylic acid monomers onto PP backbone, followed by the inclusion of the SSS monomers into the growing chains.

The total exchange capacity of the grafted fabric versus its grafting degree has been evaluated. The investigation of the synthesized CEF's ion-exchange capacity in the acidic solutions towards Cu(II) and Co(II) ions allowed us to make a conclusion that the CEF contains groups of strong (R-SO3H) and weak (R-COOH) acid in almost equal proportion and consequently it exhibits strong acid behaviour over the entire pH range in half of its total exchange capacity. The application of the CEF in Na^+ form in basic solution allows to utilize its total exchange capacity completely.

[1] H.S. Yang, J.K. Choi, Y.C. Nho, *Korean Ind. Eng. Chem.* **1999,** *10*, 477.
[2] El-S. A. Hegazy, H.A.A. El-Rehim, H.A. Shawky, *Radiat. Phys. Chem.* **2000**, *57*, 85.
[3] T. Kawai, K. Saito, K. Sugita, T. Kawakami, J.Kanno, A. Katakai, N. Seko, T. Sugo, *Radiat. Phys. Chem.* **2000,** *59*, 405.
[4] S. Tsuneda, K. Saito, H. Mitsuhara, T.Sugo, *J.Electrochem. Soc.* **1995,** *142*, 3659.
[5] S. Lacour, J.C. Bollinger, B.Serpaud, P.Chantron, R.Arcos, *Analyt.Chem.Acta* **2001,** *428,* 121.
[6] S. Shkolnik, D.Behar, *J.Appl.Polym.Sci.* **1982,** *27*, 2189.
[7] I. Kaur, S. Kumar, B.N. Misra, G.S. Chauhan, *Materials Sci. and Engineer.* **1999,** *A270*, 137.
[8] V.Viengkhou, Loo-Teck Ng, G.L.Garnett, *Radiat. Phys. Chem.* **1997,** *49*, 595.
[9] J. Chen, Y.C. Nho, J.S. Park, *Radiat. Phys. Chem.* **1998**, *52,* 201.

Macromol. Symp. **2003**, *202*, 179—187

Conformational Aspects of Segmented Poly(ester-urethanes)

S. Ioan, I. E. Cojocaru, V. C. Grigoras, D. Macocinschi, D. Filip*

"Petru Poni" Institute of Macromolecular Chemistry, Aleea Grigore Ghica Voda Nr. 41A, RO-6600 Jassy, Romania

Summary: Segmented poly(ester-urethanes) containing hard and soft segments, were obtained from aromatic diisocyanates with thiodiglycol or diethylene glycol as chain extenders, and poly(ethylene glycol)adipate usig a multistep polyaddition process. Transition temperatures by differential scanning calorimetry and thermo-optical analysis were employed to characterize polyurethane materials. Changes in the conformation of these polyurethanes were analyzed also, by viscometer measurements in N,N-dimethyl-formamide. The obtained data revealed that the thermal curves are influenced by the soft and hard segment structures in the temperature range studied.

Keywords: differential scanning calorimetry; polyurethanes; thermal properties; thermo-optical analysis; viscosity measurements

Introduction

Polyurethanes are an unique class of polymers that offer the chance to obtain the properties by a proper selection of different materials in their composition.[1] The combination of polyols, diisocyanates and low molecular chain extenders gives rise to a multitude of forms suited for extremely different practical applications as fibers, paints, foams, resins, elastomers, and many others.[2-7] Also, they are utilized as coating materials in textile industry.[8,9]

The structural and compositional diversity of polyurethane elastomers represents an useful way to study the properties and the structure of these materials. The previous publications[10-15] presented the syntheses and some properties of new segmented and crosslinked polyurethane elastomers. The influence of polymer structure on the thermal stability, the behaviour in different organic solvents, the structure and morphology of these compounds were analyzed.

The purpose of the present study is to obtain information about the conformational behavior of segmented copolymers containing aromatic diisocyanates {4,4'-methylene diphenylene diisocyanate (MDI) or 2,4-tolylene diisocyanate (TDI)}, with thiodiglycol (TDG) or diethylene glycol (DEG) as chain extenders and poly(ethylene glycol)adipate (PEGA) as a function of the temperature, and to discuss the result in function of the soft and hard segments composition.

 DOI: 10.1002/masy.200351216

The copolymers were characterized by thermo-optical and differential scanning calorimetry measurements. Also, the conformational modifications of the polyurethanes were analyzed in dilute solutions.

Experimental

Structure and Compositional Parameters of Poly(ester-urethanes)

The samples containing segmented block copolyurethanes were obtained from aromatic diisocyanates (4,4'-methylene diphenylene diisocyanate (MDI, Merck; distilled under reduced pressure) or 2,4-tolylene diisocyanate (TDI, Merck; distilled under reduced pressure), with thiodiglycol (TDG, Fluka) or diethylene glycol (DEG, Fibrex Savinesti) as chain extenders and poly(ethylene glycol)adipate (PEGA, Fibrex Savinesti, M_n = 2000 g/mol) using a two steps polyaddition process in DMF.[16] In the first stage of reaction, the NCO-terminated prepolymer was formed. PEGA was dehydrated for 3 h at 120°C followed by adding the MDI or TDI. The reaction between diisocyanate and macrodiol was kept 1 h under nitrogen atmosphere at 90°C. The amount of diisocianate and PEGA was controlled at a NCO:OH molar ratio of 3:1. The second step is the reaction of free isocyanic group with chain extenders, i.e. diethylene glycol or thiodiglycol. The reaction temperature was allowed to cool at 60°C when the chain extender was added. The polymers were precipitated in water and dried under vacuum for several days.

The purity of segmented poly(ester-urethanes) was checked by IR and ^{1}H-NMR (at 80°C in dymethyl sulfoxide – d_6) analyses.

The general chemical structure of the segmented poly(ester-urethanes) studied in this work is illustrated in Scheme 1:

Sample 1: $-(-O-R_3-O-CO-NH-R_1-NH-CO-O-R_5-O-CO-NH-R_1-NH-CO-)_q-$

Sample 2: $-(O-R_3-O-CO-NH-R_2-NH-CO-O-R_5-O-CO-NH-R_2-NH-CO-)_q-$

Sample 3: $-(O-R_4-O-CO-NH-R_1-NH-CO-O-R_5-O-CO-NH-R_1-NH-CO-)_q-$

Sample 4: $-(O-R_4-O-CO-NH-R_2-NH-CO-O-R_5-O-CO-NH-R_2-NH-CO-)_q-$

where:

$R_1 = -C_6H_4-CH_2-C_6H_4-$	derived from MDI
$R_2 = -C_6H_3(CH_3)-$	derived from TDI
$R_3 = -(CH_2)_2-S-(CH_2)_2-$	derived from TDG
$R_4 = -(CH_2)_2-O-CH_2)_2-)-$	derived from DEG
$R_5 = \left[(CH_2)_2-O-CO-(CH_2)_4-CO-O\right]_n-(CH_2)_2-$	derived from PEGA

Scheme 1. Chemical structure of segmented poly(ester-urethanes).

Table 1 presents the compositional parameters obtained for studied samples of poly(ester-urethanes).[14]

Table 1. Characterization data of segmented poly(ester-urethanes).

Sample	Soft segment	Hard segment		Weight ratio %, R_5:($R_{1\ or\ 2}$:$R_{3\ or\ 4}$)
	R_5	$R_{1\ or\ 2}$	$R_{3\ or\ 4}$	
MDITDG	PEGA	MDI	TDG	74.98:(16.67:8.35)
MDIDEG	PEGA	MDI	DEG	75.43:(16.36:8.21)
TDITDG	PEGA	TDI	TDG	79.59:(13.59:6.82)
TDUDEG	PEGA	TDI	DEG	80.11:(13.24:6.65)

Experimental Procedures

The differential scanning calorimetry (DSC) curves were determined in air atmosphere by a Mettler DSC 12E at a heating rate of 10°C/min, in temperature range from 10 to 100°C on samples of 10 mg, using the aluminum opened pans.

The thermo-optical measurements (TOA) were realized with the thermo-optical analyzer (TOA), which consists of an optical microscope equipped with a hot stage programmed by a control unit. The light that was transmitted through the sample was picked up by a photocell. The amplified photocurent was fed through a voltage divider and the voltage difference was plotted continuously against temperature (time) on a strip chard recorder. The normal heating rate was 9.6°/min in the temperature ranges 20-150°C.

The viscosity measurements were carried out in DMF in the 20-45°C temperature range (±0.01°C), by use an Ubbelohde suspended-level viscometer. All measurement was performed within one day after the samples were brought into solution in the interval of five hours; the concentration range was 0.5 to 3.5 g/dl. The kinetic energy corrections were found too negligible. The flow volume of the used viscometer was greater than 5 mL, making drainage errors unimportant. Flow times were obtained with an accuracy of ±0.035 % for different measurements of the same samples in DMF at a given temperature. Plots of η_{sp}/c vs. c were extrapolated to zero concentration to obtain intrinsic viscosity $[\eta]$ according to Huggins equation.

Results and Discussion

Conformational Transition from DSC and TOA Investigations

DSC is one of the most suitable techniques used for the testing of polymer samples having complex structures such as polyurethane elastomers. Often, the endothermic and exothermic peaks could reveal special information on thermal behaviour. Nevertheless, the shape of the DSC curves can be associated with the morphology and the chemical composition of the tested samples.

The DSC diagrams from Figure 1 were drawn for samples 1-4 based on MDI or TDI with chain extenders TDG or DEG as hard segments and PEGA as soft segments in the temperature range of 10-100°C.

All samples exhibits a large endothermic peak at about 44-52°C, corresponding to the melting point of the soft segment (T_{mss}) and a small one, as it shows in previous paper[13] at about 230°C, the melting point of a hard segment (T_{mhs}). Thus, T_{mss} is associated with the melting start of PEGA sequences (52°C) and T_{mhs} with melting start of isocyanate/chain extenders.

A different thermal behaviour of soft segments having the same chemical structure (illustrated by T_{mss}) can also be seen from DSC plots due to isocyanate/chain extenders structures. Thus, the melting temperature of PEGA as soft sequences, revealed the following order:

TDIDEG < TDITDG < MDIDEG < MDITDG

These results indicate that MDI/chain extenders segments contributed to a higher T_{mss} than TDI/chain extenders. So, it would be expected for MDI to shift T_{mss} to higher values than TDI due to a higher cohesive energy and bulkiness because of their benzene rings. Such behaviour may be attributed to the crystalline structures that could form because of the interactions (hydrogen bonding) between MDI-PEGA. In addition, for MDITDG and MDIDEG or TDITDG and TDIDEG samples we can observe the contribution of S (from TDG structure) and O (from DEG structure) atoms to the T_{mss}. It can be seen that oxygen decreases, whereas sulfur increase the T_{mss}.

It can also be noticed from Figure 1 that the endothermic effects of soft segments (ΔH_{mss}) for samples having asymmetrical isocyanate, TDI, into its structure are higher than those of samples with symmetrical isocyanate, MDI.

Literature[17] and previous our studies[13] indicate that at about 230°C the DSC heating curves of the segmented polyurethane materials exhibit an endotherm corresponding to the melting transition of the hard segments. The T_{mhs} and ΔH_{mhs} values of segmented polyurethane materials with symmetrical hard segments, MDI, exhibit a considerably higher ΔH_{mhs}, thus higher crystallinity, than the segmented polyurethane materials with asymmetrical hard segments, TDI.

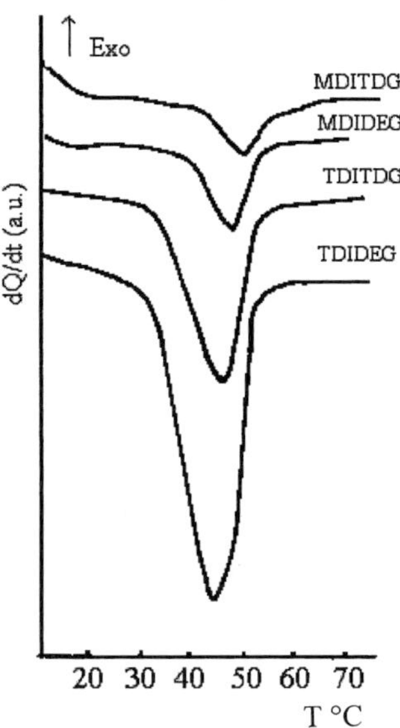

Fig. 1. DSC curves for segmented poly(ester-urethanes) in the temperature range 10-100°C:

Figure 2 presents TOA curves to detect a chain mobility transition temperature for MDIDEG, TDIDEG and TDITDG samples in the temperature range 20-100°C. The vertical scale represent transmitted light intensity by heating samples, although actually a voltage reading from the photoelectric circuit is arbitrary. It is obvious from Figure 2 that there is a close correlation between the chain mobility transition temperature observed in the temperature range studied for the specific heat increase and melting temperature of soft segments, T_{mss} from DSC data. Thus, vertical lines from Figure 2 indicate that melting temperature of soft segments are placed in the same temperature range as in DSC measurement, and revered the same order: TDITDG < TDIDEG < MDIDEG. Also, the transition phenomenon is abrupt at more flexible samples and slower at more rigid samples.

Samples with MDI presents higher chain rigidity and, thus, the higher necessary energy for the melting process of soft segments than sample with TDI. Also, the samples with DEG as chain extender are higher flexibility and lower necessary energy for the melting process of soft segments, comparatively with sample with TDG.

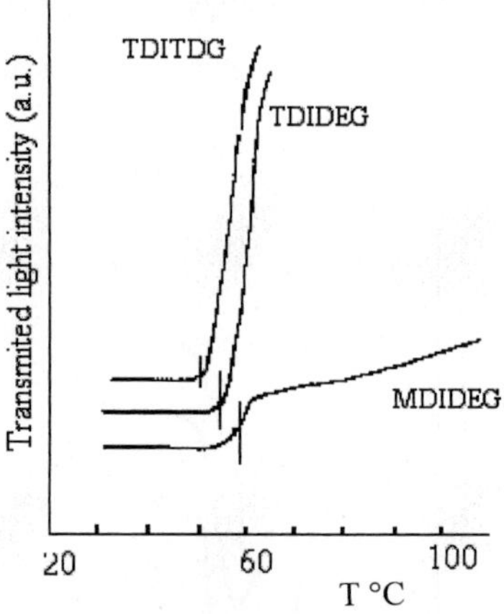

Fig. 2. TOA curves for segmented poly(ester-urethanes) in the temperature range 20-100°C. Vertical lines indicate the melting point of the soft segment, T_{mss}.

Conformational Transition from Viscosity Measurements

Dondos et al.[18,19] present the results which demonstrate that the conformational transition in the viscosity curve is generally observed for all block copolymers. They assumed that in the low temperature range the conformation is quite “segregated”, and that at temperature above the transition, the conformation tends to become closer to Gaussian.

Different properties of polyurethanes show the existence of highly flexible chains, i.e. a low degree of intermolecular interaction, and the presence of crosslinks, which can be of a chemical or physical nature. Thus, the studied polyurethanes are block copolymers, consisting of alternating rigid and flexible blocks. Due to the different polarity and chemical nature of both blocks they separate into two phases designated as soft and hard. Hard blocks also associate into domains because of rigidity and hydrogen bonding and act as physical crosslinks. Studied segmented polyurethanes are, thus, two-phase polymers and their properties in solution are strongly affected by the amount of phase separation.

The polyurethane-solvent interaction parameter can be analyzed by a simplistic two parameters model, where effective interaction parameter is supposed to be the sum of two interaction parameters, coming from solvent-soft segment and solvent-hard segment interactions, or by the complicated multiple-parameters models which take into consideration more interaction parameters.[20] In the present work the studied polyurethanes are multicomponent compounds and the determination of the interactions that may occur is very complicated, the methods of investigation of conformational transition in solution being limited to intrinsic viscosity measurements.[10]

The variation of [η] for poly(ester-urethanes) samples from Table 1 is shown in Figure 3.

All studied samples show inflexions in the viscosity in the range from 20 to 45°C. Generally, an inflexion in the [η] versus temperature plot can be seen, indicating the temperature range where the conformational changes take place. The explanation of the conformational transition is not easy, although the existence of this change is well established and reproducible. A tentative for the interpretation of these phenomena is due to differences of penetration of solvent in soft (more flexible) and hard (with physical crosslinks) segments versus temperature, which modifies the conformation of chain in solution. Moreover, there are preferential interactions between DMF and soft segment, due to the carbonyl groups.

The thermo-optical and differential scanning calorimetry measurements analysis show that the diisocyanates with methyl substituents in hard segment, such as TDI, have a higher flexibility comparatively with the samples containing MDI which are believed to possess significant chain rigidity because of the high cohesive energy and bulkiness of the benzene ring.[17] Thus, the intrinsic viscosity of samples from Table 1 with MDI is higher than that of samples with TDI not only because of the differences of molecular weights, but also because of the different flexibility.

Figure 3 shows also, that the conformation of the MDITDG polyurethane chain slightly changes to a more extended form, when goes from a lower to a higher temperature. At about 30°C this phenomenon is inverted.

The increase in viscosity, caused by an increase of the hydrodynamic volume of the polymer coils, may be the result of improved solvent power for the copolymer as a whole. The improved solubility of the copolymer could be caused by one segment or by both segments.

In this context, at lower temperature the solvent interacts especially with the soft segments not only because of the higher flexibility of the soft segments and the affinity between carbonyl groups from DMF and PEGA, but also because of a poor solvent penetration in the regions of the hard domains, that are physically crosslinked. At higher temperatures the physical crosslinks can be perturbed, thus that the modification of coil dimensions is a consequence of these phenomena.

In the MDIDEG sample (Figure 3) the transition is more evident and the transition range ΔT has a lower value.

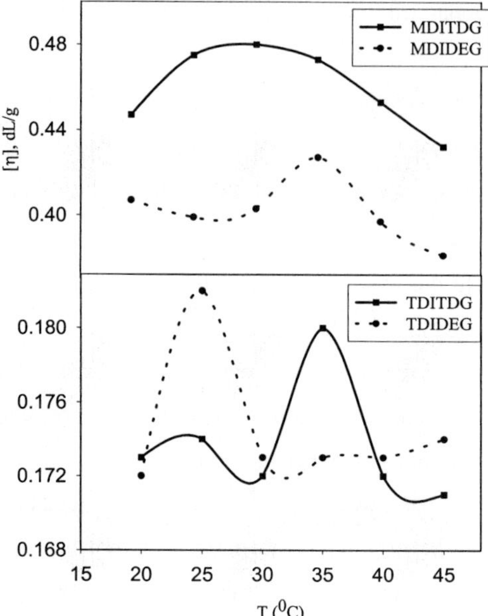

Fig. 3. Intrinsic viscosity versus temperature for segmented poly(ester-urethanes) in DMF.

Same conformational transitions occur in the polyurethanes with TDI in hard segments. The maximum of viscosity displaces from 25°C for more flexible TDIDEG sample, at 35°C for lesser flexible TDITDG sample.

It was demonstrated that the flexibility of polyurethane chains increases with the increase of the number of oxygen atoms in the backbone chain.[20] The possibility to occur the conformational transition in function of the temperature is more probable and evident in the samples with higher flexibility (e.g. MDIDEG sample with DEG in hard segments), comparatively with samples which have a lower flexibility (e.g. MDITDG sample with TDG in hard segments).

Conclusions

The segmented poly(ester-urethanes) obtained from aromatic diisocyanates (4,4'-methylene diphenylene diisocyanate or 2,4-tolylene diisocyanate) with thiodiglycol or diethylene glycol, and poly(ethylene glycol)adipate using a multistep polyaddition process were studied by DSC, TOA and viscosity measurements in N,N-dimethyl-formamide at 20-45°C.

It can be noticed that the endothermic effects of soft segments for samples having asymmetrical isocyanate, TDI, into its structure are higher than of samples with symmetrical isocyanate, MDI. Thus, samples with MDI present higher chain rigidity and, thus, the higher necessary energy for the melting process of soft segments than sample with TDI. Also, for MDITDG and MDIDEG or TDITDG and TDIDEG samples we can observe the contribution of S (from TDG structure) and O (from DEG structure) atoms to the T_{mss}. It can be seen that oxygen decreases, whereas sulfur increases the T_{mss}.

Viscosity studies show discontinuities in the intrinsic viscosity [η] as a function of temperature. This behavior is interpreted as a conformational transition of the copolymer generated to chain flexibility of soft and hard segments and modified of solvent quality versus temperature. It was demonstrated also, that the flexibility of polyurethane chains increases with the increase of the number of oxygen atoms in the backbone chain. Thus, the possibility to occur the conformational transition in function of the temperature is more evident in the samples with DEG sequences in hard segments which possess higher flexibility, comparatively with samples with TDG sequences in hard segments with lower flexibility.

[1] E. Yamazaki, H. Hanahata, J. Hiwabari, Y. Kitahama, *Polym. J.* **1997**, *29*, 811.
[2] T. Speckhard, S. L. Cooper, *Rubber Chem. Technol.* **1986**, *59*, 405.
[3] M. D. Lelah, S. L Cooper, *"Polyurethanes in Medicine"*, CRC Press, Boca Raton **1996**.
[4] T. L. Wang, T.H. Hsiech, Polym. J. **1996**, *28*, 839.
[5] J. T. Koberstein, T. P. Russel, *Macromolecules* **1986**,*19*, 149.
[6] H. S. Lee, Y.K. Wang, S. L. Hsu, *Macromolecules* **1987**, *20*, 2089.
[7] C. W. Heuse, X. Yang, D. Yang, S. L Hsu, *Macromolecules* **1992**, *25*, 925.
[8] A. Grigoriu, D. Mahocinschi, D. Filip, S. Vlad, *Eur. Polym. J.*, in press.
[9] A. Grigoriu, D. Mahocinschi, D. Filip, S. Vlad, *Rev. Rom. Text.* **2001**, *1-2*, 325.
[10] S. Ioan, G. Grigorescu, A. Stanciu, *J. Opt. Adv. Mat.* **2000**, *2*, 397.
[11] S. Ioan, G. Grigorescu, A. Stanciu, *Int. J. Polym. An. Ch.* **2002**, *7*, 221.
[12] S. Ioan, G. Grigorescu, A. Stanciu, *Polymer* **2000**, *42*, 3633.
[13] S. Ioan, G. Grigorescu, A. Stanciu, *Eur. Polym. J.* **2002**, *38*, 2295.
[14] S. Ioan, D. Macocinschi, D. Filip, A. Taranu, *Polymer Testing* **2002**, *21*, 757.
[15] S. Ioan, I. Cojocaru, D, Macocinschi, D. Filip, *J. Polym. Sci., Polym. Phys. Ed.*, in press.
[16] H. Hanahata, E. Yamazaki, Y. Kitahama, *Polym. J.* **1997**, *29*, 818.
[17] D. K. Lee, H. B. Tsai, *J. Appl. Polym. Sci.* **2000**, *75*, 167.
[18] C. Tsitsilianis, G. Staikos, A. Dondos, *Makromol. Chem.* **1990**, *191*, 2309.
[19] A. Dondos, D. Papanagopoulos, *J. Polym. Sci., Polym. Phys. Ed.* **1996**, *32*, 1281.
[20] Z. S. Petrovic, W. J. MacKnight, R. Koningsveld, K. Dusek, *Macromolecules* **1987**, *20*, 1088.

A Fluorescence Study on Critical Exponents During Sol-Gel Phase Transition in Complex Monomeric Systems

Demet Kaya, Önder Pekcan*

Department of Physics, Istanbul Technical University, Maslak, 80626 Istanbul, Turkey
Email: demet@itu.edu.tr

Summary: Methyl methacrylate (MMA), ethyl methacrylate (EMA) and various combinations of MMA with EMA were used during FCC experiments. Pyrene (P_y) was introduced as a fluorescence probe and fluorescence lifetimes from its decay traces were measured during sol-gel phase transitions. The fast transient fluorescence (FTRF) technique was used to study the critical exponents during sol-gel phase transition in free-radical crosslinking copolymerization (FCC). The results were interpreted in the view of percolation theory. The critical exponents of gel fraction, β and weight average degree of polymerization, γ were measured near the point of gel effect and found to be around 0.37 ± 0.015 and 1.69 ± 0.05 in all systems studied respectively.

Keywords: critical exponents; fluorescence; gelation; lifetimes; percolation

Introduction

A chemical gel appears during a random linking process of monomers to larger and larger molecules. The whole course of the bulk free radikal crosslinking copolymerization (FCC) is divided into three different stages: low concersion stage, gel effect stage and glass effect stage.[1] Monomer conversion first increases very slightly but then it accelerates because of the gel effect. Glass effect state occurs as the last stage of polymerization, if the reaction temperature is lower than the glass transition point of the polymer. Several theories have been developed in the past half century to describe gel formation in FCC, among which Flory-Stockmayer theory and percolation theory provide bases for modeling the sol-gel phase transition.[2-8] Flory-Stockmayer theory based on tree approximation, which are also called mean field or kinetic theory; assume equal reactivities of functional groups and the absence of cyclization reactions. In the language of percolation, one may think of monomers as occupying the vertices of a periodic lattice, and the chemical bonds as corresponding to the edges joining these vertices at any given moment, with some probability p. Then, the gel point can be identified with the percolation threshold p_c, where, in the thermodynamic limit, the incipient infinite cluster starts to form.[5-8] Identifying the weight average degree of polymerization DP_w

 DOI: 10.1002/masy.200351217

with the average cluster size and the gel fraction G with the probability of an occupied site to belong to the incipient infinite cluster, one can predict the scaling behaviour of these and related quantities near the gel point, as a function of $|p\text{-}p_c|$,

$$DP_w = A(p_c - p)^{-\gamma} \quad , \qquad p \to p^- \tag{1}$$

$$G = B(p - p_c)^{\beta} \quad , \qquad p \to p^+ \tag{2}$$

Here, β and γ are the critical exponents and A and B are the proportionality factors. The critical exponents in percolation theory, β=0.41 and γ=1.80, differ from those found in Flory-Stockmayer, β=1 and γ=1.

Gelation is a kinetic phenomenon even if it is irreversible. There are much more developments in sol-gel modeling beyond static percolation. Some realistic features like multiple bonding, reversibility, and effect of solvent are generally not considered in static percolation. In references some kinetics are included to make sol-gel kinetics different from static percolation.[7,9-11] Kinetics, reversibility for physical gels, and the quality of solvent do effect the sol-gel transition.[10] They showed that the exponents γ and β (also, ν of correlation length exponent) change considerably for various solvent conditions for physical gels. Pandey *et al* argued that sol-gel transition for chemical gelation seems also nonuniversal with respect to quality of the solvent, degree of inhomogenity depending on the quality of the solvent and rate of reaction due to interplay between the phase-separation and crosslinking.[11]

The gel effect, accompanied by an increase in both rate and degree of polymerization in the free radical polymerization of vinyl monomers, is a well-known phenomenon both for some of the linear and crosslinked bulk polymers.[12-14]

In this paper we aimed to study the free radical crosslinking copolymerization of methyl methacrylate (MMA) - ethyl methacrylate (EMA), using fast transient fluorescence (FTRF) technique. Extremely short-time response of bulk MMA-EMA crosslinked copolymer is probed by an extrinsic fluoroprobe, pyrene. The lifetime of Py displayed a sharp transition from sol to gel as a function of reaction time, which is a (nonlinear) function of the degree of conversion. Immediately after the transition point, the polymerizing system is found to be glassy. The percolation exponents β and γ near the transition point for complex monomeric systems were measured which are found to be in good agreement with the percolation results.

Fluorescence Technique

When an organic dye absorbs light, it becomes electronically excited, then fluorescence occurs from the lowest excited singlet state and decays over a time scale typically of nanoseconds.[15,16] In addition unimolecular decay pathways for deexcitation of excited state, there are a variety of bimolecular interactions, which can lead to deactivation. These are referred to collectively as quenching processes, which enhance the rate of decay of an excited state intensity. For dilute solutions of dye molecules in isotropic media, exponential decays are common. Because of these features fluorescence dyes can be used to study local environments. For about two decades the transient fluorescence (TRF) technique for measuring fluorescence decay has been routinely applied to study many polymeric systems using organic dyes.[17-20] Fast transient fluorescence (FTRF) technique, which is based on the strobe, or pulse sampling technique[21] was recently used to study gelation of styrene (S)[22] and MMA[23] in FCC. In this technique the photo multiplier tube (PMT) is gated or strobed by a voltage pulse that is synchronized with the pulsed light source. The intensity of fluorescence emission of an organic dye is measured in a very narrow time window on each pulse and saved in a computer. The time window is moved after each pulse. The strobe has the effect of turning of the PMT and measuring the emission intensity over a very short time window. When the data has been sampled over the appropriate range of time, a decay curve of fluorescence intensity versus time can be constructed.

Experimental

In this work we monitored the gelation in FCC of MMA, EMA and their mixtures with ethylene glycol dimethacrylate (EGDM) by using the in situ FTRF technique. EGDM has been commonly used as crosslinker in the synthesis of polymeric networks. The free radical copolymerization of two different monomeric systems, MMA and EMA with EGDM were separately performed in bulk in the presence of AIBN as an initiator. These experiments were carried out at four different temperatures with constant EGDM content and at constant temperature with four different EGDM contents respectively. The reaction temperatures and the EGDM contents for the MMA and EMA samples are given in Table 1. The FCC of MMA–EMA mixtures in various combinations were performed at 70 ^{0}C temperature with a single EGDM content in bulk in the presence of AIBN as an initiator. MMA and EMA contents for the mixtures are presented in Table 2. In all experiments the initiator concentration was held constant at 0.26 wt %, the P_y concentration was taken as $4x10^{-4}$ M and the samples were deoxgenated by bubbling nitrogen for 10 min.

In situ fluorescence decay experiments from which P_y lifetimes were determined were performed using Photon Technology International's (PTI) Strobe Master System (SMS). All lifetime measurements were made at 90° to the incident beam and the slit widths were kept at 10 nm. The gelation experiments were performed in a round quartz cell, which was placed in the SMS, and fluorescence decays were collected over three decades of decay time. The sample was illuminated with 345 nm excitation light and pyrene fluorescence emission was detected at 395 nm. Deconvolution of *I(t)* is performed using iterative linear-least-squared fitting technique. The uniqueness of the fit of the data to the model is determined by χ^2 ($\chi^2 < 1.20$), the distribution of weighted residuals, and the autocorrelation of the residuals.

Results and Discussion

Figure 1 presents the fluorescence decay profiles of P_y in various gelation steps for MMA6 sample. It is observed that as the gelation time, t_g is increased, excited pyrenes decay slower and slower by indicating that quenching of excited pyrenes decrease.

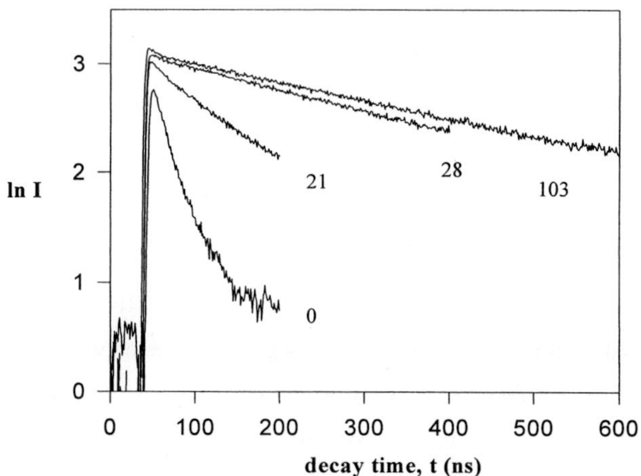

Fig. 1. Fluorescence decay profiles of excited P_y, at various gelation steps for the MMA6 sample. Numbers on each curve present the gelation times, t_g in minute.

In order to monitor gelation processes the fluorescence decay curves are measured and were fitted to the following equation.

$$I = I_0 \exp(-t / \tau) \tag{3}$$

where I and I_0 are the intensity of P_y at time t and zero and τ is the lifetime of P_y. τ values were produced at each gelation steps using the linear least square analysis. The measured τ values and their best fits during FCC are plotted versus gelation time, t_g in Figure 2a, b and c for the MMA1, MMA2 and MMA3 samples.

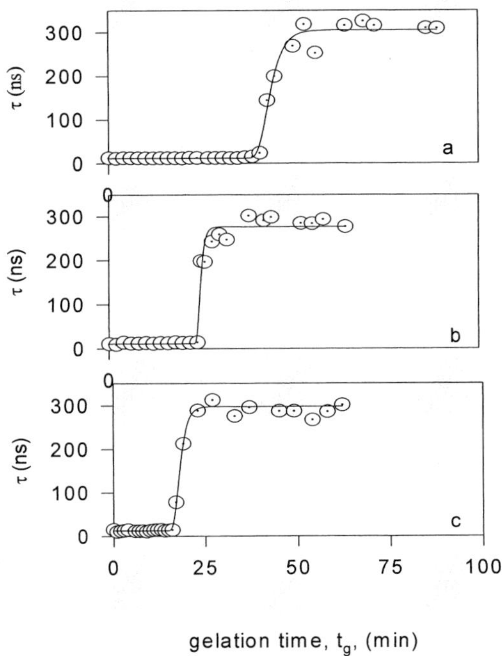

Fig. 2. Plots of the P_y lifetimes, τ , and their best fits during gelation of a) MMA1, b) MMA2 and c) MMA3 samples, respectively.

As seen in Figure 2 P_y lifetimes, τ increased drastically above the certain gelation time called the onset of the gelation time from very low values to their unquenched values during FCC for all samples. The onset of gelation time decreased as the temperature and crosslinker contents are increased, indicating early gelation process takes place at higher temperatures and crosslinker contents for MMA and EMA samples respectively.

The time derivatives of the best fitted curves of the lifetimes are plotted versus gelation time in Figure 3a, b and c for the EMA6, EMA7 and EMA8 samples. These are typical critical peaks, with rounding due to the finite size effects.[17] The maximum of the curves in Figure 3 are evaluated as the gel effect point.

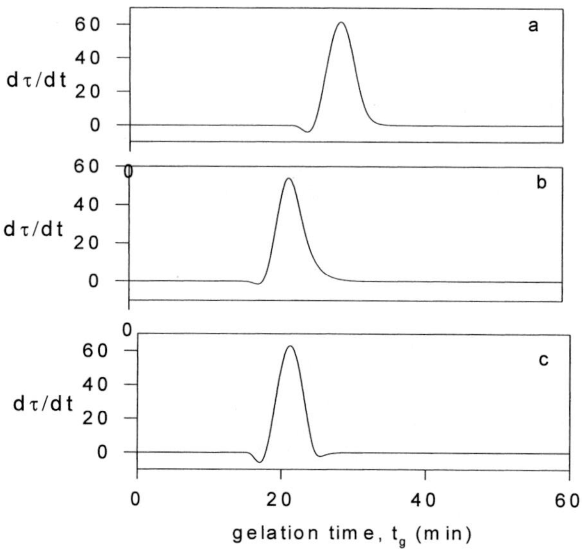

Fig. 3. The time derivatives of the best fitted curves for the lifetimes of a) EMA6, b) EMA7 and c) EMA8 against gelation time, t_g.

The position of the *point of gel effect* on the time axis, called critical time t_c , can be determined with great precision[17], assuming

$$|p - p_c| \propto |t - t_c| \tag{4}$$

at least in a narrow region about the gel point. We argued that for $t > t_c$, the fluorescence lifetimes, τ monitors the growing gel fraction, G the fraction of the monomers that belong to the macroscopic network, and for $t < t_c$ the weight average degree of polymerization, DP_w is monitored. Before the formation of a macroscopic network, i.e., for $t < t_c$, the pyrene molecules are free to interact with – and be quenched by – the sol molecules. In the sol state, the fluorescence lifetimes becomes appreciable only to the degree that the pyrene molecules find themselves surrounded by the progressively larger clusters that are formed. The number of pyrene molecules trapped in the interior of a cluster, and therefore contributing to τ, increases as the number of monomers belonging to this cluster. Thus, the average normalized fluorescent lifetimes will be proportional to the DP_w. Above the gel point, however i.e., for $t > t_c$, most of the pyrene molecules are trapped in the macroscopic network, and τ then measures the G. In summary, we have, the scaling forms for the quantities DP_w and G near the percolation threshold, p_c together with Eq. (4) yields,

$$\tau \propto DP_w \propto A'(t_c - t)^{-\gamma'} , \quad t \to t_c^- \tag{5}$$

$$\tau \propto G \propto B'(t - t_c)^{\beta} \quad , \quad t \to t_c^+ \tag{6}$$

Here, A' and B' are the new proportionally factors. Notice that we need not subtract the value of τ (t_c) from τ (t) in Eq. (6) since we are assuming that once the threshold has been crossed, the unquenched fluorescence lifetime is being contributed essentially by the monomers trapped in the incipient infinite cluster.

In this work, we fitted the double logarithmic plots of the fluorescence lifetime v.s $|t - t_c|$ for $t > t_c$ and $t < t_c$, to determine the critical exponents β and γ respectively. The critical point for each sample was determined by varying t_c in such a way as to obtain good scaling behaviour for both quantities β and γ over the greatest range in $|t - t_c|$. The results are shown in Figure 4a and b for the MMA1 and MMA2 samples.

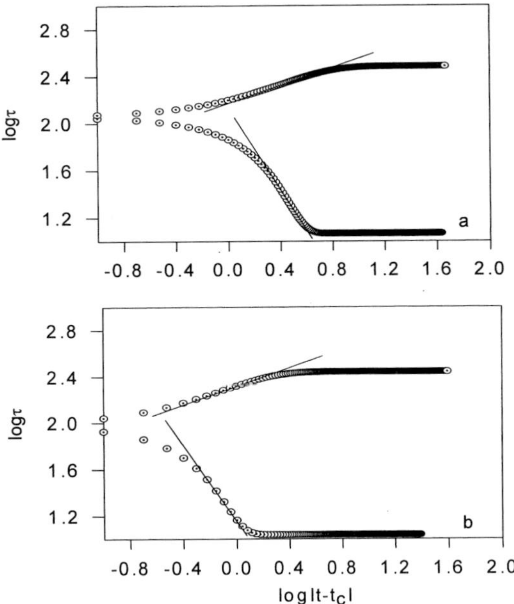

Fig. 4. The double logarithmic plots for a) MMA1 and b) MMA2 data. The β and γ exponents were determined from the slope of the straight lines.

The values of the critical exponents, β and γ which were found from the slope of log-log curves are listed in Table 1, together with the critical times t_c for all MMA and EMA samples.

The agreement with the known values (β=0.41, γ=1.80)[4,5] of the percolation exponents in three dimension is quite good. We find, β=0.37 ± 0.014 and γ=1.70 ± 0.05 as averages over sixteen experiments for two different monomeric systems.

Table 1. Experimentally measured critical exponents β, γ and critical times, t_c for the various MMA and EMA samples in various EGDM content at different temperatures, T.

	EGDM	T	t_c	β	γ
	vol %	°C	min		
MMA1	2.0	60	42.8	0.38	1.78
MMA2	2.0	65	24.5	0.40	1.63
MMA3	2.0	70	17.5	0.36	1.64
MMA4	2.0	75	20.1	0.37	1.80
MMA5	1.0	70	36.0	0.37	1.70
MMA6	1.5	70	23.5	0.40	1.70
MMA7	2.5	70	16.0	0.36	1.70
MMA8	3.5	70	13.9	0.37	1.80
EMA1	2.0	60	46.8	0.36	1.68
EMA2	2.0	65	42.0	0.37	1.70
EMA3	2.0	70	23.8	0.36	1.70
EMA4	2.0	75	22.1	0.35	1.66
EMA5	1.0	70	32.1	0.38	1.70
EMA6	1.5	70	29.5	0.36	1.66
EMA7	2.5	70	22.0	0.36	1.70
EMA8	3.5	70	20.9	0.37	1.65

P_y lifetimes, τ for MIX2, MIX5 and MIX7 are measured and the double logarithmic plots of the fluorescence lifetime v.s $|t - t_c|$ for $t > t_c$ and $t < t_c$, are fitted to determine the critical exponents, β and γ. The critical point for each sample was determined by varying t_c in such a way as to obtain good scaling behaviour for both quantities β and γ over the greatest range in $|t - t_c|$. The values of the exponents, β and γ which were produce from the slope of log-log curves, are given in Table 2 together with the critical times t_c. We find, β=0.38 ± 0.014 and γ= 1.67 ± 0.04 as averages over eight experiments for mixed polymeric system, which are in good agreement with the literature values (β=0.41, γ=1.80).[4,5] Even though the measured β and γ values are both smaller than (about ~ % 9) their theoretical values given by percolation theory, it should be expectable in real experiments.[4]

Table 2. Experimentally measured critical exponents β, γ and critical times, t_c for the various MMA and EMA samples in various EGDM content at different temperatures, T.

	EMA	MMA	T	EGDM	t_c	β	γ
	vol%	vol%	^{0}C	vol %	min		
MIX1	0.0	98.5	70	1.5	23.5	0.40	1.70
MIX2	5.0	93.5	70	1.5	32.0	0.40	1.70
MIX3	10	88.5	70	1.5	35.4	0.39	1.70
MIX4	30	68.5	70	1.5	42.5	0.38	1.60
MIX5	40	58.5	70	1.5	35.4	0.37	1.70
MIX6	50	48.5	70	1.5	39.3	0.37	1.68
MIX7	70	28.5	70	1.5	36.0	0.37	1.60
MIX8	98.5	0.0	70	1.5	29.5	0.36	1.66

In summary, this paper introduces a novel method which uses the FTRF technique to measure the critical exponents during sol-gel phase transition for different monomeric systems MMA, EMA and their mixtures near the point of gel effect. Here, one has to notice that since we measured lifetimes, no environmental corrections to the data, which are quite problematic when one uses fluorescence intensity from steady state spectrometers, are needed. In this work fluorescence lifetimes are taken proportional to the DP_w below t_c, and to the G above t_c to determine the critical exponents. In conclusion it has been shown that PMMA and PEMA gels produce same β and γ exponents even they own very different glass transition temperature. In other words fluorescence lifetimes here monitores the percolating system rather than microviscosity in the polymeric networks. This work also presents that gels made from MMA-EMA mixtures belong to the same universality class as the PMMA and PEMA systems by producing similar critical exponents.

[1] J. Qin, W. Guo, Z. Zhang, *Polymer* **2002**, 43, 1163.
[2] P. J. Flory *J.Am. Chem. Soc.* **1941**, 63, 3083.
[3] W. H. Stockmayer, *J.Chem. Phys*. **1943**, 11, 45.
[4] D. Stauffer, A.Coniglio, and M. Adam, *Adv. Polym*. Sci. **1982**, 44, 103.
[5] D. Stauffer "*Introduction to Percolation Theory*", Taylor and Francis, London 1985.
[6] D. –G. de Gennes "*Scaling Concepts in Polymer Physics*", Cornell University Press, 1988, Ithaca, p.p. 54.
[7] H. Herrmann, *J.Phys. Rev.* **1986,** 153, 136.
[8] D. Stauffer and A. Aharony " *Introduction to Percolation Theory*" Taylor and Francis, London 1992.
[9] M. Adam, D. Lairez, M. Karpasas and M. Gottlieb, *Macromolecules* **1997**, 30,5920.
[10] Y. Liu, R. B. Pandey, *J. Chem. Phys*. **1996**, 105, 825.
[11] R. B. Pandey, Y. Liu, *J. of Sol-Gel Sci. And Tech.* **1999**, 15, 147.
[12] H. K. Mahadabi, K. F. O' Driscol, *J. Polym. Sci. Polym. Chem. Ed.*. **1977**, 15, 283.
[13] J. Dionisio, H. K. Mahadabi, K. F. O' Driscol, *J . Polym. Sci. Polym. Chem. Ed.*. **1979**, 17, 1891.
[14] I. A. Maxwell, G. T. Russell, *Macromol Theory Simul.* **1993**, **2**, 95.
[15] J. B. Birks "*Photophysics of Aromatic Molecules*" J.Wiley-Interscience, New York 1971.
[16] J. R. Lakowicz "*Principles of Fluorescence Spectroscopy* " Plenum Press, New York 1983.
[17] Ö. Pekcan, M. A. Winnik and M. D. Croucher, *Phys. Rev. Lett.* **1988**, 61, 641.
[18]Ö. Pekcan, L. S. Egan, M. A. Winnik, and M. D. Croucher, *Macromolecules* **1990**, 23, 2210.
[19]M. A. Winnik in "*Polymer Surfaces and Interfacs* " J.Feast and H. Munro, Eds., Wiley, London, Ch.1. 1983.
[20] M. A.Winnik, Ö. Pekcan, L. Chen and M. D. Croucher, *Macromolecules* **1988,** 21, 55.
[21] W. R. Ware, D. R. James and A. Siemiarczuk *Rev. Sci. Instrum.* **1992**, 63, 1710.
[22] Ö. Pekcan and D. Kaya, *Polymer* **2001**, 42, 7865.
[23] D. Kaya and Ö. Pekcan, *J. Phys. Chem.B* **2002**, 106, 6961.

Macromol. Symp. **2003**, *202,* 199—219

UV Stabilising Synergies between Carbon Black and Hindered Light Stabilisers in Linear Low Density Polyethylene Films

A. Richard Horrocks, Mingguang Liu*

Centre for Materials Research and Innovation, Bolton Institute, Bolton, BL1 5AB, UK

Summary: The combined effects of selected carbon black pigments and hindered light stabilisers (HALS) on the UV stabilities of linear low density polyethylene film have been studied under UVA and UVB fluorescent radiation sources. While the presence of HALS do not change the chemistry of film photodegradation, whether they are low or high molecular variants, their presence significantly extends film lifetime relative to the sum of the effects of carbon black and HALS individually. These lifetime extensions may be defined in terms of a synergy factor defined with respect to film time to lose a specific percentage of a tensile property, namely t_{20}, the time to lose 20% of initial elongation-at-break, or the carbonyl index associated with this condition.
It is proposed that possible causes of this synergy are a result of the UV screening effect of the carbon black particles which provide lower concentrations of polymer radicals for the HALS component to interact with and/or an accompanying thermal stabilising effect by the latter as a consequence of the higher polymer local temperature during irradiation of pigmented films.

Keywords: carbon black; HALS; hindered light stabilizer; linear low polyethylene; LLDPE; photodegradation; synergy; UV

1 Introduction

In our previous paper[1] we examined the effect of carbon black variables such as particle size and structure on pigment stabilising efficiency when present in linear low density polyethylene films subjected to artificial weathering. As expected, carbon blacks having the smallest particle size and, in particular those below 20 nm, showed the greatest stabilising capacity in stabilising films to both fluorescent tube (UVA, 340 nm and UVB 310 nm) and xenon arc source radiation. However, there was no consistent influence of aggregate structure but increasing carbon black presence from 1.5 to 3.0 (w/w) % improved the stabilising power of all blacks.

The effect of pigment was not seen to be one of simple UV radiation scattering as previously proposed[2] since for UVA and UVB sources, presence of carbon black, while increasing carbonyl group generation with respect to unit loss in tensile property with respect to unfilled

 DOI: 10.1002/masy.200351218

LLDPE, also appeared to suppress Norrish Type II scissions at photochemically generated carbonyl centres in polymer chains.[1] This was especially the case for the smallest (~20nm) particle sizes, thereby suggesting that the photostabilising efficiency of carbon black is based on both physical surface-area-dependent UV absorption and photochemical activity. Under xenon arc exposure, however, this latter appeared to be minimal probably because of its generally low UV content within its overall emission spectrum.

Combinations of carbon black and photo-antioxidants have been used to give higher levels of UV and thermal stability. This is especially important because of the heat absorbing character the pigments which have been reported to produce temperatures as high as 100°C under natural sunlight exposures.[3,4] While both synergistic and antagonistic effects have been reported for polyolefin carbon black-antioxidant ternary systems, adverse effects have been attributed to adsorption and/or decomposition of the antioxidant by carbon black.[5,6] For example, Hawkins reported that phenolic antioxidants are effective in the presence of low concentrations of furnace black but are adversely effected when a channel black with higher concentrations of reactive groups present is used.[7]

The use of hindered light stabilisers during the last 10 years or so has significantly raised the weathering performance of the polyolefins in general.[8] However, the effect of combining them with carbon blacks has not been widely studied although early work demonstrated possible synergy during xenon arc weathering of polypropylene tapes.[9] However, adverse thermal degradative effects were reported if higher black loadings were used because of antioxidant adsorption.[10] Recent work by Allen et al has considered the possible interactions of carbon black and antioxidants[11] and carbon black, antioxidants and HALS.[12] Using carbonyl index as a main indicator of photo-oxidation, this latter work showed that while antagonisms were often the case between primary antioxidants and HALS, small synergies were seen between secondary antioxidants and HALS during photo-oxidation. Furthermore, the presence of HALS and carbon black alone seemed to produce an antagonistic effect which was a balance of antioxidant and adsorptive properties of the former. When HALS, antioxidant and carbon black were together, the overall effect appeared to be one of synergy with the effect of carbon black variables being restricted to variations in degrees of surface oxidation for a standard 22-25 nm particle size. The more oxidised black tended to be present in the more UV stable formulations.

This paper extends our previous work[1] and examines the effects of adding HALS to the UV stability of LLDPE films in which a standard primary/secondary antioxidant is present and which contain carbon blacks having defined particle properties.

2 Experimental

As in our previous study[1] the LLDPE powder used was supplied by Union Carbide (GRSN 7510NT) and is the copolymer of ethylene and hex-1-ene with density of 0.919g/cm^3 and MFI of 0.75 g/10 min at 190°C, 2.16kg. HPLC (carried out at Cabot Plastics (UK) Ltd.,) analysis indicated that the LLDPE powder contained 0.1% (w/w) B-blend of antioxidant system B225 (Irganox 1010 and Irgafos 168a in a 1:1 mass ratio, Ciba Speciality Chemicals) for melt stabilization during manufacture.

In order to fully understand the combined effect of carbon black and HALS stabilizers on photostabilization of the LLDPE films, 5 grades of carbon black (1.5% w/w concentration) with various properties defined in particle size and aggregate structure, and 3 different HALS stabilizers (0.3% w/w concentration) were chosen. The former selection was made in consultation with Cabot Corporation and is based on choosing extreme values of fundamental properties such as particle size, structure and surface area. The matrix of 5 carbon blacks and 3 HALS stabilizers are listed in Table 1 along with 3 unfilled, HALS-containing samples. The detailed properties of the selected carbon blacks (supplied by Cabot Corporation) have been reported elsewhere[1] and former codes are designated in parentheses for the first four blacks in Table 1. The fifth black, C5, was selected because of its lower surface area (as measured by nitrogen adsorption by Cabot Corporation) relative to C3 which has similar size and structure and yet higher surface area. Unfortunately, no data is available regarding the exact chemical characters of the various carbon black surfaces.

Table 1. Matrix of mixtures of carbon black and HALS stabilizers.

	Particle Size, nm	Structure (as DBPA, cm^3/100g)*	Surface area via nitrogen adsorption, m^2/g	HALS1 Chimassorb 944	HALS2 Tinuvin 770	HALS3 Tinuvin 765
LLDPE	-	-	-	X	X	X
C1(C1)**	60	65	30	X	X	
C2(C11)	45	121	42	X	X	
C3(C3)	18	117	200	X	X	
C4(C7)	17	68	210	X	X	
C5	19	114	140		X	X

Notes: * denotes use of dibutyl phthalate to determine surface area of carbon black aggregates;[13] ** figures in brackets denote former carbon black codes in reference 1; X denotes film samples collected before and after exposures for full characterization.

The three HALS stabilizers (supplied by Ciba) were selected based on molecular size with regard to the relatively small Tinuvin 770 (bis-2,2,6,6-tetramethyl-4-piperidyl sebacate) (HALS2) and the polymeric Chimassorb 944 (poly {[6-[(1,1,3,3-tetramethylbutyl)-imino]-1,3,5-triazine-2,4-diyl][2-(2,2,6,6-tetramethylpiperidyl)-amino]-hexamethylene-[4-(2,2,6,6-tetramethylpiperidyl)-imino]} (HALS1). The potential effect of alkylation of the secondary amine hydrogen was examined by including Tinuvin 765 (bis-1,2,2,6,6-pentamethyl-4-piperidyl sebacate) (HALS3) in the study. The matrix in Table 1 thus provides carbon black particle size/structure combinations of Large/Low (C1), Large/High (C2), Small/High (C3 & C5) and Small/Low (C4) each in the presence of two of the three HALS selected.

Masterbatches at a nominal 5% loading in LLDPE for each HALS stabiliser were prepared on a BR Brabender mixer in Cabot Corporation laboratories, USA and then diluted to a concentration of 0.3% (w/w) by mixing with LLDPE pellets or carbon black (1.5 %(w/w))-LLDPE pellets prior to film production. The concentration of carbon black in each black-HALS-LLDPE ternary becomes 1.41% (w/w) during the compounding and subsequent extrusion into 75μm films described previously.[1]

The HALS stabilizer-containing films with and without carbon blacks were exposed in the Q-Panel, QUV/se source under UVA and UVB conditions at 1.25 and 0.63 Wm^{-2} respectively and at 60°C followed by characterization using the same techniques as described previously.[1]

3 Results

3.1 Unexposed Film Properties

A summary of mechanical and physico-chemical properties of carbon black-containing LLDPE films with and without HALS stabilizers is shown in Table 2.

After addition of the HALS stabilizers, the elongation-at-break values (with an error of about ±5%) increase for all the films except C3 film which shows decreases with respect to HALS-free analogues. Furthermore, a general decrease in fusion endotherm maximum temperatures occur suggesting that HALS have encouraged growth of a population of smaller polycrystallites. This is probably because these stabilizers have polar structures[14] which modify the intermolecular force and fine structural characteristics. However, all the stabilizer-containing films filled with carbon blacks show lower retained elongation-at-breaks than those for respectively HALS-stabilized films without carbon black due to the reinforcing effect of the pigment.

Table 2. Mechanical and morphological properties of carbon black-containing LLDPE films in the absence and presence of HALS stabilizers.

Code	ε, %	BL, N	T_m, °C	H_f, J/g	T_{on},°C
LL	773	12.2	115	67.4	234
HALS1-LL	809	12.9	114	72.7	223
HALS2-LL	796	12.4	110	97.0	241
HALS3-LL	808	12.8	-	-	-
C1	717	12.1	113	75.3	234
HALS1-C1	782	11.5	115	73.3	227
HALS2-C1	801	12.5	111	84.3	241
C2	754	12.9	114	76.6	237
HALS1-C2	776	11.9	-	-	-
HALS2-C2	781	11.9	-	-	-
C3	773	12.9	115	66.5	238
HALS1-C3	729	11.0	-	-	-
HALS2-C3	743	11.1	-	-	-
C4	724	11.4	113	73.3	237
HALS1-C7	738	11.0	110	76.2	232
HALS2-C7	749	11.0	109	84.0	239
C5	771	13.2	114	60.4	235
HALS2-C5	770	11.3	-	-	-
HALS3-C5	789	11.9	-	-	-

Key : ε - Elongation-at-break.
BL - Breaking load.
T_m - Melting temperature obtained from DSC melting peak.
H_f - Heat of fusion calculated from DSC melting endothermic area.
T_{on} - Onset temperature of post-fusion oxidation.

The addition of HALS2 stabilizer results in an increase in the DSC-derived crystallinity (as heat of fusion, H_f) for the pure, C1 and C4 films while the HALS1-filled films show comparable results to the HALS-free films. The results in Table 2 indicate, therefore, that apart from a possible nucleating effect by HALS, to a first approximation, all films have very similar tensile and morphological properties.

3.2 Films Exposed to UVB (310nm) Radiation

3.2.1 Tensile Properties

The tensile properties in terms of retained elongation as a function of exposure time for the LL films containing HALS1, HALS2 and HALS3 stabilizers only are plotted in Fig.1. It can be seen that the use of HALS in LLDPE films leads to pronounced improvement in UVB stability, as expected. However, it was observed that little if any difference of photostabilizing efficiency between the low (HALS2 and HALS3) and high molecular mass HALS (HALS1) exists in these films. It has been suggested[15,16] that diffusion of low molecular mass stabilizers to the outer film surfaces subjected to photo-oxidation may lead to an enhancement of UV stability. The comparable performances of low molecular mass NH-HALS, HALS2 and the corresponding N-methyl HALS, HALS3 have been attributed to very rapid transformation of the N-methyl group into the formic acid salt of NH-HALS during initial photo-oxidation[8] and the photostabilization of films results from subsequent reaction of the NH form.[15]

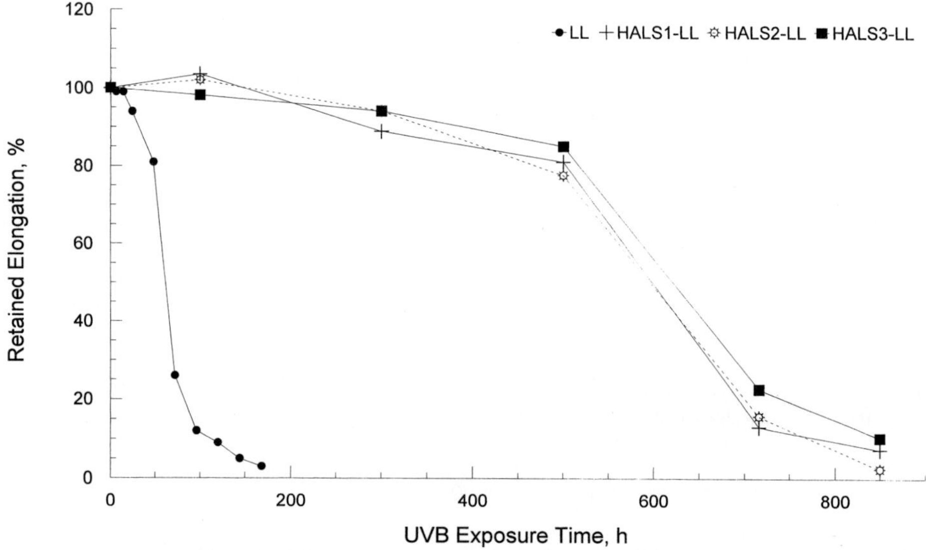

Fig. 1. Effect of UVB exposure and HALS on retained elongation of LLDPE films.

The effects of HALS1, HALS2 and HALS3 stabilizers on retained elongation for the films containing small particle-sized blacks are shown in Figs. 2 (a), (b) and (c) respectively in comparison with the HALS-free films. In all filled films, the addition of HALS stabilizers results in considerable increases in UV stability.

In terms of respective t_{20} values (the time for 20% elongation-at-break to be lost), the HALS1 and HALS2 films containing C1 and C2 blacks with large particle size show different HALS stabilizing effectiveness in contrast to the pure films (see Fig. 1 and Table 3). The HALS1 films containing either C1 or C2 black show better UV stability than respective HALS2 films. It is also noticed that all C2 films give better UV durability than similarly stabilized C1 films. These differences in behaviours may be associated with reduced migration of HALS1 coupled with varying levels of black-HALS interactions.

a)

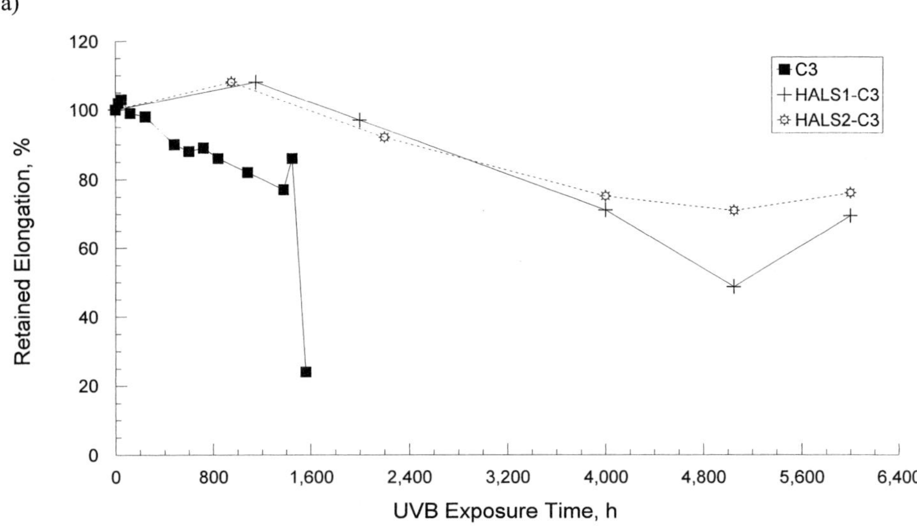

b)

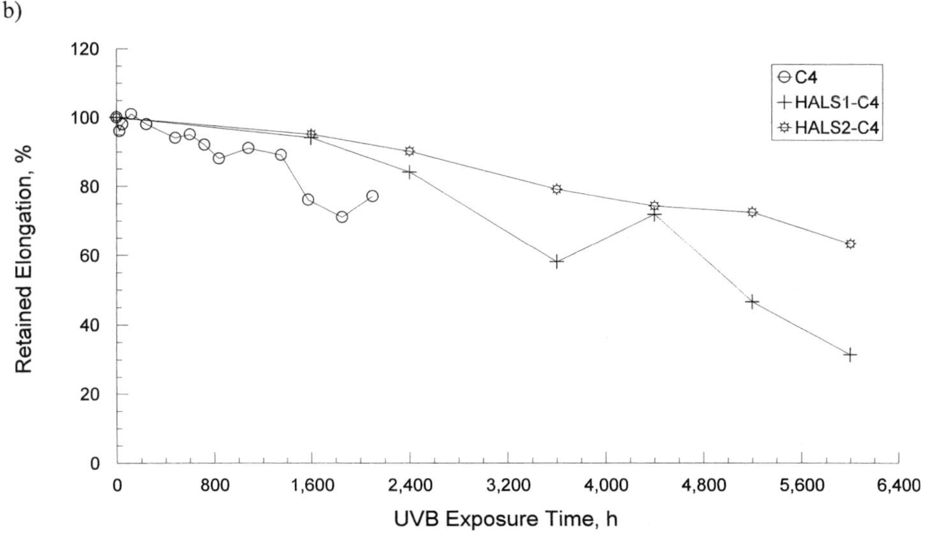

c)

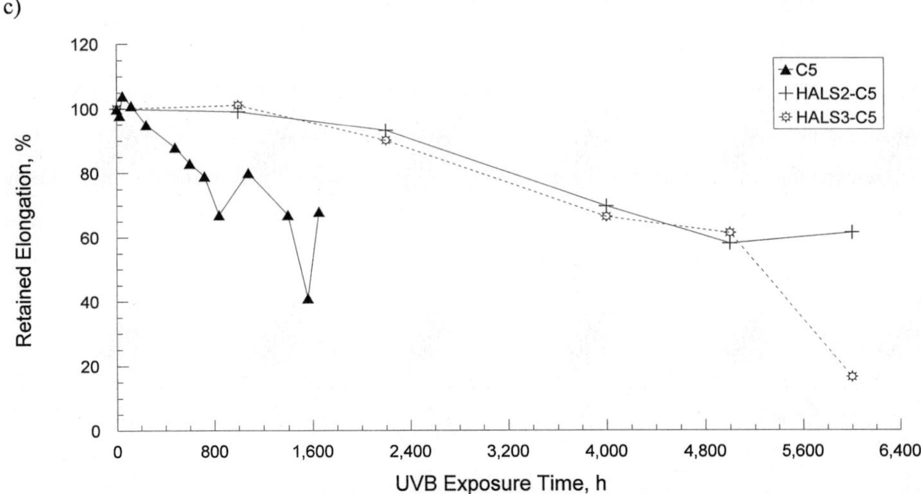

Fig. 2. Effect of UVB exposure and HALS on retained elongation of films filled with small particle-sized carbon black; (a) C3; (b) C4 and (c) C5.

Table 3. Times of 20% and 50% loss of retained elongation for HALS films in comparison with non-HALS films under UVB irradiation.

Non-HALS Films			HALS Films				
Code	t_{20}, h	t_{50}, h	Code	t_{20}, h	t_{50}, h	Synergism (%)	
						$S(t_{20})$	$S(t_{50})$
LL	50	60	HALS1-LL	510	600	-	-
			HALS2-LL	480	600	-	-
			HALS3-LL	540	620	-	-
C1	300	540	HALS1-C1	1550	2100	111	100
			HALS2-C1	1300	1650	84	56
C2	350	540	HALS1-C2	2000	3200	157	208
			HALS2-C2	1750	2450	133	134
C3	1050	1550	HALS1-C3	3400	>6000	130	>196
			HALS2-C3	3200	>6000	120	>196
C4	1450	>2200	HALS1-C4	2600	5100	37	-
			HALS2-C4	3600	>6000	94	-
C5	800	1550	HALS2-C5	3300	>6000	175	>193
			HALS3-C5	3000	5400	138	163

In C3 films, the HALS1 and HALS2 stabilizers show comparable results (see Fig. 2a) while in C4 films, the HALS2-containing film seems to be less UV degraded than the HALS1-containing film at each time (see Fig. 2b). In C5 films, both HALS2 and HALS3-containing films show similar gradual decreases in retained elongation with increasing exposure time up to 5000h after which the HALS3 film suddenly loses retained elongation-at-break compared with the HALS2-containing films which remain unchanged (see Fig. 2c).

The relative UV protective effects of the combination of HALS stabilizers and carbon blacks in comparison with the effects of HALS and blacks alone may be simply quantified in terms of changes in t_{20} and t_{50} values, the respective times of exposure required for a given film to lose 20 and 50% of elongation-at-break respectively.[1] These are given in Table 3 and show that in HALS1 films, stabilising performance order is C3>C4>C2>C1>LL and for HALS2, C4>C5>C3>C2>C1. The superior performance of the small particle-sized blacks is clearly evident. According to Al-Malaika *et al*,[17] the synergism between two stabilizers could be determined by using the following equation:

$$\% \text{ Synergism} = \frac{[E_s - E_c] - [E_1 - E_c] - [E_2 - E_c]}{E_1 - E_c + E_2 - E_c} \times 100\% \qquad (1)$$

where, E_s = Embrittlement time (t_{20}) of synergist

E_c = Embrittlement time (t_{20}) of control

E_1 = Embrittlement time (t_{20}) of stabilizer 1 (carbon black)

E_2 = Embrittlement time (t_{20}) of stabilizer 2 (HALS)

If we replace "Embrittlement time" by $t_{20,}$ then the calculated results based on this equation are shown in Table 3. These indicate that the mixtures of the three HAL stabilizers and any grade of carbon black used give synergistic photostabilizing effects in the LLDPE films during UVB exposure. In large particle-sized black films, the C2 films show higher values of synergism than the C1 films after the addition of HALS1 and HALS2 stabilizers.

3.2.2 FTIR Spectroscopic Studies

The main changes that have occurred in IR spectra for HALS-filled films with and without carbon black are characteristic of those relating to hydroperoxide and carbonyl groups in the absorption regions of 3100-3700cm^{-1} and 1500-1850cm^{-1}. Other IR peaks identified previously,[1] such as the 887 cm^{-1} vinylidene band were also monitored but found to be of low intensity during UV ageing of HALS only films and almost absent in HALS-carbon-containing films (see Figs. 3 and 4).

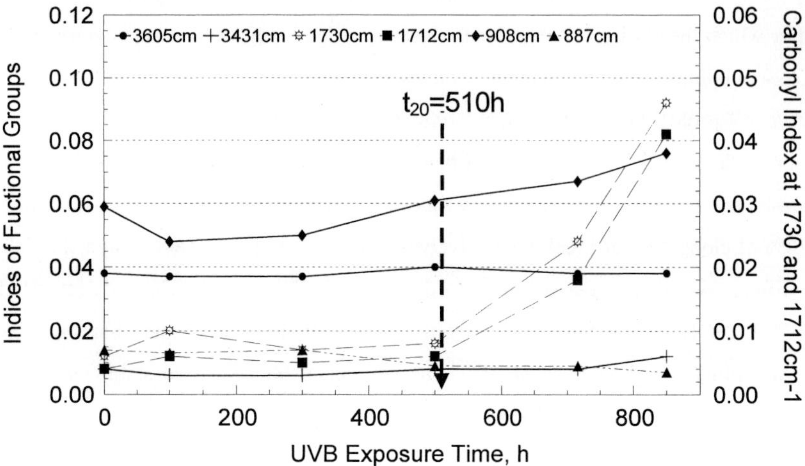

Fig. 3. Changes in absorbance index of functional groups with UVB exposure for LLDPE films containing HALS1.

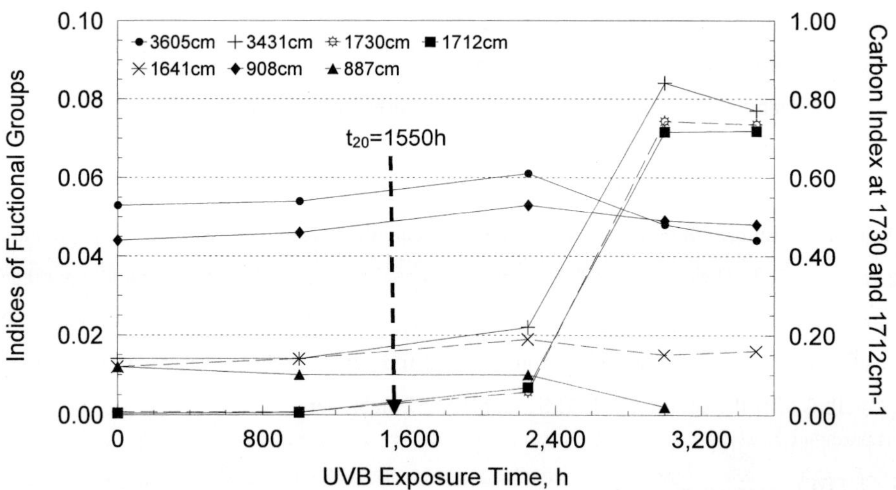

Fig. 4. Changes in absorbance index of functional groups with UVB exposure for C1 films containing HALS1.

With regard to the former, no independent hydroperoxide analysis has been carried out and so the assignment is an assumption and the conclusions drawn could be influenced by this. The control, C1 and C4 black containing films in the presence of HALS1 and HALS2 stabilizers have IR absorption spectral shapes in the region of 3100-3700cm^{-1} that are quite similar to their analogues without HALS where a broad absorption peaking at 3431cm^{-1} from associated hydroperoxide and a sharp peak at 3605cm^{-1} from free hydroperoxide were observed.[1] The former is substantially suppressed in the clear HALS films compared to the clear HALS-free films.

It is interesting that the absorption spectral detail in the region of 1500-1850cm^{-1} for the clear HALS films are quite different from the clear HALS-free films, but the growth of the most typical broad carbonyl band absorbing between 1712-1732cm^{-1} was noticed with increasing UV exposure time for all HALS-containing films. The C=C stretching band at 1641cm^{-1} present in vinyl and vinylidene groups was observed for all films before and after exposures. For the HALS1 films with and without carbon black prior to UV irradiation, two unique peaks were observed at 1531cm^{-1} and 1568cm^{-1} which could be associated with nitroxide derivatives[15] generated in the thermal extrusion process. These two peaks were not present after subsequent exposure. However, in the unexposed HALS3 films with and without carbon black, a sharp peak at 1738cm^{-1} associated with ester from the stabilizer itself was noticed which decreased in intensity with photo-ageing.

To illustrate changes in functional groups with exposure time for control and C1 (large particle size) films containing HALS1, the plots of functional group indices as a function of exposure time are given in Figs. 3 and 4 respectively. Respective t_{20} values are indicated on these curves. For all films with and without carbon black, after a certain induction period, the significant increases in index with exposure time were observed from ketone carbonyl at 1712cm^{-1} and ester at 1730cm^{-1} which were accompanied by increases in associated hydroperoxide at 3431cm^{-1}. However, the gradual growth of vinyl index at 908cm^{-1} observed in the carbon black-free films is not seen in the carbon black-filled films as noted previously.[1] Induction times, t_i, for carbonyl group formation are listed in Table 4.

Comparison of Tables 3 and 4 shows that respective t_{20} and t_i values are similar in the absence of carbon black. However, for C1 $t_i > t_{20}$, for C2 $t_i \sim t_{20}$ and for all small-particle sized black-containing films, $t_i < t_{20}$. It is interesting to note that the growth of carbonyl index stopped after 3000 h for HALS1-C1 (see Fig. 4) and after 4000 h exposure for HALS1-C4 films possibly because of lower diffusional mobility of the reactive centres in the polymeric HALS1 as previously suggested.[15]

Table 4. Induction time and Carbonyl Index at t_{20} and t_{50} under UVB irradiation.

Non-HALS Films			HALS Films			Synergism
Code	CI_{20}	t_i, h	Code	CI_{20}	t_i, h	$S(t_i)$,%
LL	0.047	72	HALS1-LL	0.006	500	-
			HALS2-LL	0.034	720	-
			HALS3-LL	0.012	720	-
C1	0.048	360	HALS1-C1	0.03	2250	204
			HALS2-C1	0.03	3000	213
C2	0.062	360	HALS1-C2	0.028	2000	169
			HALS2-C2	0.050	2000	106
C3	0.13	480	HALS1-C3	0.16	2200	155
			HALS2-C3	0.19	2200	102
C4	0.175	500	HALS1-C4	0.1	1600	79
			HALS2-C4	0.35	2400	116
C5	0.25	500	HALS2-C5	0.30	2200	98
			HALS3-C5	0.14	2200	98

Notes: t_i - Induction time.
CI_{20} - Carbonyl index at the time of 20% loss of retained elongation.

In the large particle-sized black HALS films, the rates of formation of associated hydroperoxide and carbonyl group in terms of index with exposure time for HALS2-C2 film are the fastest. For small particle-sized blacks, changes in these two functional groups with exposure time are very similar, although the rates of formation of these groups in HALS2-C5 film tend to be the fastest. It is further noticed that at longer exposure times, the rates of formation of associated hydroperoxide in the C3 and C5 films containing HALS tend to decrease and the rates of formation of carbonyl group in the C3, C4 and C5 films containing HALS tend to level off.

Using equation (1), percentage synergism factors, $S(t_i)$ may be calculated and these are listed in Table 4. These indicate that for suppression of carbonyl index, the larger carbon blacks are more effective than the smaller blacks unlike the results in Table 3, where no great size dependence was seen for either $S(t_{20})$ or $S(t_{50})$. Table 4 also lists, in addition to the induction times, t_i values, the carbonyl indices at t_{20} , namely CI_{20}. Assuming that surface chemistries are similar and that surface area of carbon black is a significant factor in determining photostabilising efficiency, possible relationships between t_{20}, t_i and CI_{20} values for HALS1 and HALS2 films and increasing carbon black surface area are depicted in Fig. 5. With increasing surface areas of carbon black, the values of t_{20}, t_i and their corresponding carbonyl index increase up to the surface area value of 140m^2/g after which t_{20} seems to remain constant, and CI_{20} tends to decrease, except for the value for HALS2-C4 film.

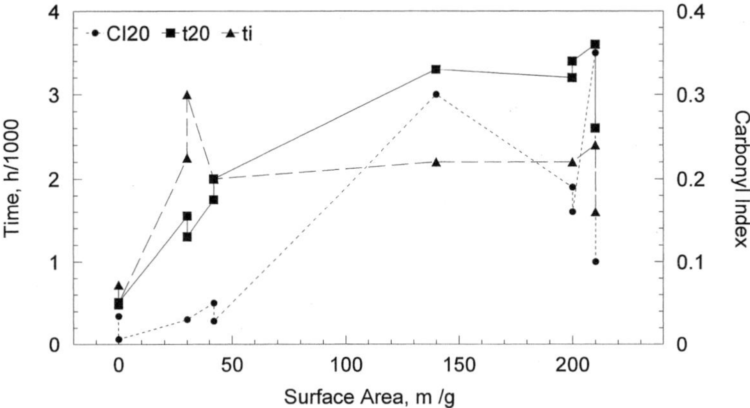

Fig. 5. Relationships of t_{20} , CI_{20} and t_i for HALS1 and HALS2 films versus carbon black surface area under UVB exposure.

3.2.3 DSC Studies

The melting temperatures, crystallinity and onset temperatures of post-fusion oxidation were obtained from the DSC tests as shown in Table 5. The melting temperatures for all HALS films tested show little change with increasing exposure time. The crystallinities derived from the DSC measurements with exposure time for both HALS1 or HALS2-containing LL films show continual rises similar to trends seen for all non-HALS-containing films with and without carbon black.[1] However, those containing carbon blacks initially show a similar increasing trend with exposure time as carbon black-free films up to 1000 h but then pass through a maximum. Thus the addition of HALS has no effect on the nucleating effect of carbon black, however, at longer exposure times, the consequences of chemical oxidative chain scissions dominate thus resulting in decreases in crystallinity.[18]

Table 5. DSC morphological properties of HALS filled films with and without carbon black after UVB exposure.

Code	UVB Time, h	T_m, ^0C	H_f, mJ/mg	T_{on}, ^0C
HALS1-LL	0	114	73	223
	100	110	109	205
	300	107	117	211
	716	111	125	210
	850	111	124	207
HALS2-LL	0	110	97	241
	100	110	107	204
	300	109	112	203
	716	109	124	207
	850	110	128	205
HALS1-C1	0	115	73	227
	1000	110	109	213
	2250	109	119	209
	3000	111	54	201
	3500	112	64	201
HALS2-C1	0	111	84	241
	1000	110	127	210
	2250	111	120	208
	3000	111	96	199
	3500	115	62	187
HALS1-C4	0	110	76	232
	1600	107	104	208
	3600	113	70	202
	4400	112	90	198
	6000	115	70	204
HALS2-C4	0	109	84	239
	1600	107	109	202
	3600	110	88	181
	4400	112	67	184
	6000	114	82	184

Notes: T_m - Melting temperature obtained from melting peak.
H_f - Heat of fusion calculated from melting endothermic area.
T_{on} - Onset temperature of post-fusion oxidation.

Generally, all T_{on} values decrease with time with asymptotic trends being shown by clear HALS films and almost linear behaviour when carbon blacks are present. These trends suggest that thermally reactive species produced as a consequence of UV exposure are significantly reduced in concentration in the presence of HALS and carbon black. Table 5 shows that HALS2-C1 and HALS2-C4 films yield lower temperatures than other films at longer exposure times. While little differences in T_{on} occurred between HALS1 and HALS2 stabilized controls, the latter effect suggests that HALS2 may be a less effective photostabiliser than HALS1. It is interesting to note that this difference between HALS1 and HALS2 is seen in Table 4 where CI_{20} figures are greater for HALS2-containing films.

3.3 Films Exposed to QUV/se, UVA Radiation Only

3.3.1 Tensile Properties

Under the UVA exposure conditions, the use of HALS significantly improves the UV stability of the films and the HALS1 film shows better photostabilizing efficiency than the HALS2 and HALS3 films. The superior photostability of HALS1 may be attributed not only to its stable polymeric structure, but also to the longer wavelength radiation emitted from the UVA source which favours photo-oxidation as opposed to photolysis.

Table 6. Times of 20% and 50% loss of retained elongation for HALS films in comparison with non-HALS films under UVA exposure.

Non-HALS Films			HALS Films			Synergism, %	
Code	t_{20} ,h	t_{50} ,h	Code	t_{20} ,h	t_{50} ,h	$S(t_{20})$	$S(t_{50})$
LL	110	120	HALS1-LL	650	1050	-	-
			HALS2-LL	400	600	-	-
			HALS3-LL	550	650	-	-
C1	450	800	HALS1-C1	1350	2200	41	29
			HALS2-C1	800	1550	10	23
C2	500	900	HALS1-C2	2000	4000	103	134
			HALS2-C2	2000	2800	178	122
C3	2100	2800	HALS1-C3	5400	>6300	109	>71
			HALS2-C3	4400	>6300	88	>96
C4	2200	3200	HALS1-C4	3700	6300	39	54
			HALS2-C4	5000	7200	105	99
C5	1800	2800	HALS2-C5	5200	>6300	157	>75
			HALS3-C5	4700	>6300	115	>99

A pronounced improvement of UVA durability was also seen in the carbon black-filled films after addition of HALS in a manner similar to UVB-exposed films. In the large particle sized-black films, the HALS1 stabilizer provides better protection to photodegradation than the HALS2 stabilizer and the C2 films have better UV durability than the C1 films; this difference was observed as well during UVB exposure. From Table 6, the UV stability of the films containing HALS1 stabilizer can be described as the order of C3>C4>C2>C1>LL and in the presence of HALS2, there is little distinction between the small particle sized-black films giving the durability order of C4 ~ C5>C3>C2>C1>LL films.

By using equation (1), synergism factors between carbon black and HALS during UVA exposure were calculated as shown in Table 6 for t_{20} and t_{50} conditions for HALS films in comparison with HALS-free films for the same UVA exposure. These results again indicate that there exists synergism between carbon black and HALS under UVA exposure conditions as well. Compared to Table 3 for the UVB-exposed films, the values of synergism at t_{20} and t_{50} in Table 6 are generally lower. This may indicate that the addition of HALS results in greater improvement of UV durability during UVB exposure than during UVA exposure.

3.3.2 FTIR Spectroscopic Studies

Overall, the changes in the IR spectra for the selected films during UVA exposures are similar to the same films exposed to the UVB source (see Figs. 3 and 4). Again, the formation of associated hydroperoxide and carbonyl groups in the clear HALS films was significantly suppressed by the presence of HALS. Compared to the HALS-free films, the absorbance indices of the two groups are relatively low even after complete loss of elongations-at-break.

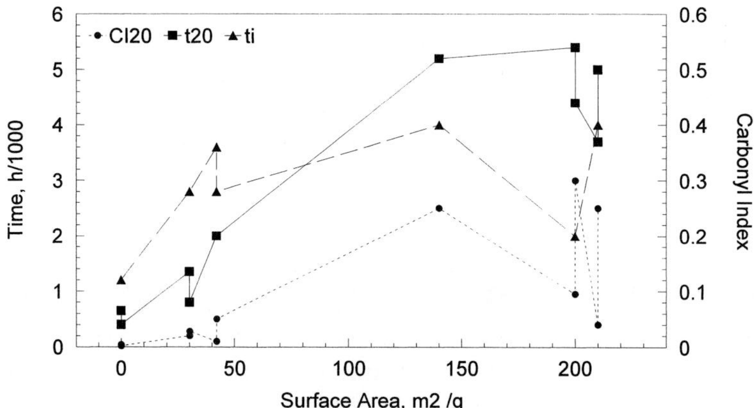

Fig. 6. Relationships of t_{20}, CI_{20} and t_i for HALS1 and HALS2 films versus carbon black surface area under UVA exposure.

Induction time values t_i, for carbonyl group generation are listed in Table 7 along with respective CI_{20} values. Generally when compared with similar results for UVB exposure in Table 4, respective t_i values are longer and CI_{20} values less for UVA-exposed samples. For clear LL HALS-containing films and large particle-sized black-containing films, $t_i > t_{20}$. For small particle-sized blacks, t_i < t_{20} possibly indicating their superior property of reducing Norrish II scissions which would otherwise accelerate carbonyl group formation. Percentage synergism factors, $S(t_i)$, calculated from equation (1) are also listed in Table 7 and apart from C5 films, tend to exhibit respectively lower values than shown in Table 4 for UVB-exposed samples. It is impossible to discern a simple carbon black property relationship here, especially since the C3 films indicate antagonism (with negative $S(t_i)$ values), and the similar small particle-sized, high structured black in C5 films showing very high levels of synergy. Furthermore, they do not match the synergism values based on tensile loss times in Table 6. The plots of t_i, and t_{20}, each as a function of surface area of carbon black are shown for HALS1 and HALS2 films in Fig. 6. In general, t_i and t_{20} values increase with increasing surface area of carbon black up to the value of 140m^2/g regardless of the types of HALS, as observed in the same films exposed to UVB exposure conditions (see Fig. 5). Above this, t_{20} tends to gradually increase further before dropping at 200 m^2/g in contrast to the t_i value, which has achieved its maximum. Fig. 6 also shows that, as seen for the HALS films exposed to the UVB source, t_i are greater than respective t_{20} values for the control and C1 (30m^2/g) films containing both of HALS1 and HALS2 stabilizers. However, with increasing surface area of carbon black, t_i < t_{20} except for the t_{20} value of HALS1-C4 film.

Table 7. Induction time and carbonyl indices at t_{20} under UVA exposure.

Non-HALS Films			HALS Films			Synergism
Code	CI_{20}	t_i, h	Code	CI_{20}	t_i, h	$S(t_i)$,%
LL	0.04	150	HALS1-LL	0.004	1200	-
			HALS2-LL	0.002	1200	-
			HALS3-LL	0.013	800	-
C1	0.05	700	HALS1-C1	0.02	2800	66
			HALS2-C1	0.028	2800	66
C2	0.10	700	HALS1-C2	0.01	3600	116
			HALS2-C2	0.05	2800	66
C3	0.22	1200	HALS1-C3	0.095	2000	-12
			HALS2C3	0.30	2000	-12
C4	0.22	1200	HALS1-C4	0.04	4000	83
			HALS2-C4	0.25	4000	83
C5	0.22	1200	HALS2-C5	0.25	4000	175
			HALS3-C5	0.23	4000	285

Notes: t_i - Induction time.
CI_{20} - Carbonyl index at the time of 20% loss of retained elongation.

3.3.3 DSC Studies

With exposure, there is little change in the melting temperatures for all the exposed HALS films tested. The DSC-derived crystallinities are very similar to those for UVB-exposed films. In contrast to the control films with HALS that show only an initial increase in crystallinity and then remain constant, those with carbon black and HALS show initial increases to respective maxima with increasing exposure time.

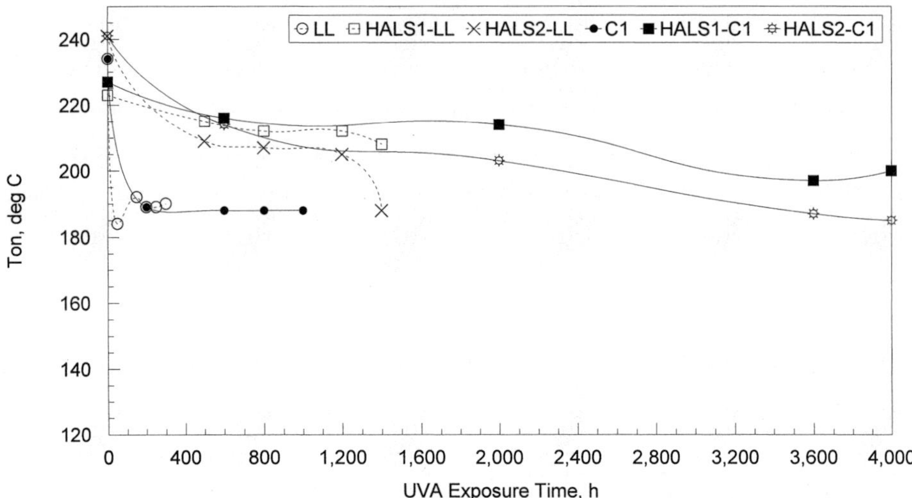

Fig. 7. Effect of HALS on oxidation onset temperatures of the LLDPE, C1 and C4 films after UVA exposure.

T_{on} values as a function of exposure time for the films selected for DSC tests are plotted in Fig. 7. As seen for UVB-exposed films previously (Table 5), under UVA exposure T_{on} versus time trends for HALS films asymptote becoming almost linear when carbon is present (note the HALS-C1 and HALS C2 trends). In the latter, higher T_{on} values may be due to the presence of HALS trapping free radicals thus inhibiting photo-oxidation and the generation of oxidation sensitive groups as mentioned above. T_{on} of the HALS1-filled films have higher values at all exposure times than those of the HALS2-filled films in agreement with the time-dependence of retained elongation results. Thus the relative tensile and physicochemical (by FTIR and DSC) property changes demonstrate consistently the superior photostabilising effect of HALS1 HALS.

4 Possible Mechanism of the Combined Effect of HALS and Carbon Black

The combined photostabilizing effect of carbon black and HALS may be summarized from various points of view:

(i) Elongation-at-break Retention: In terms of retention of elongation-at-break, UV durability of carbon black-HALS-containing films follows the order of small>large particle size of carbon black regardless of types of HALS as seen for the HALS-free films.[1] The HALS1-stabilized films show better UV resistance than the HALS2-stabilized films in large particle-sized black (C1 and C2) films. However, in small particle-sized black (C3 and C4) films, there is superior UV stability of the HALS2-filled compared to the HALS1-filled film containing C4 black in contrast to those containing C3 black. These latter show little if any differences between the HALS1 and HALS2-stabilized films. The percentage synergism values (see Tables 3 and 6) support the above observation.

(ii) Carbonyl Group Formation: With regard to carbon black variables, the large particle-sized black films show faster rates of formation of carbonyl group compared to those containing small particle-sized blacks. But induction times (t_i) show significant differences only under UVA conditions for small particle-sized blacks where $t_i < t_{20}$ holds for both exposure conditions. As previously observed from the HALS-free films,[1] the carbonyl index values at t_{20} and t_{50} increase with increasing specific surface areas (which relate inversely to particle size) up to a surface area value of 140 m^2/g (equivalent to 19nm particle diameter). This suggests that the presence of HALS may not suppress the formation of carbon black-initiated carbonyl groups from the larger surface area of carbon blacks having particle diameter >19nm while generating high t_{20} and t_{50} values. Furthermore, since it has been proposed[1] that carbon blacks may generate carbonyl groups by reducing Norrish II type chain scissions in UV-exposed HALS-free films, so this effect continues and is little affected in the HALS-containing films.

However, if we consider the stabilizing efficiencies of HALS1 and HALS2 stabilizers in terms of their effects on t_i in filled films, the lower molecular weight HALS2 stabilizer is no less effective than the HALS1 stabilizer (see Tables 4 and 7), in spite of its higher diffusional mobility. In addition, the use of HALS2 stabilizer slightly enhances the formation of carbonyl group in small particle-sized black films.

(iii) DSC-derived T_{on} trends show that after longer exposures, values for the HALS2-stabilized films are lower than those for the HALS1-stabilized films. These results correlate with the higher values of carbonyl index CI_{20} values for the HALS2-stabilized films (see Tables 4 and 8), indicating that the extent of photodegradation may be measured by the magnitude of depression of T_{on}.

In summary, the mechanism of synergism between HALS and carbon black does not seem to involve any new chemical effects. The effect of carbon black may be essentially physical where the screening effect also protects the HALS from photolysis, because the intensity of UV radiation is reduced by its presence. Hence the observed synergistic effects between HALS and carbon black are the greatest for smaller particle sizes with UVA conditions showing greater differences. It is likely that the reduced UV radiation intensity is not intense enough to decompose the light stabilising component and its derivatives giving rise to lower degradation and hence HALS consumption thus extending the induction time for photostabilisation of the carbon black-HALS films. Furthermore, the participation of carbonyl groups generated or present on surfaces of carbon black particles may also be involved in these photostabilizing reactions. This appears to be especially so under UVA conditions where CI_{20} values are generally greater for non-HALS than for HALS-containing black films (see Table 7).
One final consideration, of course, is the possible thermal antioxidant effect of the HALS present which will especially be evident if the local carbon black-polymer surface interface is higher in temperature than is the matrix in clear films. In the present experiments possible resolution of the observed synergistic effects into respective photo- and thermal stabilising effects is not possible. However, it might be assumed in the carbon black only-containing tapes, that increases in t_{20} and t_i values will be sum of a positive photodegradative stabilising effect and a negative enhanced thermal degradative one. Addition of HALS will have a positive effect on both these and so contribute to the S(t) values defined in Tables 3, 4, and 7. That the $S(t_{20})$ and particularly $S(t_i)$ values were greater for UVB than for UVA degraded films may also be partly credited to this thermal stabilising effect, although further work involving local polymer temperature measurement during irradiation would be required to more fully understand its importance.

[1] M. Liu, A. R. Horrocks, *Polym Degrad Stab* **2002**, 75, 485. see also M Liu, PhD Thesis, University of Manchester, UK, 1997.
[2] W. L. Hawkins, M. A. Worthington, F. H. Winslow, *Rubber Age* **1960**, 88: 279.
[3] W. L. Hawkins, F. H. Winslow, in: „*Reinforcement of Elastomers*", G. Kraus, Ed., Interscience Publishers, New York, **1965**, 566.
[4] H. Schneider, in: „Geotextiles, geomembranes and related products" G. Den Hoedt, Ed., Balkema, Rotterdam **1990**, 723.
[5] E. Kovacs, Z. Wolkober, in: „*Degradation and stabilization of polyolefins*", B. Sedlacek, C. G. Overberger, H. F. Mark, T. G. Fox, Eds., J Polym Sci. Polymer Symposia **1976**, 57, 171.
[6] W. L. Hawkins, V. L. Lanza, B. B. Loeffler, W. Matreyek, F. H. Winslow, *J. App Polym Sci* **1959**, 1, 43.
[7] W. L. Hawkins, in: „*Degradation and stabilization of polyolefins*", B. Sedlacek, C. G. Overberger, H. F. Mark, T. G. Fox, Eds., J Polym Sci. Polymer Symposia **1976**, 57, 319.
[8] F. Gugumus. in: „*Oxidation inhibition in organic materials*", .J Pospisil, P. P. Klemchuk, Eds., CRC Press, Florida, **1990**, 29.
[9] R. L. Gray, *Geotechnical Fabric Repor,t* **1990**, (Nov), 22.
[10] E .Stengrevics, P. Horng, *Plast Compd,* **1987**, 10(35), 35.
[11] J. M. Pena, N. S. Allen, M. Edge, C. M. Liauw, B. Valnage, *Polym Degrad Stab,* **2001**, 72, 163.
[12]] J. M. Pena, N. S. Allen, M. Edge, C. M. Liauw, B. Valnage, *Polym Degrad Stab,* **2001**, 72, 259.
[13] W. M. Hess, C. R. Herd, in: „*Carbon Black*", 2nd ed., J. Donnet, R. C. Bansal, M. Wang, eds., Marcel Dekker Inc, New York, **1993**, 116.
[14] Anon. Ciba Geigy Ltd. Publication, No. 28474/e, Sept. **1986**.
[15] D. K. C. Hodgeman, in: „*Developments in polymer degradation*", volume 4, N. Grassie, ed., Applied Science Publishers Ltd, London, **1982**, 189.
[16] F. Gugumus, *J. Polym Degrad Stab,* **1995**, 50, 101.
[17] S. Al-Malaika, G. Scott, in: „*Degradation and stabilization of polyolefins*", N. S. Allen, ed., Applied Science Publishers Ltd, London, **1983**, 283.
[18] M. Liu, A. R. Horrocks, M. Hall, *J Polym Degrad Stab,* **1995**, 49, 151.

Evaluation of the Fire-Retardant Properties of New Modifiers in Unsaturated Polyester Resins Using the Cone Calorimetric Method

Ewa Kicko-Walczak

Industrial Chemistry Research Institute, Department of Ecological Polymers and Materials for Medicine, Technology and Modification of Polyester Resins Laboratory; Rydygiera 8, 01-793 Warszawa, Poland
Email: Ewa.Kicko-Walczak@ichp.pl

Summary: An investigation of the burning behaviour of a series of unsaturated polyester resin formulations has been carried out using the Cone calorimeter technique. A complex analysis of the pyrolysis of selected glass-reinforced polyester (GRP) laminates was made based on fire-nonretarded, fire-retarded and smoke suppressed compositions of unsaturated polyester resins (UPR) by the use of mixtures of zinc hydroxystannate $ZnSn(OH)_6$, (ZHS), zinc stannate $ZnSnO_3$ (ZS) and zinc borate or zinc molybdate. The heat release (average) rate (HRR and HRR_{av}), time required to ignite a GRP, the mass loss rate and (average) surface extinction area (SAE and SAE_{av}) were studied in relation to thermal irradiation power and to presence or absence of a fire retardant. The cone calorimeter is shown to enable all the stages of pyrolysis of GRP involving combustion and smoke emission to be analyzed, and confirmed the good effects of Zn/Sn systems as a fire retardant additives for UPR.

Keywords: Cone calorimetry; polyesters; pyrolysis; reinforcement, unsaturated polyester resins (UPR)

Introduction

Fire retardancy of polymeric materials has long been studied at several national centers and by numerous European and American companies. These studies have been focused primarily on the mechanism of the polymer combustion phenomena, on new fire retardants, toxicity, improvement of flammability tests and evaluation of the degree of fire-retardancy and standardization of products.

An essential problem encountered in conflagrations involving plastics is the emission of smoke. This hazard is particularly imminent when curable polymers of the unsaturated polyester resin (UPR) type are used and less so when thermoplastics are involved. I have already presented[1] a

 DOI: 10.1002/masy.200351219

general characteristics of the course of combustion of UPR and of glass-reinforced polyester laminates (GRP) and also described the principles underlying the fire retardancy of such products. Related information has also been presented elsewhere.[2-6]

Generally, flame retardancy is achieved by modification of polyester resins involving introduction of additive fire-retardants.[3,4] The most effective modifiers should change the course of the chemical reactions occurring on pyrolysis of UPR or GRP so as to reduce the content of aromatic hydrocarbons in the products which, on burning, produce soot and to increase the content of aliphatic hydrocarbons. This way increases the amount of coal passing to the solid residue, known as the „coke".

Emission of smoke and toxic gases is related to the chemical composition of a burning UPR. It should be emphasized, however, that even a specially designed structure of the polyester backbone does not guarantee that the resulting product will be difficult to burn to. Therefore, it was deemed to be advisable to use modifiers whose relatively low percentages reduce combustion and emission of smoke.[6]

Two inorganic tin compounds, zinc hydroxystannate (ZHS) and zinc stannate (ZS), have recently been introduced as novel fire-retardant additives for polymer formulations. Previous work at ICHRI[1] has shown that these essentially non-toxic compounds are effective flame retardants and smoke suppressants in a range of polymeric substrates, including halogenated and non-chlorinated polyesters.

To characterize the flammability and to class the polymeric materials into suitable categories of fire resistance requires the combustion factors to be fully established, including the ignition temperature, the time to start burning, the rate of burning, the time of spontaneous burning, the scope of burning and the percentage of char.

TGA, DTA, gas chromatography and mass spectrometry are helpful in following the combustion phenomena.

The most objective and representative of the behavior of polymers on burning are now considered to be studies carried out by the use of cone calorimetry,[8-11] which involve the measurement of the rate of heat evolution and generation of smoke released from various materials.

Experimental

Materials

- Polimal 120 (Z.Ch. Organika-Sarzyna, Sarzyna, PL), isophthalic-maleic-propylene polyester resin: styrene, 35%: no halogens.
- Polimal 161 (Z.Ch. Organika-Sarzyna, Sarzyna, PL), a maleic-phthalic-epichlorohydrin polyester resin with an incorporated bromine compound: styrene 35%; 9,3% Cl + 6,7% Br.
- Polimal HET (Z.Ch. Organika-Sarzyna, Sarzyna, PL) a polyester resin prepared from maleic anhydride, hexachloroendomethylenetetrahydrophthalic (HET) acid and diethylene glycol; styrene, 33%; Cl 21%.
- A borosilicate glass roving, surface weight 450g/m^2.
- Zinc hydroxystannate (ALCAN Chemical Europe, UK)
- Zinc stannate (ALCAN Chemical Europe, UK)
- Zinc borate, $2ZnO \cdot 3B_2O_3 \cdot 3H_2O$ (ALCAN Chemical Europe, UK)
- Zinc molybdate, $ZnMoO_4$ (Sherwin-Williams Chemical, UK)
- Ketonox, methyl ethyl ketone peroxide, a 40% solution in dimethyl phthalate
- Cobalt naphthenate, a 1% Co solution in styrene.

Preparation of Samples

Three GRP samples 100 x 100 x 100 mm in size were examined, each representing a polyester composition selected as a result of preliminary fire-retardancy study and each containing or not containing selected fire-retardants. The samples were conditioned at 23 ± 2° C, at R. H. 50 ± 5%, to constant weight (± 1%). Before being examined, the sample was wrapped into an aluminum film and screened at the back with a ceramic blanket and then fixed in a holder inside the calorimetric chamber. The samples were exposed in a horizontal position to thermal radiation of a selected power equal to 25, 35 or 50 kW/m^2. A sparking igniter was used to ignite gases.

Methods of Study

Earlier studies on flame retardancy of selected polyester materials carried out by different techniques and reported elsewhere[11] served us to select the most efficient fire retardants

constituting primarily the Sn/Zn system for the present cone calorimetry studies (''Cone 2" type calorimeter, ATLAS Electronic Devices Co.). Two GRP laminate types were chosen, based on the following polyester compositions;
(A) chlorinated Polimal HET and a nonhalogenated Polimal 120 containing $ZnSn(OH)_6$ + $2ZnO \cdot 3B_2O_3 \cdot 3H_2O$ as fire-retardant, smoking intensity 12 980 Lx·s and oxygen index 39,7%;
(B) halogenated (Cl + Br) Polimal 161 and nonhalogenated Polimal 120 containing $ZnSnO_3$+ $ZnMoO_4$; smoking intensity 13 459 Lx.s, oxygen index 38,7%. The laminates were prepared as described elsewhere.[11]

The principle of operation of the cone calorimeter is based on the laws of consumption of oxygen, i.e., the heat evolved on combustion can be determined quantitatively in terms of the oxygen consumed. With most flammable materials including high-M plastics, organic liquids, and wood, the heat released is 13,1 MJ per 1 kilogram of oxygen consumed. Deviations from this average value encountered with various materials do not exceed ± 5%. This value should not be confused with the heat of combustion which is defined as the quantity of heat released per 1 kilogram of the material burnt.
A cone calorimeter measures the concentration of oxygen in the gases evolved and the flow rate of these gases in the exhaust. The combination of the quantity of oxygen consumed and the constant 13,1 MJ allows to establish the amount of the heat released and the rate of heat release.[8,9] A considerable progress in the method is that results can be presented in the form of physical data characterizing the flammability properties of materials.[10]
Smoke evolution was studied with the aid of a laser photometer. Results are given in rational units of "specific extinction area" which is the area (m^2) of extinction produced by a unit mass (kg) of the sample. Although the scientific principles of such measurements have long been known, the cone calorimeter is the first device in which the above unit is consistent with the theory. This is the counterpart of smoke evolution.
The decrease in sample weight is expressed in $kg \cdot m^{-2}$ and the average weight loss rate per sample's unit surface area ($g \cdot m^{-2} \cdot s^{-1}$) is calculated over the experimental time from the moment of ignition until the end of the test. The sample is held on a balance which allows to measure the rate of weight loss during the test.

Results and Discussion

The HRR curves are presented (Fig. 1) for the GRP samples containing and not containing the Sn/Zn fire-retarding system in relation to thermal radiation power. Regardless of the polyester type and irradiation power, the curves ascertained allow to distinguish the following stages. The initial segments correspond to the period of heating of the polyester, volatilization of volatile parts and evolution of gases. The length of the section is related to the power of thermal irradiation; the higher the power, the shorter the section is (cf. Figs. 1a, c).

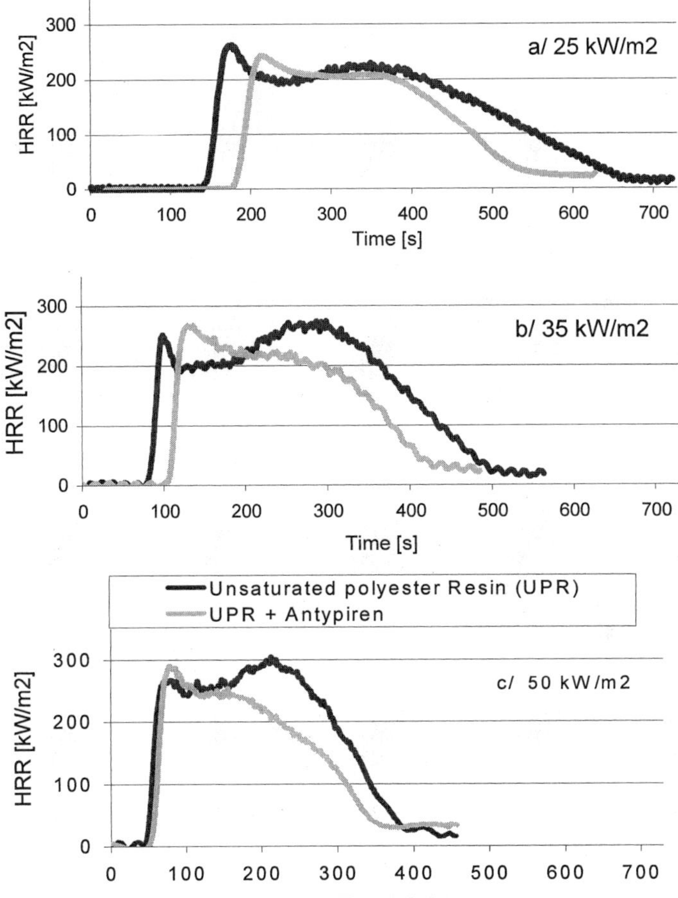

Fig. 1. The heat release rate (HRR) in relation to thermal irradiation power (kW/m^2): a-25, b-35, c- 50; 1 - fire-nonretarded, 2- fire-retarded GRPL.

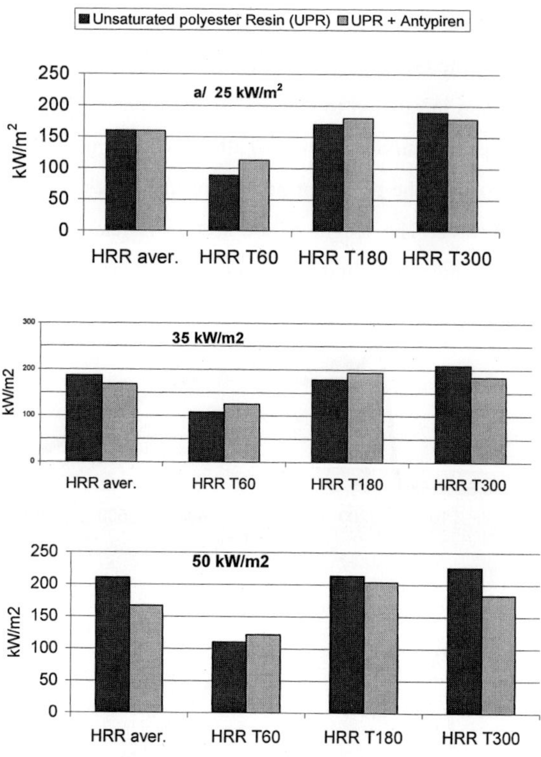

Fig. 2. The average HRR (HRR_{av}) and the HRR at the 60th, 180th and 300th second reckoned from ignition in relation to thermal irradiation power (for other designationss see Fig. 1).

The evolving gases become ignited. Immediately past the ignition point, the HRR curve sharply increases to produce a peak corresponding to the combustion of pyrolyzates and evolution of a high amount of heat (ca. 250kW/m^2). The time of decomposition an combustion is related to the power of thermal radiation; the higher the power, the shorter the time is. In case of fire-nonretarded GRP, the heat release rate (HRR) increased as the power of thermal radiation was increased. The curves (Fig.1) show that, at 25 kW/m^2 used as the power of thermal irradiation, the polyester was burnt in a more mild manner and unlike the case at 35 and 50 kW/m^2, the second sharp peak (at 270 to ~`300 kW/m^2) was missing. In case of fire-retarded GRP samples, the rate of heat release did not fall past the first maximum (first HRR peak), i.e., there was no other HRR peak. This fact is due to the insulating effect of the carbon layer which stabilizes the release of the heat within the period of time that is related to the power of irradiation. The higher the power of irradiation, the more pronounced is the peak.

The data in Fig. 2 allow to see that, with fire-nonretarded GRP, the average heat release rate rose from 160 to 210 kW/m^2 as the power of irradiation was raised. With fire-retarded GRP, the rise was insignificant, 159 to 166 kW/m^2. The effect of the fire retardant is most eminent at the 300th second, unlike that at the 60th and the 180th second.

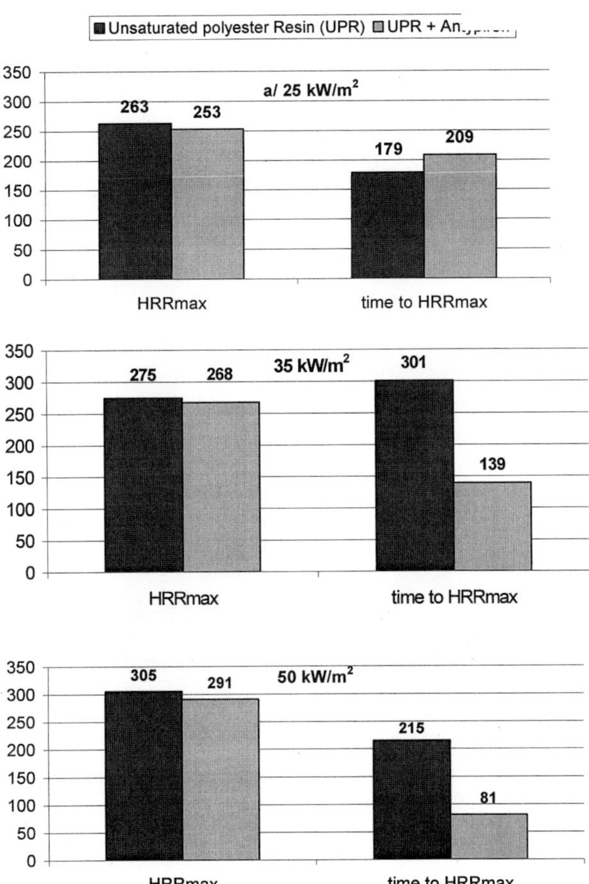

Fig. 3. The maximum HRR (HRR_{max}) and time required to attain HRR_{max} in relation to thermal irradiation power (for other designations see Fig. 1.).

The maximum heat release rate (HRR_{max}) is seen (Fig. 3) to increase as the power of thermal irradiation is increased, regardless of the GRP type. The fire retardant is seen to have reduced the HRR_{max} only slightly: on the other hand, the time elapsed to achieve HRR_{max} varies considerably. The presence of fire-retardants has affected the intensity of burning at the second stage of the fire. With fire-retarded GRP, the intensity of the second stage of the fire has been reduced. When the power of thermal irradiation used was 35 or 50 kW/m^2, the polyester burned more intensely and the time to reach HRR_{max} was longer, but within this period of time the fire-retarded polyester is seen to release heat at a reduced rate.

The time to ignite (τ_{ign}) is presented (Fig.4) in relation to thermal irradiation power (Fig.4); the higher the power, the shorter is the time to ignite the polyester. The fire retardant added to the polyester has invariably protracted the time required to ignite the polyester . The higher the difference between the powers of irradiation, the more pronounced is the effect of the fire retardant.

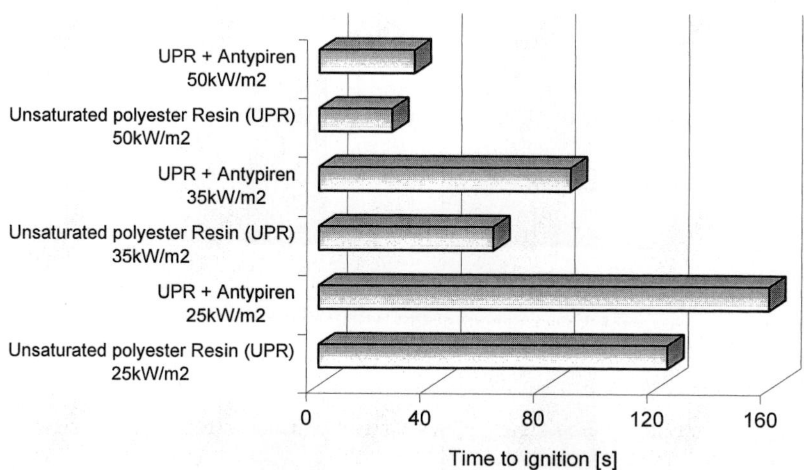

Fig. 4. Time required to ignite (τ_{ign}) in relation to thermal irradiation power: 1 - fire-retarded, 2 - fire nonretarded GRP.

The total heat released (THR) by the polyester is expressed in MJ/m^2. In presence of a fire retardant (Fig.5), the polyester's THR is reduced (Fig.5), viz., 56-58 MJ/m^2; and there is no large difference between the heat released and the thermal irradiation. With fire - nonretarded GRP, the THR is higher by 30 to 40%, viz., 72-83 MJ/m^2.

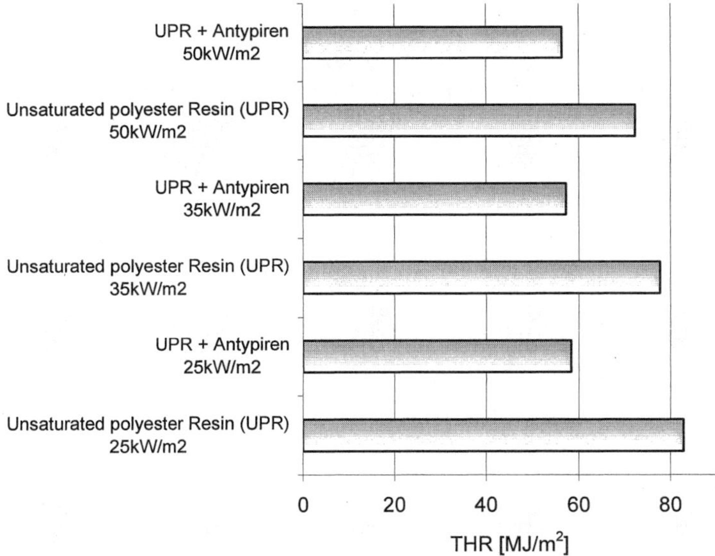

Fig. 5. Total heat released (THR) by: 1 - fire retarded, 2 - fire-nonretarded GRP in relation to thermal irradiation power.

The specific extinction area (SEA) curves are presented (Fig. 6) in relation to thermal irradiation power. In presence of a fire retardant, the amount of smoke is considerably smaller. The fire-retarded samples have invariably shown suppressed smoke evolution.

The difference is particularly well visible in the average extinction area (SEA_{av}) that has been measured over the entire duration of the test starting with the moment of ignition (Fig.7).The fire-retarded GRP show SEA_{av} of 471-541 m^2/kg v.s. the 1023-1037 m^2/kg for the non-retarded GRP.

The mass loss rate (MLR) curves, which are presented in relation to thermal radiation power (Fig.8), are seen to follow courses analogous to those of the HRR curves (Fig.1). Starting with the moment of ignition, the MLR curve rises to reach a certain level and then a maximum within a period of time that is related to the heat flux. The mass losses are the largest when the entire polyester surface area exposed to thermal radiation is burning. In presence of a fire-ratardant, the second maximum in the MLR curve does not occur.

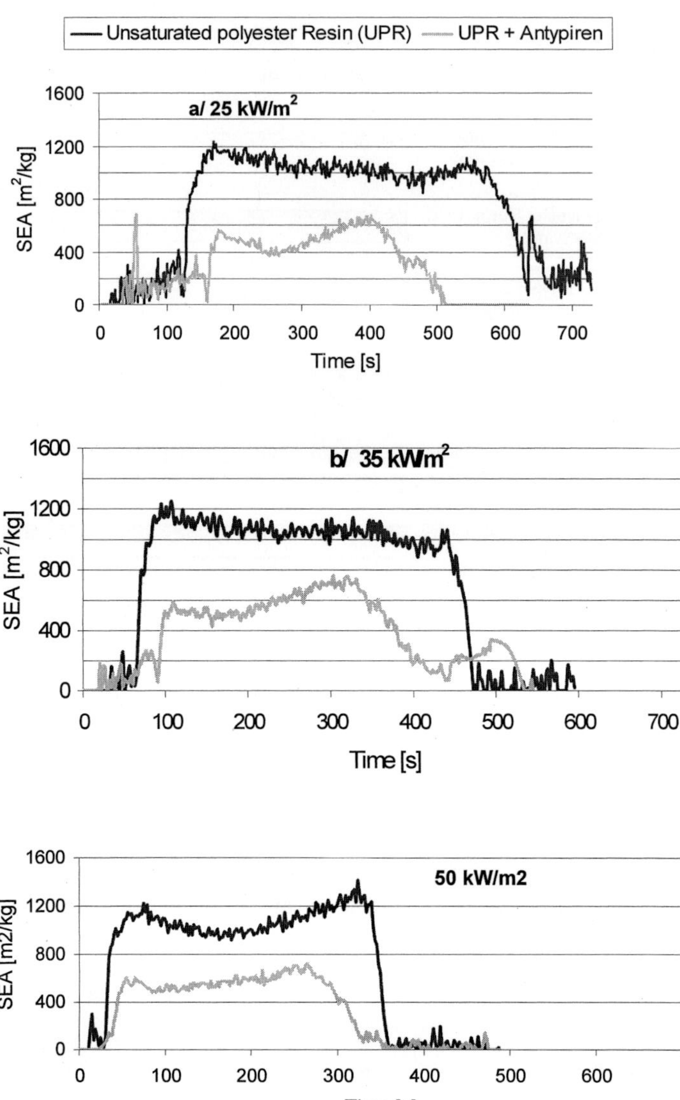

Fig. 6. The specific extinction area (SEA, m^2/kg) of: 1 - fire-nonretarded, 2- fire-retarded GRP, in relation to thermal irradiation power: a - 25, b - 35, c - 50 kW/m^2.

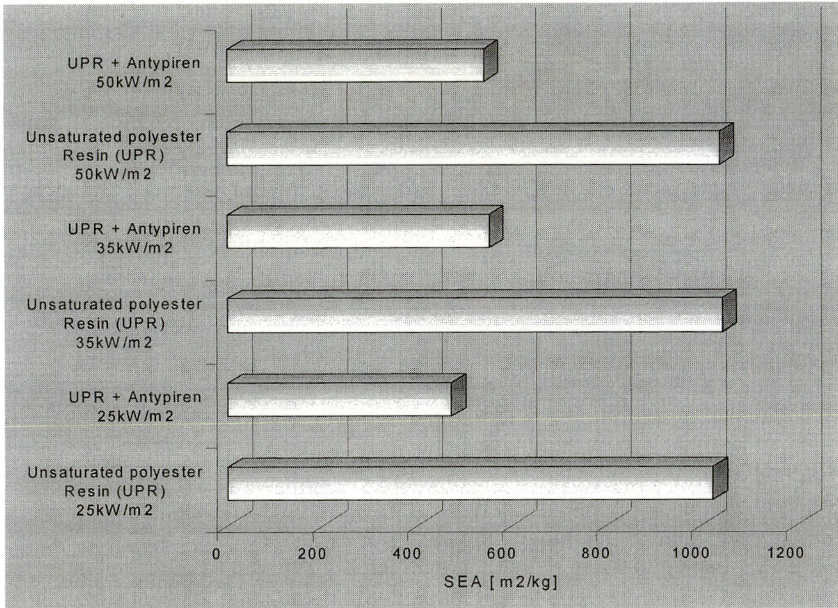

Fig. 7. The average extinction area (SEA_{av}) of: 1 - fire-nonretarded, 2 - fire-retarded GRP, in relation to thermal irradiation power indicated on ordinates.

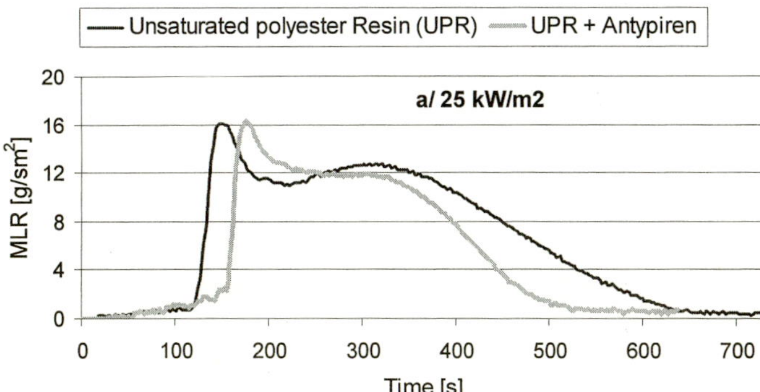

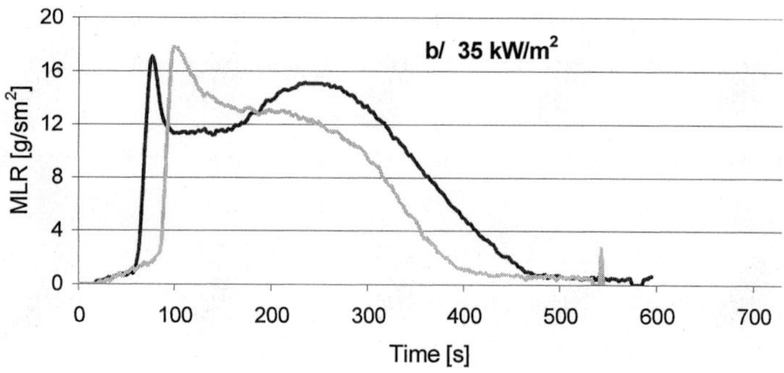

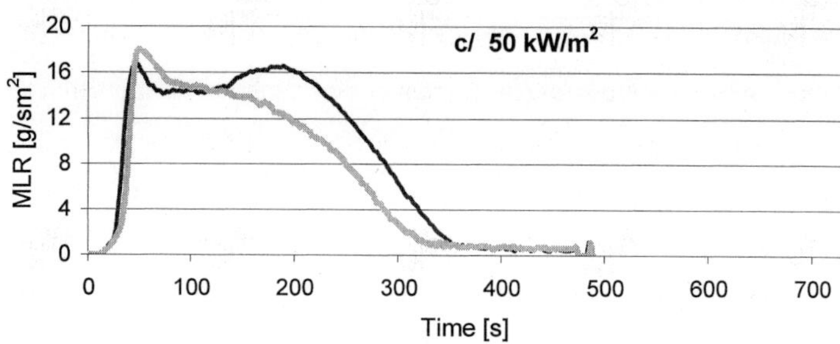

Fig. 8. The mass loss rate (MLR) in relation to thermal irradiation power: a - 25, b - 35, c - 50 kW/m^2 (for other designations see Fig. 6).

In the calorimetric tests, samples were weighed before and after the test. The total heat released was divided by the mass decrement to achieve the average heat of combustion. Most tests are discontinued soon after the flame has disappeared, and before oxidation of carbon becomes predominant. The affective heat of combustion is believed to be constant over this period of time.

Conclusions

The cone calorimeter allows to perform a complex analysis of all the stages of combustion and of smoke evolution during pyrolysis and to fully evaluate the level of fire-retardancy of GRP materials based on fire-retarded polyester as compared with the materials that have not been fire retarded.

The cone-calorimetric analysis performed on GRP showed the tin-zinc systems to be effective fire retardants for polyester resins, which reduce their flammability level and essentially suppress smoke evolution essentially.

The GRP examined are seen to be applicable as materials for widespread use, particularly in the production of boats, baths, shower units, automobile parts, railway carriage panels and electrical components.

[1] E. Kicko-Walczak, *Polimer,y* **1995**, 40, 665.
[2] P. .A. Cusack, A. W. Monk, S. J. Reynolds, *Fire and Materials* **1989**, 14, 23.
[3] V. Babrauskas, Ibid, **1994**, 18,289.
[4] E. Kicko-Walczak, *Polimery* **1992**, 37,527.
[5] P. A. Cusack, M. S. Heer, A. W. Monk, *Polym. Degrad. Stabil.* **1991**, 32, 17.
[6] E. D. Weil, M. M. Hirschler, N. G. Patel, M. M. Said, S. Shaikr, *Fire and Materials* **1992**, 16, 159.
[7] M. Władyka-Przybylak, M. Helwig, R. Kozłowski, Conf. Proc., XVIII Symp. „Protection of Wood" (In Polish) Jachranka, 4-16 September **1996**, 112-116.
[8] ISO Standard 5660-1 „Rate of heat release from building products (Cone calorimeter method)"
[9] E. Mikkola, *J. Fire Sci.*, **1990**, 9, 276.
[10] M. Władyka-Przybylak, M. Helwig, R. Kozłowski, *Nat. Fibers* (Włókna Naturalne) **1996**, 40, 145.
[11] E. Kicko-Walczak, *J. Appl. Polym. Sci.* **1999**, 74, 373.

Surface Modified Aluminium Hydroxide in Flame Retarded Noise Damping Sheets

Sándor Keszei,[1] *Péter Anna,*[1] *Györgi Marosi,**[1] *Andrea Márton,*[1]
Györgi Bertalan,[1] *Ferenc Valló*[2]

[1] Organic Chemical Technology Department, BUTE, H-1111 Műegyetem rkp.3, Budapest, Hungary
Email: gmarosi@mail.bme.hu
[2] MAL Corp. 8401 Ajka, Gyártelep P.O.Box. 124, Hungary

Summary: Flame retarded polyethylene compounds were prepared using a series of aluminium hydroxide of different particle size applying a milling processes and special precipitation technologies. The processability and flame retardant efficiency of the flame retarded systems were compared. The effects of various surface modifications were analysed in case of one selected type of aluminium hydroxide. A silicone terminated reactive surfactant promoted not only the processability but also the flame retardant efficiency. Noise damping sheets were prepared by simultaneous application of aluminium hydroxide and barium sulphate in an elastomer blend matrix. V0 flame retardant grade could be achieved this way accompanied with improvement in the acoustic properties and maintenance of the mechanical properties.

Keywords: aluminium hydroxide; flame retardancy; noise damping; surface modification

Introduction

Metal hydroxides are used for a long time in huge quantities as flame-retardant additives because of their minimal smoke emission and corrosivity. Still a continuous development exists in this field. New surface modification methods[1,2] for improving the compatibility and processability, and new production technologies for modifying the morphology are reported in the literature, for example bulk manufacturing of synthetic hydrotalcite,[3] application of new milling and precipitation technologies for influencing the size distribution[4] and application of synergists such as expandable graphite,[5] boroxo-siloxane,[3] zinc-borate,[6] lignin[7] and compound like zinc hydroxyl stannate[8] for increasing the efficiency of the additives. The main application field of the metal hydroxide containing elastomeric compounds is the preparation of flame retardant cable insulation with high loading of additive. Changes in compounding techniques and formulation technology as well as the use of processing aids and modification of the metal hydrate are needed for incorporating these flame retardants into polyethylene matrix to form better products.[9]

 DOI: 10.1002/masy.200351220

polyethylene matrix to form better products.[9]

In this work the processability and flame retardancy of polyethylene compounds containing variously prepared aluminium hydroxide of different particle sizes are compared. The influence of different surface treatments on the processability and flame retardance are also discussed, as well as the noise damping acoustic character of barium hydroxide/aluminium hydroxide filler-containing compounds.

Experimental

The materials used in the experiments are as follows: medium density polyethylene (PE): Tipelin K 341 94 (TVK, Hungary); thermoplastic elastomer blend (TEB): PEMÜBLEND (PEMÜ, Hungary); aluminium hydroxide of various grades (ATH): Alolt types (MAL Corp. Hungary), the average particle sizes and the SEM pictures of the particles are shown in Figure 1; surface treating agents: various derivatives of stearic acid of low cationic activity (LIS), of medium cationic activity (MIS), of high cationic activity (HIS), of non-ionic character (NIS) and silicone-containing reactive surfactant (RS) (The agents were prepared in the laboratory); barium sulphate $BaSO_4$ (Barite, Merck).

The PE-based compounds, containing 60 % ATH, were homogenised in the mixing chamber of a Brabender Plasti-Corder (PL2000 type, Brabender) equipment for 12 min at 170°C with rotor speed 50 rpm. Sheets (100 x 100 x 3 mm) were prepared by compression moulding in a Collin P200E press at 170°C and a pressure of 3 MPa. The surface modifying additives were applied in 0.1%, related to the amount of the ATH.

The reference noise damping compound consisted of 64.3% Barite in PEMÜBLEND. The flame retardant noise damping compounds were prepared by gradual replacement of Barite by ATH.

The bending loss factor was measured on a small, narrow band with bending wave excitation.[10] The applied sizes of samples, length: 300 mm, width: 20 mm, thickness: 0.68 mm, cover the most important acoustic frequency range. The samples were excited by an electrodynamics exciter. The samples were fixed to the exciter by a thin layer of adhesive.

The flammability properties were characterized by limited oxygen index (LOI) and UL 94 tests. Mechanical properties were characterized by determination of tensile strength and elongation at break, applying Instron 1195 testing machine.

Fig. 1. SEM micrographs of various $Al(OH)_3$ types produced by MAL Corp. Hungary.

Results and Discussion

Effect of Particle Diameter on the Processability and Flammability

The average diameter of the various grades of ATH used for the test varies in the range of 1.1 to 5 μm. The decrease of filler diameter generally increases the melt viscosity of the polymer compound, which increases the mixing torque of melt during polymer compounding. Consequently, very small filler diameter is not favourable for the processing. Regarding the flame retardant properties, however, the smaller diameters are favourable. This antagonistic impact is reflected in Figures 2 and 3. Figure 2 shows the diagram of the steady state kneading torques of the PE compounds containing 60% of various grades of ATH. A dynamic increase in the processing torque can be observed with decreasing particle size, getting very high below 1.5 μm.

Figure 3 shows the LOI and UL 94 flammability characteristics of PE compounds containing 60% of various grades of ATH. No strict correlation can be observed between the particle size and the flammability, which can be explained by the uncompletedispersion of the small size ATH. However, as a general trend, compounds with smaller particles have better FR characteristics. The best flame retardancy could be achieved by the application of ATH having an average diameter of 1.5 μm.Thus ATH of 1.5 μm was applied for the further works.

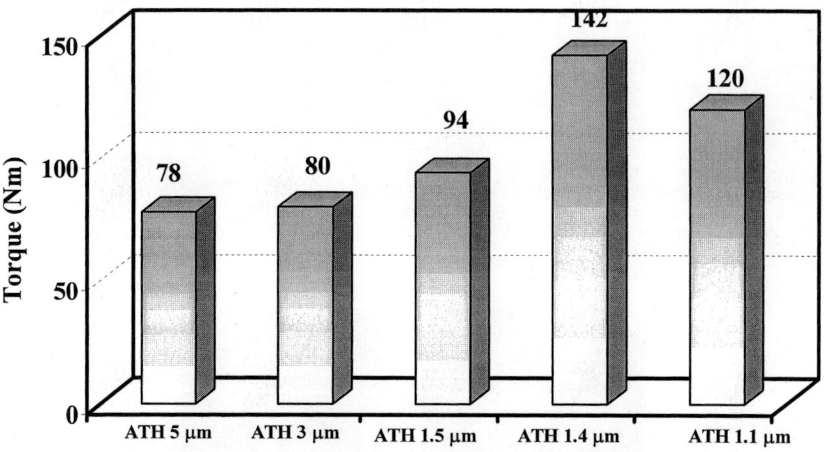

Fig. 2. Processing torque of PE compounds containing 60% of various grades of ATH.

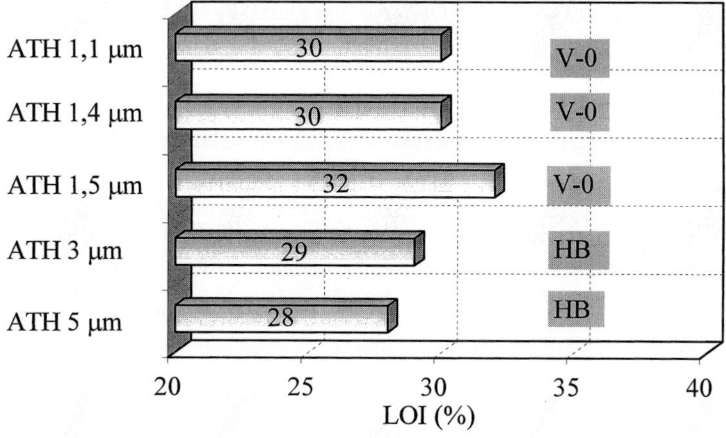

Fig. 3. Flammability of PE compounds containing 60% of various grades of ATH.

Surface Modification of ATH

A series of stearic acid derivatives, of different ionic activity, that is of low-, medium- and high cationic activity (LIS, MIS, HIS), and of non-ionic character (NIS) and silicone containing reactive surfactant (RS) were applied for surface modification of the ATH. The influence of the various modifying agents on the mixing torque can be seen in the Figure 4. The additives in general reduce the mixing torque considerably. The most favourable effect is produced by the additives LIS and RS.

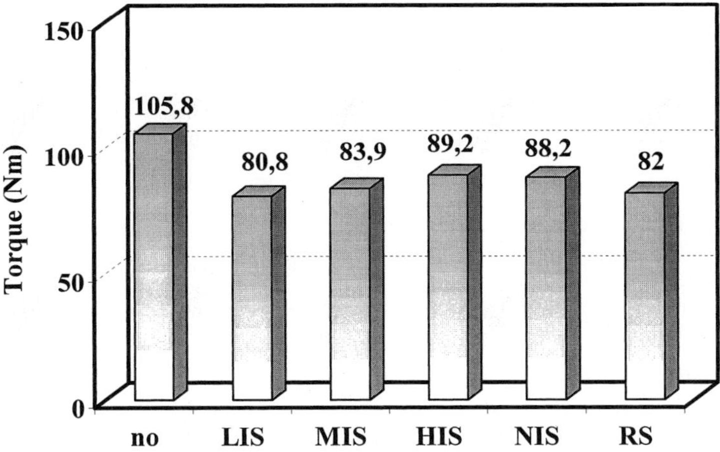

Fig. 4. The equilibrium mixing torque of ATH-PE compounds containing different surface modifiers.

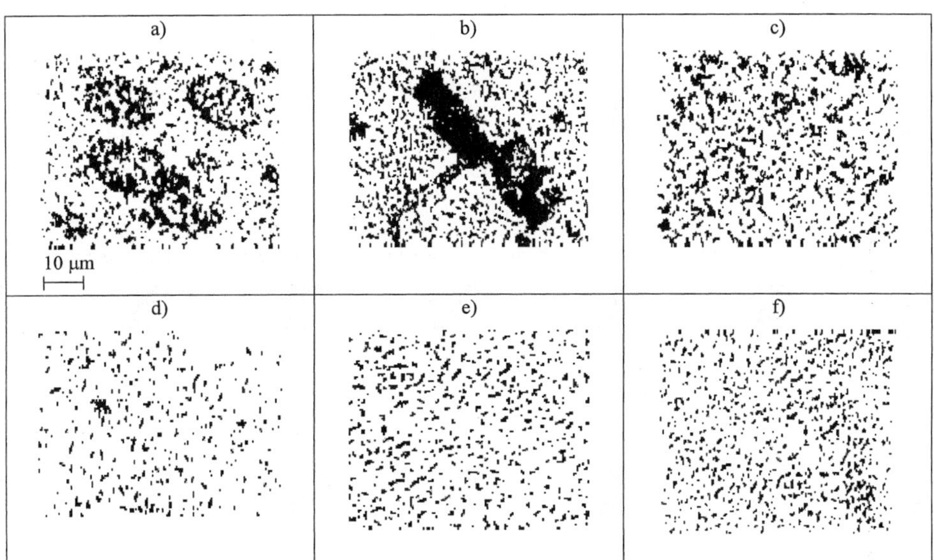

Fig. 5. Influence of various surface modifying additives on the distribution of ATH in PE compounds a) no modification, b) LIS, c) MIS, d) HIS, e) NIS, f) RS

The effect of the additives on the distribution of ATH is demonstrated by the optical microscopic picture of the thin film of the compounds in Figure 5 (enlargement: 100X). Concerning the homogenization effect of the additives, substantial differences can be observed. Without surface modification, picture "**a**" in Figure 5, the ATH is not distributed homogenously, a part of them agglomerates in globular islets. The additive of low ionic activity, picture "**b**" in Figure 5, does not improve the distribution, only the shape of the agglomerates change. The homogeneity is improved by the use of MIS and HIS additives (pictures "**c**" and "**d**" in Figure 5). The best distributions, however, could be achieved by application of NIS and RS additives (pictures "**e**" and "**f**" in Figure 5). Based on the favourable effects of the additives on the mixing torque and mainly on the homogeneity of the compounds, in the further experiments NIS was applied as surface modifying agent.

Noise Damping Properties

The floor assembly and some of the walls in public transport vehicles are prepared using a special sandwich structure, having a noise absorbing interlayer. This interlayer is a polymer compound consisting in general of a thermoplastic elastomer blend (TEB) and of barium sulphate (Barite). In spite of the high loading of Barite, above 60%, the compound is flammable. A flame retardant noise damping compound was formulated by partial substitution of Barite by ATH.

The influence of gradual substitution of Barite on the physical properties of the noise absorbing compound is illustrated in the Figure 6. Increase of the ATH content, on the expense of the Barite content, favourably influences the strength and the modulus. However, above a certain ATH content the elongation at break is reduced considerably. The specific density (a factor effecting the noise absorption) was reduced gradually with the increasing share of ATH.

The acoustic properties, that is the bending loss factor of the compounds consisting of TEB as matrix and a mixture of Barite/ATH in various weight ratios, can be seen in Figure 7. This series demonstrates the influence of the ATH contents on the acoustic properties. It can be observed that the trend of transfer damping in the low frequency range (up to 100 Hz) is slightly reduced in the presence of the ATH, while in the high frequency range (above 600 Hz), all of the compounds show a similar acoustic character. In the frequency range from 110 Hz to 700 Hz, however, the compound in which 66.6% of the Barite have been replaced by ATH, shows a better noise damping than the reference compound, containing only Barite. Thus, the partial replacement of Barite may result in improved acoustic character, in spite of the reduction of the specific density. The only question is whether the applied amount of ATH is satisfactory for achieving the necessary flame retardancy.

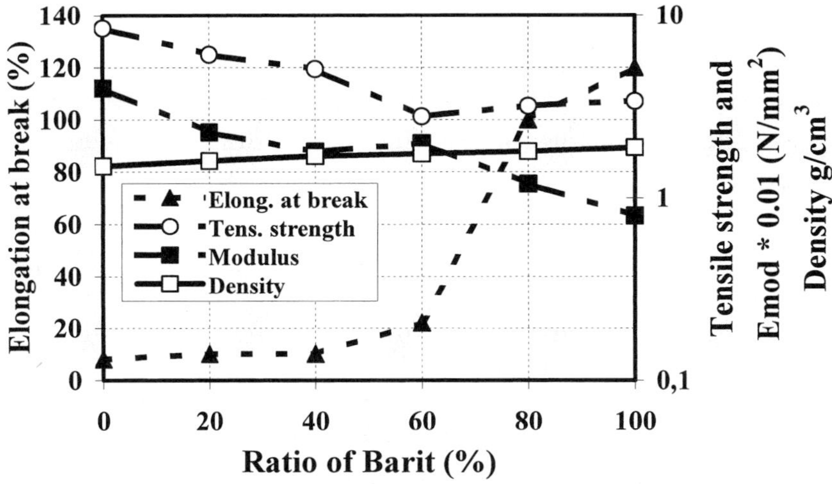

Fig. 6. Physical properties of noise damping compounds containing 64.3% additives (Barite+ATH) plotted against weight ratio of Barite.

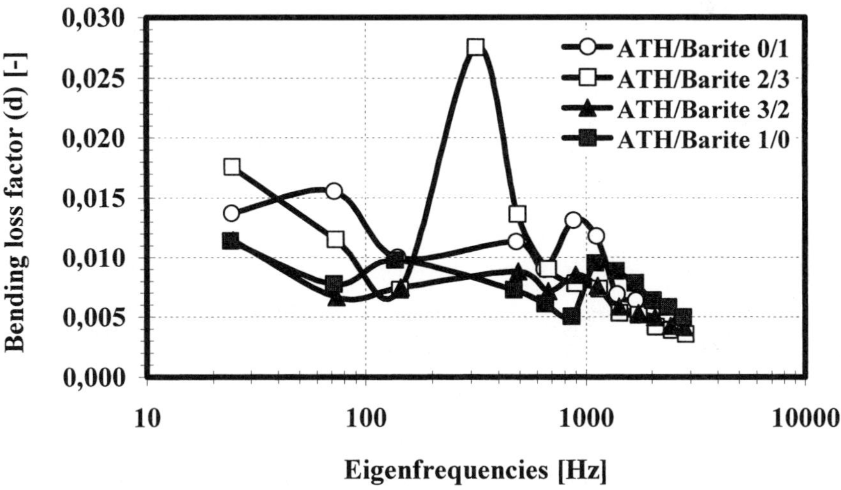

Fig. 7. Bending loss factor of different TEB containing 64.3% (Barite+ATH) additives (for bending wave excitation).

Figure 8 shows the LOI and UL 94 flammability properties of the compounds containing ATH/Barite in various weight ratios. As is seen, that the gradual replacement of Barite by ATH increases the flame retardant character. Above the ATH/Barite weight ratio of 3/2 the compounds achieves the self extinguishing V0 UL 94 flammability level, and the LOI also achieves or exceeds the 30% value as a sign of good flame retardancy. Based on Figures 7 and 8 it can be concluded, that a partial or a total substitution of Barite by ATH in the noise damping compound, having a TEB matrix, provides an improved flame retardant character. Simultaneously, at a particular ATH/Barite weight ratio, it also improves the sound absorbing character.

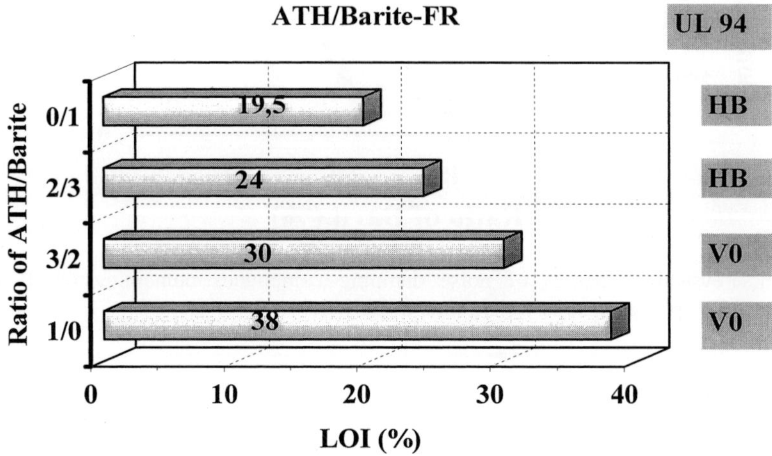

Fig. 8. Flammability of noise damping compounds containing ATH/Barite in various weight ratios.

Conclusion

In the 5 to 1.1 μm diameter range the decrease in the particle size of ATH, increases the mixing torque necessary for the preparation of the PE compounds. No direct correlation exists between the particle size and flame retardant effect of ATH. The best flame retardancy was obtained when the ATH of 1.5 μm particle size was applied. The application of stearic acid derivatives of different cationic activity, non-ionic and reactive silicon derivatives (especially the last two mentioned) considerably reduces the mixing torque necessary for compounding and improves the dispersion of ATH. A partial substitution of barium sulphate by aluminium hydroxide in the thermoplastic elastomer blend compounds provides a flame retardant character and improves the noise damping.

Acknowledgement

This work has been financially supported by the Commission of the European Communities through contract No G5RD-CT-1999-00120 and by the Hungarian Research Fund through projects OTKA T32941 and T026182.

We express our thanks to I. Dombi for the bending wave loss factor measurements in the Acoustic Laboratory of the Institute of Quality Control in Construction, Budapest, Hungary.

[1] H. D. Metzemacher, R. Seeling, "*Surface Modified Filler Composition*", US patent, US005827906, 1998.
[2] Gy. Marosi, P. Anna, Gy. Bertalan, Sz. Szabó, I. Ravadits, J. Papp, "*Role of Interface Modification in Flame-Retardant Multiphase Polyolefin Systems*", in *Fire and Polymers, Material and Solutions for Hazard Prevention*, ACS Symposium Series 797 (Ed.: Nelson G., Wilkie CA.) **2001**, 161.
[3] K. Koyama, K. Murakami, M. Ueda, H. Matsuda, "*Polyolefin-based resin composition*", Japan Patent, JP2000103912, 2000.
[4] Gy. Marosi, P. Anna, A. Márton, Gy. Bertalan, A. Bóta, A. Tóth, M. Moha, I. Rácz, *Polym. Advanced Technol.* **2002**, *13,* 1.
[5] W. Zhengzhou, Q. Baojun, F. Weicheng, H. Ping, *J. Appl Polym Sci,* **2001**, *81*, 206.
[6]S. Bourbigot, F. Carpentier, M. Le Bras, "*Thermal Degradation and Decomposition Mechanism of EVA-Magnesium Hydroxide-Zinc Borate*", in *Fire and Polymers, Material and Solutions for Hazard Prevention,* ACS Symposium Series 797 (Ed.: Nelson G., Wilkie CA.) **2001,** 173.
[7]A. De Chirico, M. Armanini, P. Chini, G. Cioccolo, F. Provasoli, G. Audisio, *Polym. Degrad. Stability*, **2002**, *79(1),* 139.
[8] P. R. Hornsby, *International Materials Reviews*, **2001**, *46(4),* 199.
[9] J. Innes, A. Innes, *Plastics Additives & Compounding,* **2002**, *15(5),* 28.
[10] "*Acoustic insulation*", EN ISO 14-3:1995

Surface Treated Cellulose Fibres in Flame Retarded PP Composites

*P. Anna, E. Zimonyi, A. Márton, A. Szép, Sz. Matkó, S. Keszei, Gy. Bertalan, Gy. Marosi**

Department of Organic Chemical Technology, Budapest University of Technology and Economics H-1111 Műegyetem rkp. 3, Budapest, Hungary
Email: gmarosi@mail.bme.hu

Summary: Polypropylene-based composites were prepared containing non-treated and various treated cotton fibre and wood flakes. A correlation was observed among the fibre treatment and compounding parameters, mechanical and discoloration properties. The structural changes in fibres were demonstrated by Raman spectroscopic and DSC measurements. The possibility for forming cellulose fibre containing flame retardant composites was also investigated. The efficiency of various treatments on compounding, discoloration and mechanical properties enhance in the following order: no treatment < non ionic surfactant < reactive silicone segment containing non ionic surfactant < special silylation treatment. The best results obtained with the special silylation treatment were explained with the more organophilic character and by the thermal stability of the treated fibres. Cellulose fibre as a polyol-charring component and ammonium-polyphosphate together constitute a high performance intumescent flame retardant system in the PP matrix.

Keywords: cellulose fibres; DSC; fire retardant; polypropylene composites; Raman spectroscopy; surface modification; wood flakes

Introduction

In modern polymer composite technology there is a great demand for environment friendly materials completely adapted to the environment, i.e. they should be biodegradable and should originate from renewable resources. One possible approach is the embedding of natural reinforcing fibres, e.g. flax, ramie, hemp, etc. into biopolymeric matrices made of derivatives from cellulose, starch, lactic acid, etc. New fibre reinforced materials called biocomposites have been created and are still being developed with the unique benefit, of the total biodegradable character.[1,2] The bio-fibres, however, do not possess the necessary thermal and mechanical properties desirable for the compounding of engineering plastics. On the other hand, best engineering plastics are obtained from synthetic polymers, but they are not biodegradable. The composites of natural fibres and non-biodegradable synthetic polymers may offer a new class of materials but are not completely biodegradable.

 DOI: 10.1002/masy.200351221

Government regulations and growing environmental awareness throughout the world have triggered a paradigm shift towards designing materials compatible with the environment[3] such as the combination of natural fibres and non-biodegradable synthetic polymers. Advantages of bio-fibres over traditional reinforcing materials such as glass fibre, talc and mica are their low cost, low density, high toughness, acceptable specific strength.[4,5] The main drawback of these reinforcing materials is their hydrophilic nature, which lowers their compatibility with hydrophobic polymer matrices. The other disadvantage is the relatively low processing temperatures necessary to avoid fibre degradation and/or volatile emission that can deteriorate composite properties. Several approaches are described in the literature aiming at the improvement of the interfacial adhesion between organic fillers and polymer matrix, such as addition of maleic anhydride polymers as compatibiliser[6] or by combined application of maleic anhydride containing compatibiliser and treatment of the cellulose based filler with a silane coupling agent.[7,8] These methods improve the compatibility of cellulose fibres, but do not enhance the thermal stability of the fibres. The thermal stability of cellulose based fibres could be improved by a special silylation method,[9] which also resulted in enhanced mechanical properties of the polymer composites. Both effects were explained by structural changes in the amorphous phase of the cellulose fibres. It seems to be a promising way to use this special silylated cellulose fibre as reinforcing additive.

In the present work polypropylene composites were prepared using non-ionic surfactant, reactive silicone containing surfactant, a special silylated cotton fibre and wood flake. Inter-relations among fiber treatment, compounding efficiency, mechanical and discoloration properties were demonstrated. Structural changes in fibres were demonstrated with Raman spectroscopic and DSC measurements. The possibility of preparing cellulose fibre containing composite was also investigated.

Experimental

Materials: Tipplen H535, polypropylene homo-polymer (TVK, Hungary), MFI 4 g/10 min (230°C, 2.16 kg). Wood flake: 1.2 mm chips of pinewood. Fibre: natural cotton fibre of 30 mm. A special silylated product of natural cellulosic fibre was prepared with 20% tetraetoxy silane TES in presence of catalytic amount of dibutyl tin dilaurate (DBTDL) by heating them to 140°C and keeping at constant temperature until ethanol development was observed. The excess of TES was removed by distillation. ESTOL 1474 (Unichema International): glycerol-monostearate (GMS), a non-ionic surfactant, white powder, melting temperature 60°C. Silicone containing reactive surfactant (SRS): a glycerol mono oleate-triethoxy silane, an experimental product, prepared from 1 mol tetra ethoxy silane and 1 mol glycerol mono oleate in presence of DBTDL by heating at 120°C until development of ethanol was observed, molecular mass was 1498 g/mol, as determined by the method of boiling point elevation. Tetra ethoxy silane (TES) (Wacker, Germany). EXOLIT 422 (HOECHST, Germany), ammonium-polyphosphate (APP), white powder, average degree of polymerisation 700, acidity of suspension (1 % in water) pH 5.5.

Sample preparation: Compounding was performed in the mixing chamber 350 of a Brabender Plasti-Corder PL2000 (Brabender, Germany) at a rotor speed of 50 rpm, at 200°C, with 10 min homogenisation time. Fibres added a rate of 30 wt% were previously homogenised physically with surface treating additives, (2 % of the fibre weight), then were fed into the PP melt. Sheets (100×100×3 mm) were formed using a laboratory compression moulding equipment.

Measurements: Tensile strength, and elongation at break were measured on a Textenser instrument (Textile Research Institute, Hungary), at an elongation rate of 10 mm/min.

Micro-Raman measurements were performed by a Labram Raman Microscope system (Jobin Yvon Horiba, France), excitation at 632,81 nm by HeNe laser. Thermo gravimetric measurements (TG) were performed using Setaram Labsys equipment (Setaram, France), with 10 mg test material, at heating rate of 7,5°C/min, in air atmosphere. Colour measurements with MOMCOLOR-100 device (MOM, Hungary) by using white reference, for colour characterisation the ΔL^* value, the lightness-difference between the sample and the reference was used according the CIE recommendation.[10] The flammability was characterised by LOI and UL 94 tests.

Results and Discussion

Effect of Surface Treatment on Compounding

The influence of various surface treatments on cellulose fibre manifested in the compounding, surface discoloration and mechanical properties of PP compounds. The extent of the effect was characterised by measuring the torque during the compounding, the colour of surface, the tensile strength and elongation at break of the plates prepared from the compounds. The influence of surface treatment on the processing characteristics of the wood flake containing composite is demonstrated in Figure 1. The homogenisation of the non-modified wood flake (NoM-W) containing PP composite requires a high torque (nearly 40 Nm) and a long mixing time, above 7 min. Addition of a non-ionic surfactant (NIS-W) reduces the torque maximum considerably (below 30 Nm), but the required homogenisation time could not be reduced. The homogenisation takes place in a multi-step process. The mixing torque as a measure of the power consumed during the compounding is related to the melt viscosity of the mixed system. The application of silicone containing reactive surfactant (SRS-W) results in a much more intense change on the torque curve. The maximum value reduces below 20 Nm, the homogenisation requires more than 7 min. The most intensive effect on the compounding characteristics can be observed in the preparation of the composite containing the special silylated wood flakes (Sil-W). In contrast to the former composites, the torque begins to increase immediately after feeding the wood flakes. The maximum of the torque reaches only 17 Nm, and the homogenisation is completed after 3 min. That means, the most favourable effect with respect to compounding can be achieved by the special Sil treatment.

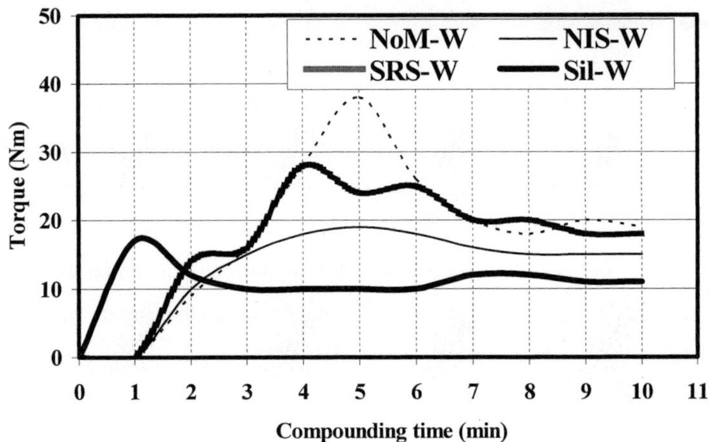

Fig. 1. Torques during the preparation of PP compounds containing various surface-treated wood flakes.

Figure 2 shows the torque values measured during the preparation of PP composites made with various treated cotton fibres..The observed tendencies are similar to those observed with wood flakes, but not so pronounced. Each modification reduces the torque maximum, and modifies the compounding time necessary for the homogenisation, similarly to the compounds discussed above,the differences are, however, not so pronounced. With respect to the homogenisation time the most favourable effect, similarly to the case of wood flakes, can be obtained by application of the special Sil treatment.

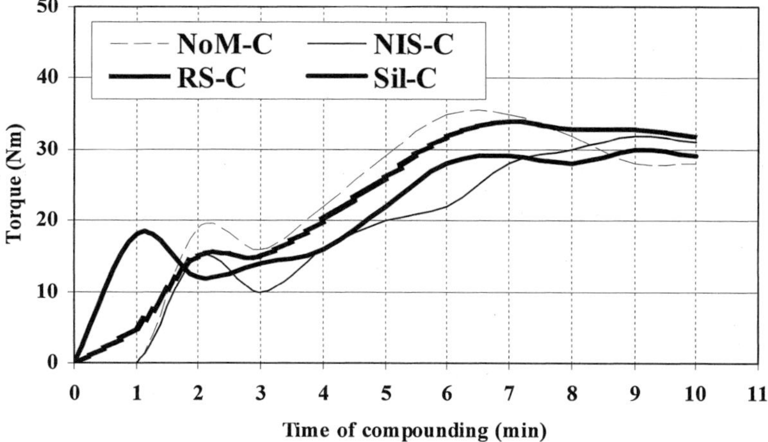

Fig. 2. Torque as function of compounding time of PP composites containing variously surface treated cotton fibres.

Considering the mixing curves discussed above one can conclude that the treatments, which render the surface of cellulose based fillers organophillic, promote the homogenisation during the compounding, making the surface more compatible with the matrix of the composite. The favourable effect is minimal for NIS and maximal for the special Sil treatment.

Effect of Surface Treatment on the Mechanical Properties

The beneficial effects of treatments are also reflected in other properties. Figure 3 shows the relative mechanical properties (related to the NoM-W sample) of the PP composites containing various modified wood flakes. NIS and the SRS treatments have a negligible, while the special Sil modification has a considerable effect on the mechanical properties. The relative strength of the Sil-W sample increased nearly by 100%, the elongation at break by more than 200%.

The effects of various treatments are similar in the case of cotton fibre reinforced PP composites, as demonstrated in Figure 4, only the improving effect on the characteristics has a reverse direction, i.e. the changes in the strength and elongation at break are reversed. This inverse improvement is most characteristic and most intense in the case of the special Sil treated sample, where the strength increases by 150%, and the elongation only by 50%, that is the improvement is higher for the strength. The improvement of the mechanical properties in both types of the fibre reinforced composites can be explained by the increased organophilic character of the fibre surfaces, which leads to an enhanced interaction between the originally hydrophilic fibre surface and the hydrophobic matrix. The higher increase of the strength of the special Sil treated cotton fibre reinforced composite, compared to the wood flake reinforcement, is in connection with the longer size of the cotton fibres. The longer fibre size, on the contrary, prevents the elongation, resulting in a lower elongation at break. The higher elongation at break of the wood flake-containing composite is due to the shorter fibre size of the wood flakes, which allows the orientation and movement of the matrix polymer chains during the tensile test.

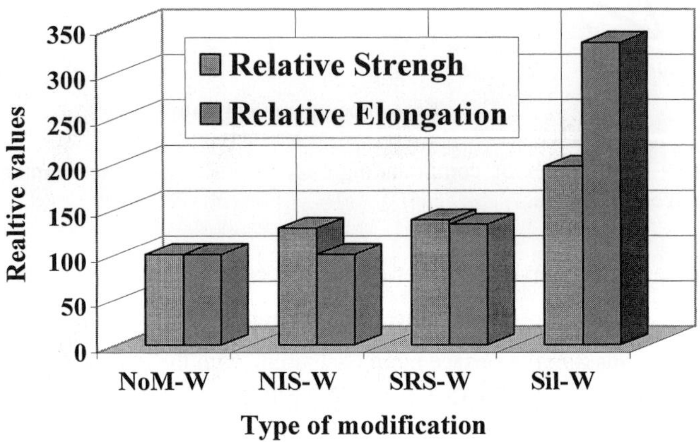

Fig. 3. Relative tensile strength and elongation at break of PP composites containing various treated wood flakes.

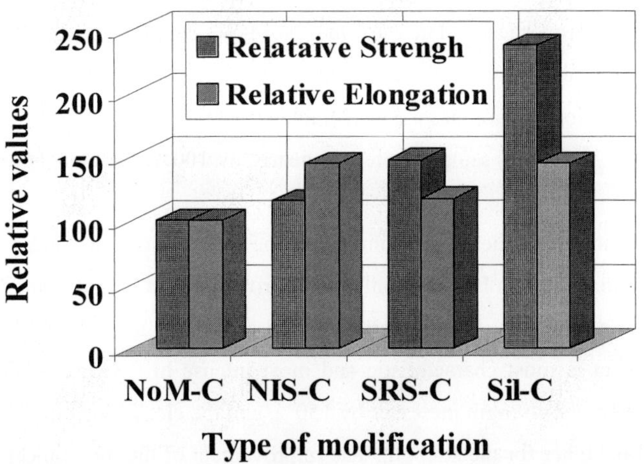

Fig. 4. Relative tensile strength and elongation at break of PP composites containing various treated cotton fibre.

Effects of Surface Treatment on the Discoloration of Composites

The various modifications manifest themselves also in the different discoloration of the PP composites, as demonstrated by Figure 5, where the lightness differences, ΔL*, as compared to the white reference, are presented for the compounds containing wood flakes and cotton fibre.

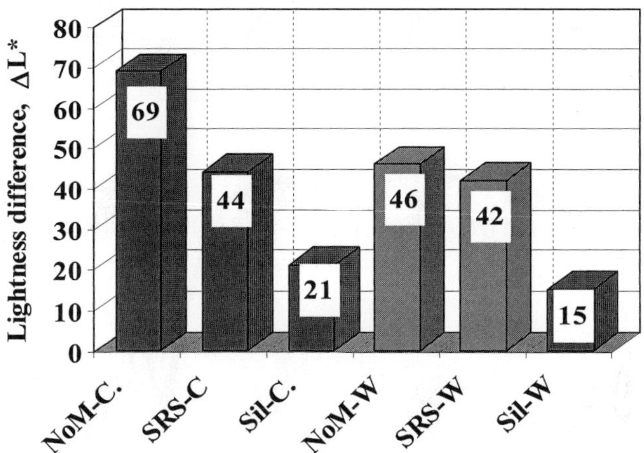

Fig. 5. Lightness differences of PP composites containing various treated cotton fibre and wood flakes.

The larger the lightness difference, the higher the discoloration of the sample. As it can be seen, the cotton fibre is more susceptible to discoloration. Each treatment reduces the lightness difference. The Sil treatment is most effective in preventing discoloration.

The mechanical properties, mainly the tensile strength and the discoloration measurements suggest, that the Sil treatments improve the thermal stability of cellulose fibres. This assumption was confirmed by thermo-gravimetric analysis. The DTG curves of the non-treated and Sil treated cotton fibres are illustrated in Figure 6. The DTG curve of non-treated (NoM-C) cotton fibre shows a starting degradation already above 180°C, while the degradation of the Sil treated fibre, Sil-C begins only above 200°C. The degradation of the non treated fibre is negligible with respect to weight loss, but considerable with respect to discoloration.

For the explanation of the improved thermal stability, Raman spectra, illustrated in Figure 7, of the non-treated and Sil treated cotton fibre samples were studied. Comparing the spectra of the non treated (NoM-C) and the Sil treated (Sil-C) fibres considerable differences can be observed is some region of the spectra:[11]

- in the range of 1000 cm^{-1} characteristic for the cyclic axial secondary alcoholic -OH-groups,
- around 1340 cm^{-1}, which band can be attributed to $-CH_2-$ deformational vibration and
- in the region 3200 to 3500 cm^{-1} characteristic for the stretching vibration of the $-OH \rightarrow O$ bond system.

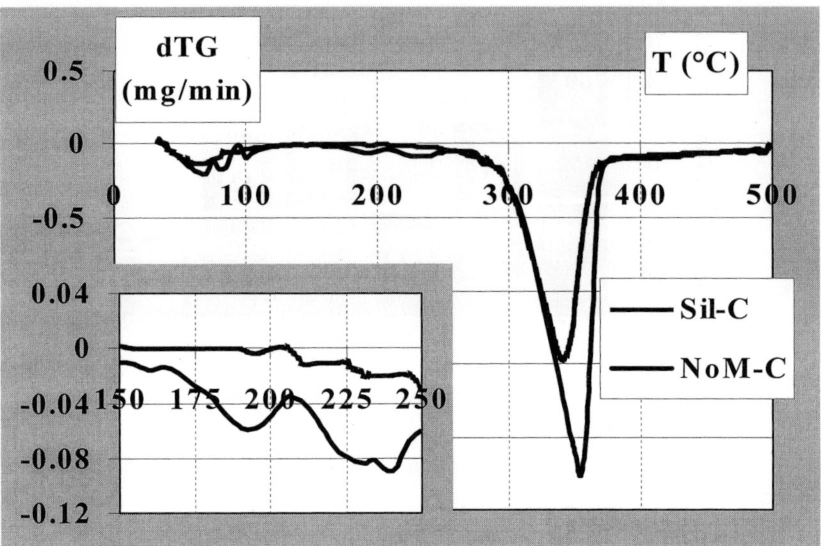

Fig. 6. DTG curves of natural and Sil treated cotton fiber.

Typically, the intensity of the peaks mentioned above increase after Sil treatment. The complex peak system in the 3200 to 3500 cm^{-1} region becomes not only more intense, but also narrower. These changes indicate an apparent increase in the concentration of the -OH groups in the ordered phase, an increase of crystallinity and a reduction of the amorphous phase. The reduction of -OH groups in the amorphous phase is an evident result of the Sil treatment, as the silylation reaction can not take place in the crystalline phase. The increase of the concentration of the -OH groups in the crystalline phase is a result of post-crystallization, which is the consequence of the easier segmental movement, which was facilitated by the reduction of secondary chemical bonds in the amorphous phase. These structural changes explain the reduced discoloration and the higher thermal stability of the Sil treated fibre containing PP composites. The -OH groups in the amorphous phase have higher susceptibility for easy thermal dehydration, as their concentration was reduced, the thermal stability of the system increased.

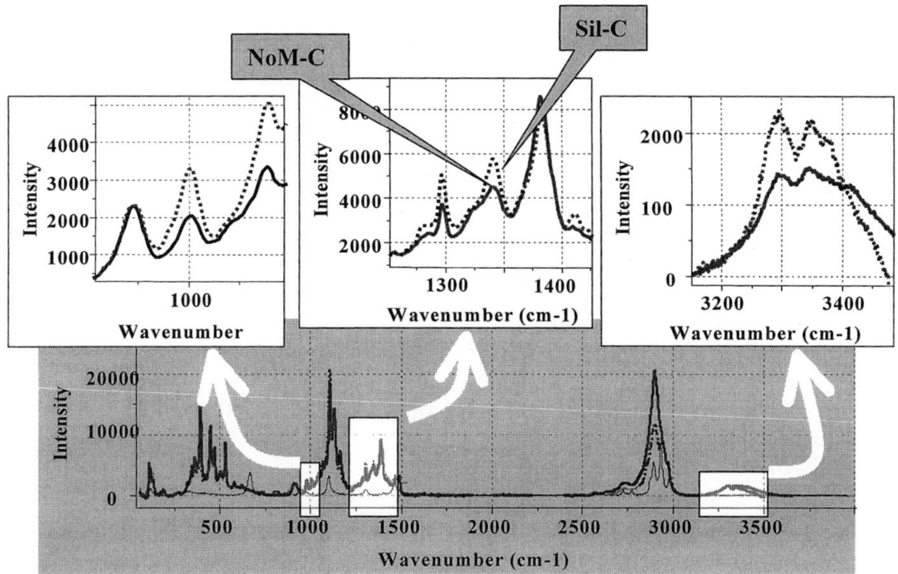

Fig. 7. Overview Raman spectra of non-treated and Sil treated cellulose and TES, and magnified spectra sections of non-treated and Sil treated cellulose.

Flame Retarding of Cellulose Fiber Containing PP Composite

The cellulose fiber containing composites contain a large amount of -OH groups. The polyols are effective char forming components in intumescent flame retardant polyolefin compounds.[12,13,14] It is an obvious suggestion to combine the -OH groups of cellulose-PP composite as char former with an acid source (APP) to form natural fiber reinforced composite in flame retarded form. The Figure 8 shows the LOI values of the SRS modified wood flake containing PP composites prepared with 10% APP. As it can be seen, LOI hardly changes between 10 and 40% wood flake content. Increasing the fiber content to 50% the flammability strongly changes and reaches 30%. With 50% wood flake content the composite achieves the V0 flammability grade in the UL 94 test. The combination of cellulose fiber reinforced PP composite with APP acid source results in flame-retardant compound.

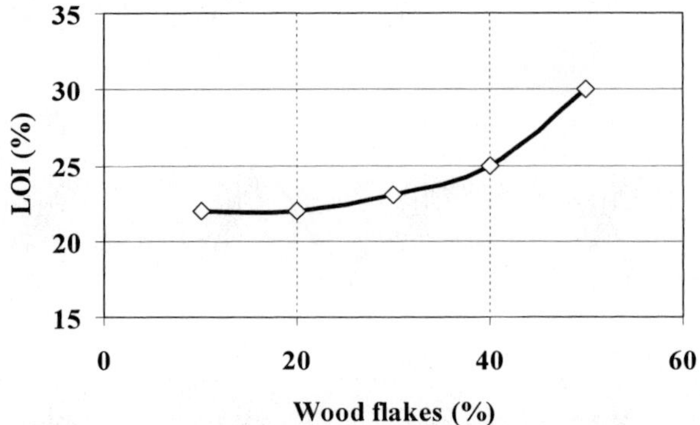

Fig. 8. LOI values of PP composites containing various amounts of SRS treated wood flakes and 10% ammonium polyphosphate.

Conclusion

The pre-treatment of cellulose fibres with non-ionic, ionic, reactive non ionic surfactants and by silylation, improves the compounding, thermal stability, mechanical properties and reduces the discoloration when used in PP compounds. Silylation has the most favourable effect, owing to the organophilisation and modification of crystalline structure, that is to the reduction of –OH groups in the amorphous phase. The addition of ammonium polyphosphate to the cellulose fibre containing composite results in flame retardant compound.

Acknowledgement

This work has been financially supported by the Hungarian Research Fund through projects OTKA T32941, NKFP 00169/2001 and NKFP 3A/0036/2002.

[1] A. K. Mohanty, M. Misra, G. Hinrichsen., *Macromol. Mater. Eng.*, **2000**, *276-277*, 1.
[2] A. S. Herman, J. Nickel, U. Riedel, *Polym. Degrad. Stab.*, **1998**, *59*, 251.
[3] R. Karnani, M. Krishnan, R. Narayan, *Polym. Eng. Sci.*, **1977**, *37*, 476.
[4] N. Hoogen, *Kunststoffe*, **1999**, *89(3)*, 103.
[5] R. Karnani, M. Krishnan, R. Narayan, *Polymer Engineering and Science*, **1997**, *37(2)*, 476.
[6] C. .Joly ,M. Kofman, R. Gauthier, *J. Macromol. Sci., Pure Appl. Chem. A*, **1996**, *33*, 1981.
[7] H. Nitz, P. Reichert, H. Römling, R. Mülhaupt, *Macromol. Mater. Eng.*, **2000**, *276/277*, 51.
[8] P. W. Balasuriya, L. Ye, Y.-W. Mai, J. Wu, *J Appl Polym Sci.*, **2002**, *83*, 2505.
[9] A. Borbély-Kuszmann, J. Nagy, E. Zimonyi-Hegedűs, *Macromol.Chem.*, **1976**, *177*, 947.
[10] CIE Publication 15 Supp. 2, "*CIE Recommandation on Uniform Color Spaces, Color-Difference Equations, and Metric Color Terms*", Bureau Central de la CIE, Paris, Fr. 1978.
[11] J. H. Wiley, R. H. Atalla, *Carbon hydrate Research*, **1987**, *160*, 113.
[12] G. Bertelli, G. Camino, E. Marchetti, L. Costa, R. Locatelli, *Angewandte Macromol. Chemie*, **1989**, *169*, 137.
[13] S. Bourbigot, M. Le Bras, R. Delobel, P. Breant, J. M. Trmilolon, D. Price, *Polym. Degr. Stab.*, **1996**, *54*, 275.
[14] Gy. Marosi, R. Lágner, Gy. Bertalan, P. Anna, A. Tohl, *Journal of Thermal Analysis*, **1997**, *48*, 717.

Use of Reactive Surfactants in Basalt Fiber Reinforced Polypropylene Composites

Sz. Matkó, P. Anna, Gy. Marosi, A. Szép, S. Keszei, T. Czigány, K. Pölöskei*

Budapest University of Technology and Economics, Műegyetem rkp. 3., 1111 Budapest, Hungary
E-mail: gmarosi@mail.bme.hu

Summary: Basalt fibers, similarly to other silicate fibers, can be introduced into both thermoplastic and thermosetting polymer matrices. In this work some basalt fiber reinforced polypropylene composites were investigated. The fiber-matrix adhesion was improved by commercial and non-commercial maleic anhydride derivatives. The latter types, called reactive surfactants, were prepared in laboratory scale and the progress of the syntheses was determined by Raman microscopy. The additives allowed performing reactive interface modification during the compounding process. Due to the interface modification with the additives in low concentration the mechanical properties improved. The boundary layers on the surface of the reinforcing fibers were observed using scanning electron microscopy (SEM).

Keywords: basalt fiber; interface modification; reactive surfactant; reinforced polypropylene; scanning electron microscopy (SEM)

Introduction

The effectiveness of maleic anhydride derivatives as interfacial additives (coupling agents) in e-glass fiber reinforced polyolefin composites is well known. Using these compounds as modifiers, the mechanical properties improve.[1-5] The similarities between glass and basalt fibers suggest that widely used interface modifiers, like maleic anhydride derivatives, can be advantageously applied in the lower cost basalt fiber reinforced polymer composites as well. Reactive surfactants, however, being able to reach the interfaces and reacting there with both phases within the short residence time spent in a compounding machine, have not been studied in such systems.

Basalt based silicate fibers can be produced from the naturally occurring igneous basalt rock, which consists of a high proportion of basic magnesium silicates such as horn-blende, pyroxene, olivine and some feldspar.[6] The fiber is formed at high temperatures by different

 DOI: 10.1002/masy.200351222

technologies, such as drawing and melt-spinning. The mechanical properties of basalt fibers span over a wide range depending on their diameter, chemical composition, structural homogeneity, surface porosity and temperature.[7-10]

These fibers can be introduced into both thermosetting[11-16] and thermoplastic[17-20] polymer matrices as well as into rubbers.[21] Interfacial contacts formed spontaneously between basalt and most of polymers are released easily if thermal or mechanical effects initiate local stresses. The poor fiber-matrix adhesion can be improved with different types of coupling agents such as silane compounds[11-14] or maleic anhydride (MAH) derivatives. In polypropylene composites the most widely used compound is polypropylene-g-maleic anhydride (PP-g-MAH).[17,22] The key question is how to find a method for the modification of the interfacial structure that is both efficient and economic. In case of short fibers this method is undoubtedly reactive compounding/processing wherein the surface modifier molecules have to reach the interfacial region within a short residence time. The most active molecules, designed to find interfaces rapidly, are surfactants, which, however, cannot bind phases together. Recently developed reactive surfactants are designed to combine the advantages of surfactants and coupling agents, facilitating the efficient in-line reactive interface modification.[23] One of the proposed ways to synthesize reactive surfactants, being able to react with both of the neighboring phases, is the introduction of MAH into hydrocarbon molecules containing one or more C=C bond. MAH derivative type reactive surfactants could be effectively used in polypropylene (PP) composites.[24] The molecules containing maleic anhydride or maleic acid (MA) group and double bond may couple to the surface of polar silicate via their MAH or carboxyl group by ionic or secondary chemical bonds. Furthermore the long carbon chain may be grafted onto the PP-chains by means of its double C=C bond in the presence of peroxide initiators.

In this work various reactive interface modifier additives were compared to each other in short basalt fiber reinforced PP composites.

Experimental

Materials

Basalt fibers: chemical composition: SiO_2: 46.2, Al_2O_3: 13.0, Fe_2O_3: 12.0, MgO: 10.0, CaO: 10.0, TiO_2: 2.0, K_2O: 1.8, Na_2O: 3.5, others: <1%. Manufacturer: Toplan Bazaltgyapot Kft.,

Tapolca, Hungary. The length of fibers, which were produced by the Junkers melt spinning technology from natural basalt rock between 1300 and 1500°C ranges from 5 to 10 cm. As a result of the aerodynamic conditions occurring during the process large ball-like heads are formed at the end of the fibers. Figures 1 and 2 show the model of the melt spinning and the SEM photograph of a fiber-end respectively.

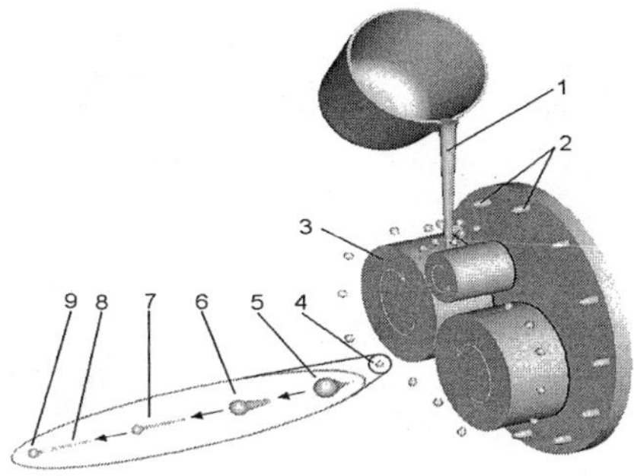

Fig. 1. The model of melt spinning.

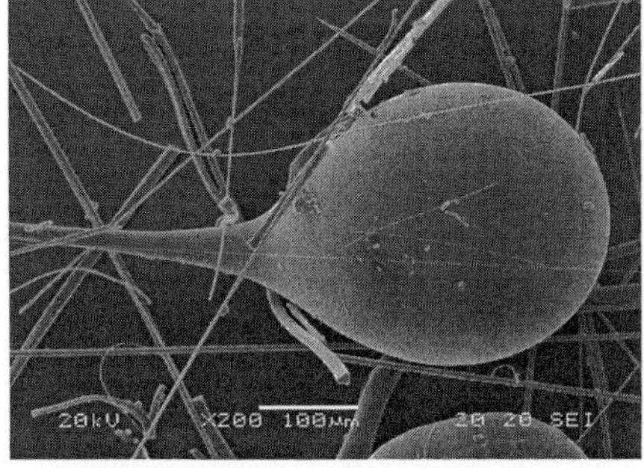

Fig. 2. SEM photograph of a large ball.

The short fibers with diameters ranging from 2 to 30 µm were made from these long fibers by milling in a hammer mill. As a result of stresses arising during grinding the fiber length decreased drastically and the balls of fiber-ends broke down. The average fiber length after the milling was in the range of 2-10 mm. In order to investigate reinforcing effect the short fibers' alone the balls were removed from the milled product in water. The balls, because of their lower hydrodynamic resistance, settled faster than the fibers. After the sedimenation of the balls the floating fibers were filtered from the water. Repeating this process four times the whole fraction of balls could be separated from the milled product. Then the clean milled product was dried in an oven between 80 and 100°C.

Polypropylene (PP): Tipplen H384F (Tiszai Chemical Works (TVK) Co., Tiszaújváros, Hungary), melt flow index (230°C, 21.6 N): 12.0 g/10 min. Maleic anhydride grafted polypropylene wax (PP-g-MAH): Licomont AR 504 (Clariant GmbH., Germany). Peroxide initiator: Luperox F90P (Elf Atochem Italia S. r. I.), $C_{20}H_{34}O_4$ (1,3-1,4 bis-ter-buthylperoxyisopropyl benzene), molecular weight: 338.49 g/mole.

Methods

Preparation of reactive surfactants was carried out in laboratory scale by means of Diels-Alder and esterification reactions.

The LIN-MAH compound was formed by the reaction of maleic anhydride and 9,11-linolic acid (LIN) in the presence of $5 \cdot 10^{-3}$% hydroquinone inhibitor for avoiding the polymerization of the LIN; initial mole ratio was 1:1, temperature was 120°C, duration of the reaction was 90 min.

COOH + O O O →

COOH

O O O

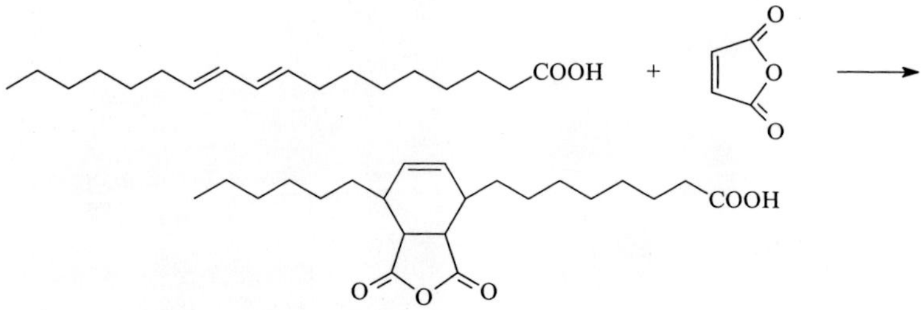

Scheme 1

SFO-MAH additive was prepared from maleic anhydride and sunflower oil (SFO) containing $5 \cdot 10^{-3}$ % hydroquinone, initial mole ratiowas 1:1, temperature was 120°C, duration of the reaction was 90 min. (SFO is a triglyceride of unsaturated (9,11-linolic acid: 26-34%, 9,12-linoleic acid: 26-34%, oleic acid: 22-39%) and saturated (5-7%.) fatty acids.)

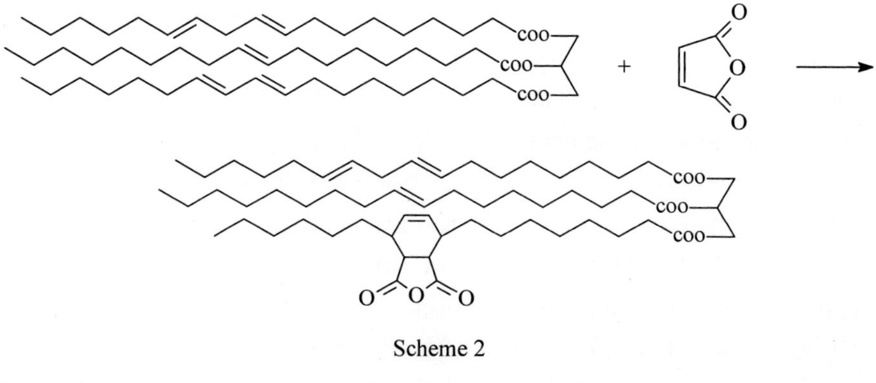

COO COO COO + O O O →

COO COO COO

O O O

Scheme 2

GMO-MAH additive is the diester of the glycerine monooleate (GMO) and maleic anhydride. The initial mole ratio in esterification reaction was 1:2, the temperature was 80°C, the duration of the reaction was 90 min.

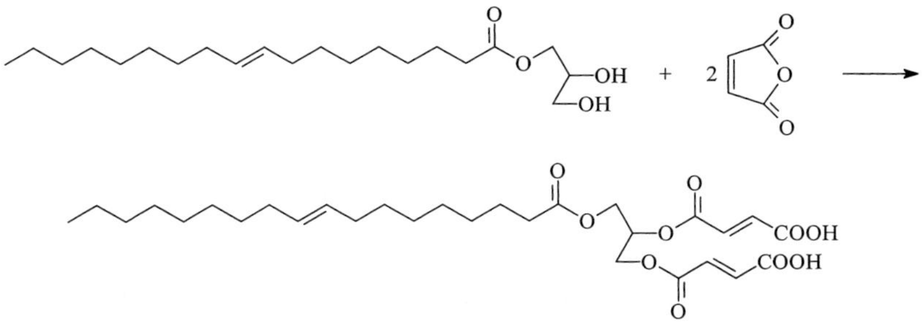

Scheme 3

In order to remove the excess of solid maleic anhydride grains at the end of the reactions from the products the liquid was dissolved in toluene and filtered on a glass filter funnel. The solvent was distilled from the filtrate at atmospheric pressure (bp.: 110.6°C).

Raman microscopy was applied for monitoring the reactions, using a Jobin Yvon "LabRam" dispersion Raman microscope equipped with a CCD detector and three changeable excitation light sources: a He-Ne laser (632 nm), a frequency-doubled Nd:YAG laser (532 nm) and a diode laser (785 nm). The optical resolution of the instrument (the size of the examined area) was adjustable by means of objectives according to the features of the sample. The best available optical resolution for microscopic analysis is 0.709 μm (exciting line at 532 nm, 100× magnification, NA = 0,9 objective), while the resolution of the objective attached to an optical fiber, called "Super Head" is 0.5 mm. In the applied system the Super Head operates at the frequency of Nd:YAG laser. The laser power at the output was 50 mW.

Compounding was carried out in a Brabender Plasti Corder PL 2000 internal mixer. Polypropylene was fed into the equipment first, and immediately after the melting of the whole mass of basalt fibers and interface modifier were added. The temperature of compounding was relatively high (230°C) the processing time (15 min) and the rotation speed (15rpm) were relatively small, in order to reduce the fragmentation of the fibers. The LIN-MAH, NFO-MAH and GMO-MA additives were grafted onto the polypropylene chains by adding a calculated amount of Luperox F 90 P peroxide initiator ($5 \cdot 10^{-3}$ weight% of the composite). The peroxide, the surface treating compounds and the basalt fibers were mixed before adding them to the molten polypropylene. In the case of interface modification by PP-g-MAH no peroxide initiator was used. Each composite contained 20.0 weight% basalt fibers. The amount of interfacial additives was 2.0 weight% relative to the mass of fiber which means 0.4 weight% of the whole composite. Compression moulding of 160×160×4 and 160×160×2 mm size plates was performed in a Collin P200E press at 230°C applying 20 bar pressure and 15 min pressing time. Differential Scanning Calorimetry measurements were performed in Setaram DSC 92 equipment. The non-isothermal crystallization behavior of the virgin PP H384F and the composites was determined by heating the samples at first were up to 230°C at a rate of 20K/min, and then the crystallization curve was recorded at a cooling rate of 10K/min. The curves were evaluated with the Setsoft 2000 1.6.3 software.

Mechanical testing was carried out at room temperature: Charpy fracture tests were performed using a Ceast-DAS 8000 equipment according to the ISO 179 1:2000(E) standard. The notched specimens of 80×10×4 mm size had an „A" type slot and were broken with a 2J energy hammer on a 62 mm thick span holder. Three point bending tests were carried out using a Zwick Z020 equipment according to ISO 178:2001(E) standard: the 80×10×4 mm size specimens were bent on a 62 mm thick span holder with 5 mm/min cross-head speed, the Young modulus values were calculated from the initial segments of the curves. Tensile characteristics were determined by the same equipment according to the ISO 527-1 standard, the results were evaluated by the Testxpert 6.01 software.

Scanning electron microscopic (SEM) observation of the fiber-matrix boundary layers of the composites was made using a JEOL 5500 LV instrument.

Results and Discussion

The synthesis of reactive surfactants was carried out at first and the reactions were controlled using the fiber optic head of Raman microscope. The results of the synthesis of a Diels –Alder adduct is shown here as an example. Before Raman microscopic monitoring of the synthesis the spectra of the row materials (LIN and MAH) and the product (LIN-MAH) were determined, as shown in Figure 3.

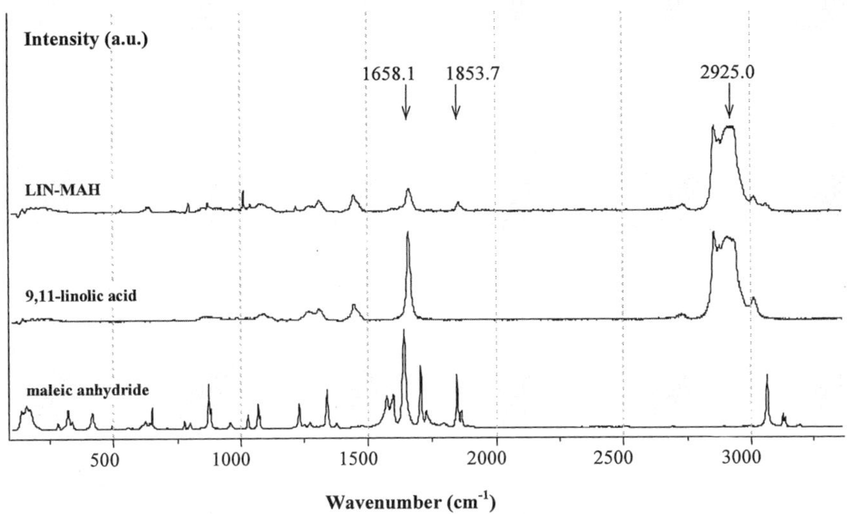

Fig. 3. The Raman spectra of linolic acid (LIN), maleic acid (MAH) and LIN-MAH compound.

The series of spectra recorded during the preparation of the LIN-MAH additive is shown in Figure 4.

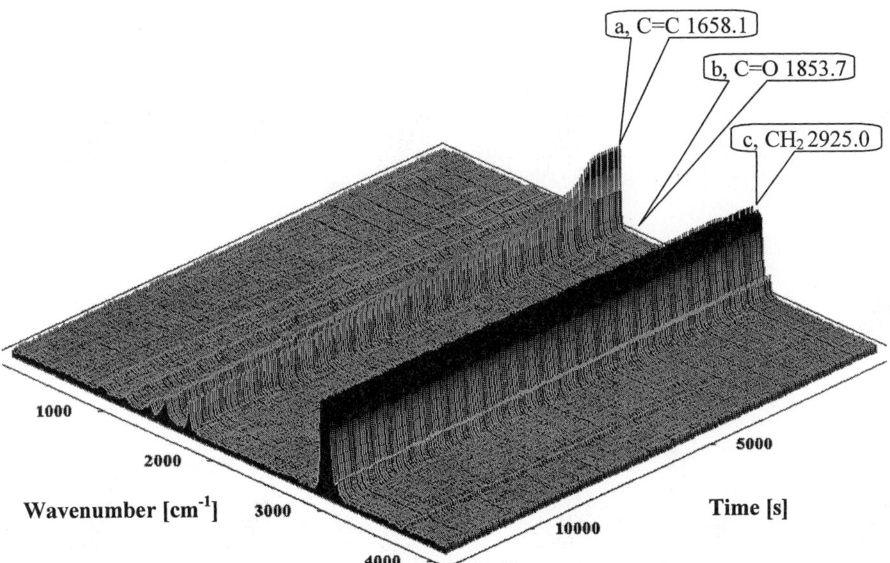

Fig. 4. Series of spectra registered by Raman microscope during monitoring the Diels-Alder synthesis showing the changing intensity of the characteristic stretching vibrations of the C=C (a), C=O (b) and CH_2 (c) groups.

As shown in Scheme 2, 9,11 linoleic acid contains two conjugated C=C bonds. The characteristic symmetric stretching vibration of these bonds appears at 1658.1cm^{-1} and its intensity decreases in time. The symmetric stretching vibration of the C=O bond in the maleic anhydride appears at 1853.7cm^{-1}. Its intensity increased after feeding then decreased due to the reaction. The intensity of the stretching vibration of the CH_2 group appears at 2925.0cm^{-1}. The intensity of this group slightly increased in time.

The prepared additives were used at compounding of PP and basalt fibers as reactive interface modifiers. After compounding the crystallization characteristics were analysed.

Differential Scanning Calorimetric measurements were made to detect the possible nucleation effect of basalt fibers on PP. DSC curves in Figure 5 show that the non-treated basalt fibers (Curve 2), similarly to the virgin glass fibers[25] have no influence on the crystallization of the polypropylene matrix (Curve 1). Crystallinity, calculated from the heat of the crystallization, decreased by 5%. The fibers at their surface probably hinder the ordering process of the polymer chains during the crystallization of the matrix.

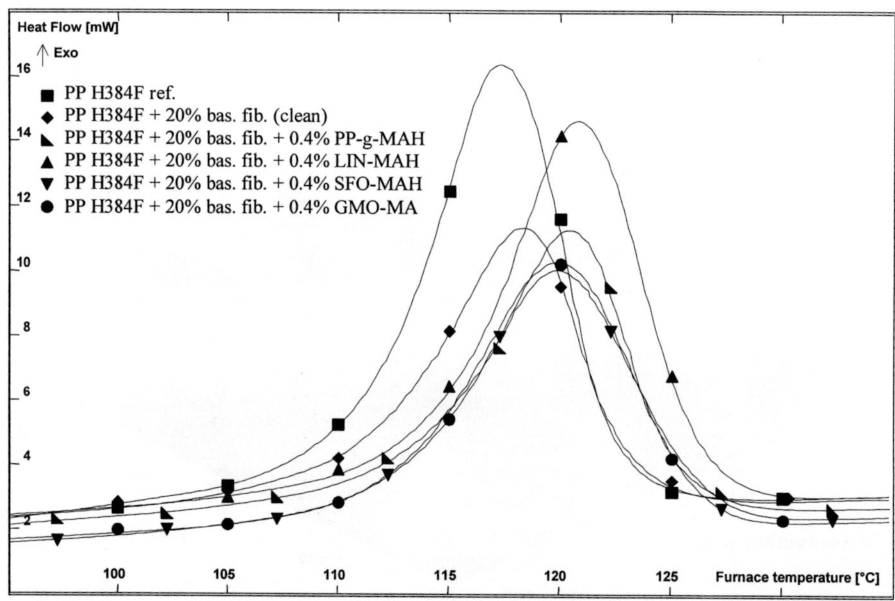

Fig. 5. The DSC curves of the virgin polypropylene and the composites.

It can be seen from the Curves 3, 4, 5 and 6 that the surface treated basalt fibers exert an influence on the crystallization process of the PP H384F matrix. The T_{cp} (crystallization peak) values shift toward higher values by 2.6-3.5°C. Crystallinities increased as compared to the non-treated case by 4-9%. The reactive surfactant molecules adsorbed on the surface of the fibers may promote the ordering of the polyolefin chains during the crystallization, thus overcompensating the blocking effect of the fibers.

Table 1. The results of the DSC measurement.

Sample	Peak of the curve of crystallization	Crystallization heat	Crystallinity*
	°C	J/g	%
1. PP H384F ref.	117.3	- 72.43	52
2. non treat. clean fibers	118.3	- 52.53	47
3. PP-g-MAH	120.5	- 59.37	53
4. LIN-MAH	120.8	- 62.80	56
5. SFO-MAH	119.9	- 61.30	55
6. GMO-MA	119.9	- 56.73	51

* Calculated with the using of the crystallization heat of the 100% crystalline PP (- 138 J/g).[26]

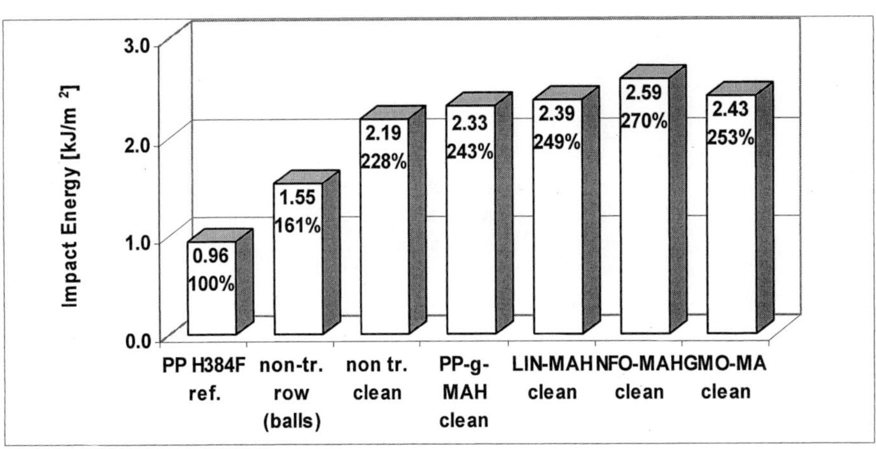

Fig. 6. The results of the Charpy fracture tests.

Mechanical testing results show the effect of removing the fiber ends (cleaning) and modification of the interfaces The results of the mechanical tests are shown in Figures 6, 7 and 8.

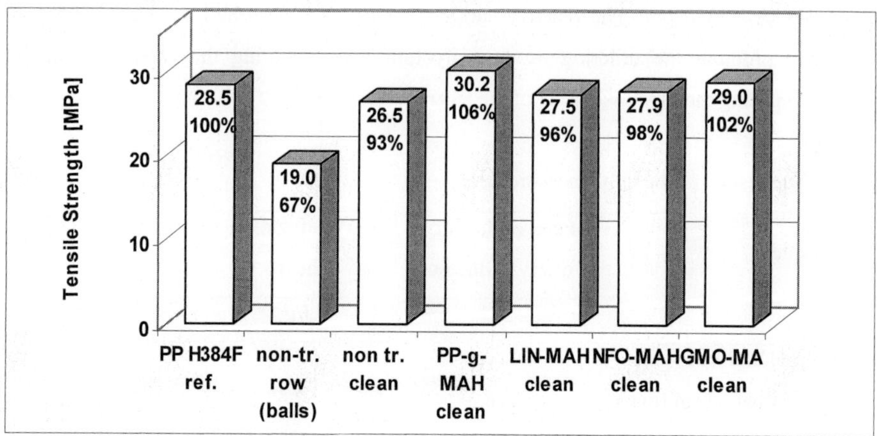

Fig. 7. The results of the tensile tests.

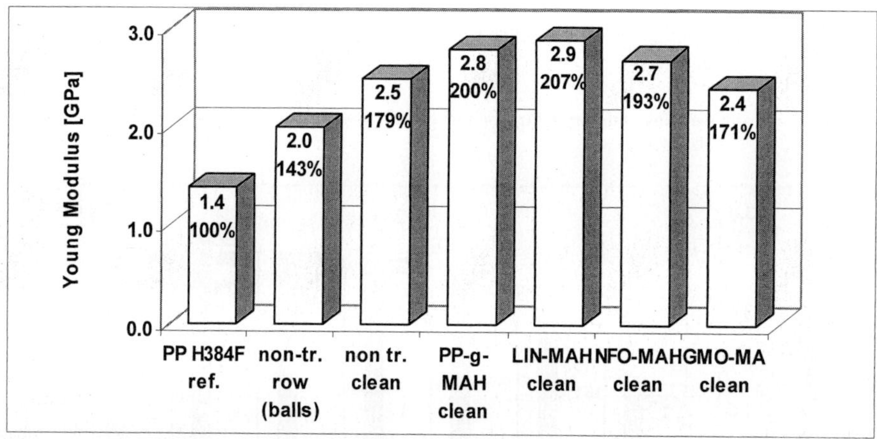

Fig. 8. The Young modulus values calculated from the bending curves.

Introducing the row basalt fibers (milled without removing the ball fraction) into the PP H384 matrix increases both impact resistance and Young modulus significantly. The improvements of these values are 61 and 43% respectively. At the same time the tensile strength is reduced by 33% as compared to the virgin PP H384F sample.

Cleaning the milled product from the broken fiber-ends and mixing it into the polypropylene matrix improves each mechanical property. The impact resistance improves further by 67% and becomes more than twice higher than the value of the PP H384F. The increase of the tensile strength is 26% which means that it is just slightly lower than that of the matrix. The toughness of the composite increased further by more than 30%.

The surface treatment of the fibers yields further improvement in the mechanical properties. The impact resistance improves by 15-42%. The increase of the tensile strength in the case of LIN-MAH and SFO-MAH are modest (3 and 5%). The tensile strength values of these composites are comparable to that of the matrix. In the case of the GMO-MA and PP-g-MAH additives the improvement is about 10%. The Young modulus values generally increase. The only exception is the GMO-MA treatment where moderate decreasing was found.

The improved mechanical properties can be explained by increased fiber-matrix adhesion. The SEM photograph in Figure 9 shows that the absence of interlayer modification yields weak fiber-matrix adhesion. In this case the polymer matrix and the fiber are separated, due to the local stresses initiated by mechanical or thermal effects. In contrast, Figure 10 shows that in the system containing surface treating additive the adhesion of the fibers to the PP increases, therefore the polymer matrix adheres to the surface of the reinforcing fibers and does not separate from it easily.

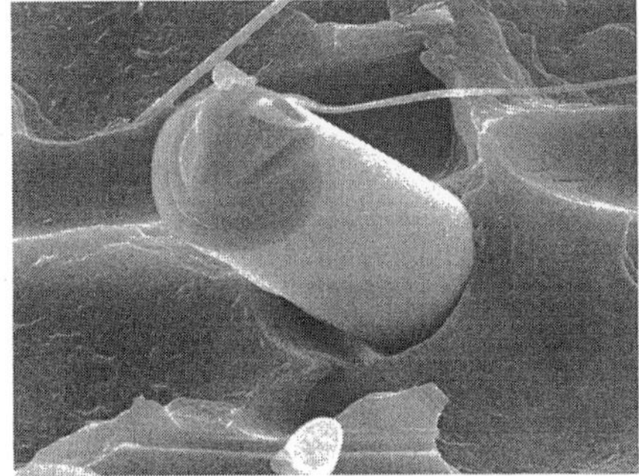

Fig. 9. The SEM photograph of the fracture surface without interface modification.

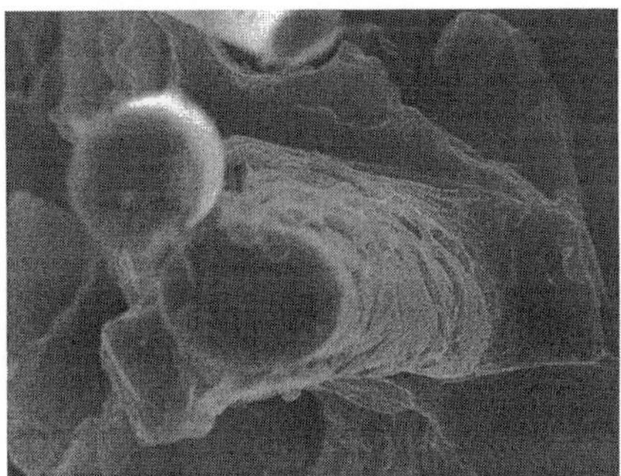

Fig. 10. The SEM photograph of the fracture surface in presence of the PP-g-MAH.

Conclusion

New type of reactive interface modifiers could be prepared by means of Diels-Alder reaction and esterification. These additives combining the advantages of surfactants and coupling agents facilitate the reactive compounding/processing. They reach the interfacial region within the short residence time and bond phases together by reacting with both of them. Raman microscopy was found to be a suitable method to follow the progress of Diels-Alder reaction. The results of the DSC measurements show that the basalt fiber is inactive, but the interfacial additives have a nucleating effect on the polypropylene matrix.

The fiber-ends exert a weakening effect on the mechanical properties of the composites. Applying a cleaning process on the fibers, and reactive interface modification improved the mechanical properties significantly. The results show that both the commercial and the prepared non-commercial interfacial additives are effective at low concentration. The bonding to the PP matrix is promoted in case maleated PP wax by penetration PP chains into the matrix and in case of reactive surfactants by means of radical reactions. These compounds increased the fiber-matrix adhesion as confirmed by scanning electron microscopy (SEM) images.

Acknowledgment

This work has been financially supported by the Hungarian Research Fund through projects OTKA T32941, NKFP 00169/2001 and NKFP 3A/0036/2002.

[1] A. P. Gupta, U. K. Saroop, G. S. Jha, M. Verma, Polym.-Plast. Technol. Eng. 2003, 42, 297.
[2] A. Zheng, H. G. Wang, X. S. Zhu, S. Masuda, Compos. Interf. 2002, 9, 319.
[3] J. Y. Yu, X. M. Kong, L. Q. Wei, H. Zhou, L. Wang, J. Wuhan Univ. Technol-Mater. Sci. Edit. 2002, 17, 62.
[4] G. Jannerfeldt, R. Tornqvist, N. Rambert, L. Boogh, J. A. E. Manson, Appl. Compos. Mater. 2001, 8, 327.
[5] H. A. Rijsdijk, M. Contant, A. A. J. M. Peijs, Compos. Sci. Technol. 1993, 48, 161.
[6] W. E. Worrall, *"Clays and Ceramic Raw Materials"* Elsevier Applied Science Publishers Ltd., 1986, p.49.
[7] T. Jung, R. V. Subramanian, *Scripta Metallurgica et Materialia* **1993,** *28*, 527.
[8] I. Wojnarovits, *Glass Sci. Technol.* **1995,** *68*, 360.
[9] V. Kovacic, J. Militky, *Fibres Text. East. Eur.* **1996,** *4*, 72.
[10] M. A. Sokolinskaya, L. K. Zabava, T. M. Tsybulya, A. A. Medvedev, *Steklo i Keramika* **1991,** *10*, 8. (in Russian)
[11] J. M. Park, R. V. Subramanian, *J. Adhes. Sci. Technol.* **1991,** *5*, 459.
[12] J. M. Park, R. V. Subramanian, A. E. Bayoumi, *J. Adhes. Sci. Tecnol.* **1994,** *8*, 133.
[13] J. M. Park, R. V. Subramanian, A. E. Bayoumi, *J. Adhes. Sci. Tecnol.* **1994,** *8*, 1473.
[14] J. M. Park, W. G. Shin, D. J. Yoon, *Compos. Sci. Technol.* **1999,** *59*, 355.
[15] R. A. Veselovskij, V. V. Efanova, I. P. Petukhov, V. N. Kestelmen, *Int. J. Polym. Mater.* **1994,** *23*, 139.
[16] J. Militky, V. Kovacic, J. Rubnerova, *Int. J. Polym. Mater.* **2000,** *47*, 527.
[17] M. Botev, H. Betchev, D. Bikiaris, C. Panayiotou, *J. Appl. Polym. Sci.* **1999,** *74*, 523.
[18] M. S. Batiashvili, N. D. Kheladze, Soviet Patent 1666480.
[19] J. S. Szabó, T. Czigány, *Polymer Testing* **2003,** *22*, 711.
[20] J. Rohrmann, Eur. Pat. Appl. EP 634451.
[21] E. N. Makarycheva, G. I. Kostrykina, *Kauch Rezina* **1991,** *9*, 18. (in Russian)
[22] M. Xantos, *Polym. Eng. Sci.* **1998,** *28*, 1393.
[23] Gy. Marosi, Gy. Bertalan, P. Anna, Hungarian Patent, 218 016.
[24] Gy. Marosi, P. Anna, I. Csontos, A. Márton, Gy. Bertalan, *Macromol. Symp.* **2001,** *176*, 189.
[25] D. M. Dean, A. A. Marchione, L. Rebenfeld, R. A. Register, *Polym. Adv. Tecnol.* **1999,** *10*, 655.
[26] J. G. Fatou, *"Differential Thermal Analysis and Thermogravimetry of Fibers. Applied Fiber Science"* ed. By F. Happes, Academic Press, London, 1979

Analysis of Multicomponent Polymer Systems by Raman Microscopy

*A. Szép, B. Marosfői, Gy. Bertalan, P. Anna, Gy. Marosi**

Budapest University of Technology and Economics, H-1111 Budapest, Műegyetem rkp. 3, Hungary
E-mail: gmarosi@mail.bme.hu

Summary: The Raman microscope is one of the most convenient instruments for analyzing the structural characteristics and changes in the interfacial region of multicomponent systems. This is confirmed by the results obtained in the field of packaging materials, nanocomposites, and basalt fibre reinforced composites. In the course of this study, the chemical character of the surface and interfacial regions were investigated and, in addition, the local characteristics of the crystallization process of the polymer matrix could be determined.
Keywords: crystallization; interfaces; Raman microscopy; surfaces

Introduction

The scope of an analytical process applicable for characterizing polymer systems is basically determined by the sensitivity of the method to the physical and chemical differences between the samples and by the modifying or destructive effects caused during sample preparation or analysis.

In order to answer the current questions emerging in materials science, especially in the area of multi-component polymer systems, the applied analytical methods have to meet more and more complex requirements. These are, for example, high sensitivity to the chemical and physical differences, good localizability, possibility for depth profiling, high lateral and depth resolution. In most cases, macroscopic samples are available for the examinations and it is a basic requirement to avoid structural changes of the sample during the measurement. The development of multicomponent polymer systems requires a comprehensive knowledge about the interfaces.

The importance of vibrational spectroscopic methods has been increasing during the recent decades. The conventional macroscopic spectroscopic methods are usually suitable for

 DOI: 10.1002/masy.200351223

description of the chemical and physical state of the sample providing overall information on the content of the sample but they do not provide local information at microscopic level. Furthermore, the examined sample may be damaged due to the operations during sample preparation and measurement of the studied material. The follow-up of the transformation and transport processes is often excluded from the scope because of the presence of a component disturbing or even impeding the analysis. For instance the biomedical application of IR spectroscopy is significantly restricted by the strong absorption by water. Raman spectroscopy, however, being much less sensitive to water, enables the examination of morphological differences and requires considerably simpler sample preparation compared to other vibrational spectroscopic methods. In the Raman spectrum there are fewer overlapped, compound band and furthermore, with certain limitation, the intensity of the bands is directly proportional to the concentration of material. Due to differences in selection rules, Raman and IR spectroscopy are complementary analytical methods; the examined wavenumber ranges largely overlap, thus both of them provide information on the vibrational or vibrational-rotational states of the molecules.[1]

A Raman microprobe, combining a spectroscope with an optical microscope, enables one to make local measurements at microscale as well.[2,3] Regarding the optical resolution, the Raman microprobe allows analysis of a sample area of 500 nm in diameter and the examination of the layers under the surface can be carried out without damaging the surface.[4]

Raman microscopy can be applied in various fields of material science, pharmaceutical industry, biology, and chemistry to local qualitative and quantitative analyses, process control,[5] determination of thermodynamic quantities and equilibriums and to studies of isomeric transformations. In biology and medicine Raman spectra can be recorded even under in vivo circumstances.[6,7] Further important examples of application are the assessment of the age and origin of paintings based on the pigment type used and forensic examinations (e.g. authentication of signatures or detection of drugs).[8] Raman micro-spectroscopy is an excellent method for detecting local morphological differences.[9,10] Owing to the recently used low power laser beam, the examination can be repeated on the same sample any number of times, because the sample does not undergo a structural change even during prolonged measurements.[11] By means of the visual image projected with the mediation of the microscope on the computer screen, the exact fixation of the examined detail of the sample is simple. Local characteristics like contamination, defect of crystallization, etc. can be characterized easily this way.[12] For industry, this method offers the possibility of quality control during the manufacturing process without any waste product.[13]

As we have shown above, the application range of Raman microscopy increases rapidly and is becoming a very important method for material analysis especially in polymer laminates, surface-treated and multicomponent systems. In multicomponent polymer systems, the method may give valuable information about the individual layers, their surfaces and interfaces, i.e. of polymer-polymer or polymer-substrate interfaces,[14] inter-diffusion and inter-reaction effects,[15] etc.

The aim of this paper is to demonstrate the versatility of the application of Raman microscopy in polymeric material investigation. The main examples are: multilayered materials, surface-modified polymer, effects of flame and heat treatment on chemical structure and transport processes, the structure of nanocomposites and fibre reinforced polypropylene.

2 Experimental

2.1 Sample Preparation

2.1.1 Preparation of Multi-layered Packaging Material Sample

A multilayer packaging material including an aluminum layer of 40 μm thickness and a polyamide layer of 25 μm thickness, with an adhesive promoter between them was investigated "as received". The quality and thickness of the adhesive layer was examined using depth profiling and on the cross-section of the multi-layered package.

2.1.2 Preparation of Surface Coated and Nanocomposite Containing Samples

Compounds were prepared in a Brabender mixing chamber at 200°C using the following components: polypropylene (PP, Tipplen H 535, TVK Co, Hungary, density: 0,9 g/cm^3, melt index: 4g/10min at 21.6 N, 230°C); polyboroxosiloxane (BSil), prepared by reacting silicon oligomer (Tegomer 2111, Th. Goldschmidt AG) with boric acid[16] and montmorillonite of Benton SD-1 type (MMT, nanoclay organophillized with cationic surfactant, Reox Inc., particle size <1μm). Coating of the particles was performed in-line during compounding.

2.1.3 Preparation of a Sample for Crystallization

PP with basalt fibre (Toplan Ltd, Hungary) was prepared in a Brabender mixing chamber at 200°C.

2.1.4 Raman Spectroscopic Analysis

A LabRam confocal Raman microscope (Jobin Yvon) was used for the examinations. The optical resolution of the instrument, that is the size of the examined area was adjustable by means of changeable objectives according to the features of the sample. The instrument was operated with three changeable excitation light sources: a He-Ne laser (632 nm), a frequency doubled Nd-YAG laser (532 nm), and a diode laser (785 nm). A good spatial resolution was ensured by means of a confocal aperture at a back-focal image plane and by using a charge-coupled device (CCD) detector as an "electronic" aperture. The location of the accumulation of a spectrum was chosen on the optical image enlarged by the microscope and projected on the monitor, and in the selected spots or places the accumulations were performed automatically according to a predetermined program.

For examination of a multilayered packaging material, a section of the sample was made and on it a line-mapping spectral image was accumulated along a 50-micrometer long line with 1-micrometer resolution. The He-Ne laser and a 50× objective were used in this type of examination. Recording of the spectra composing the Raman mapping image required an acquisition time of 45 seconds per spot with double accumulation for elimination of noises.

For determination of the heat-induced structural changes in the coating layer of polypropylene, the spectra were accumulated using He-Ne laser excitation, but with a shorter, 20-second acquisition times and double accumulation. In favour of the improvement of the precision of the focus in depth, the 100x objective was chosen with a confocal hole diameter of 200 μm, in accordance with the results described by Everall.[17,18]

The Raman signals of nanocomposites were detectable only with the 532 nm laser-excitation (owing to the fluorescence at other frequencies) during a short acquisition time of 20 seconds and double accumulation. The heat treatment of the samples was carried out by a heating-cooling stage that could be directly connected to the Raman microscope.

The study of crystallization of polymer composites was also carried out on the heating-controlling stage. Raman signals were detected in 60 seconds using the He-Ne laser and a 50× objective.

3 Results and Discussion

In this paper we discuss examples of the application of the Raman microscope for investigation of multicomponent polymer systems. Further details of the research work carried out on the studied materials are given in the cited papers of the authors.

3.1 Examination of Multilayered Packaging Materials

In case of multilayered packaging material developed for ensuring improved gas barrier properties, the purpose of the analysis was to explore the internal structure of the combined material. Earlier it was stated for multilayer materials (in which the refractive index of the layers are similar) that a cross-sectional profile obtained after sample sectioning shows very good agreement with a confocal depth profile determined using immersion lens.[19] In the studied case, however, the Al layer acts as a mirror decreasing the accuracy of the depth profile measurement therefore the films were investigated by cutting the sample in cross section direction to obtain spectra of the individual layers. The Raman spectra were collected using a step size of 1 μm along a line perpendicular to the layers on the section. The optical image of the cross-section taken by an optical microscope can be seen in Figure 1a. The thickness of the composite film layers can be easily determined by visual examination, but in fact the adhesive layer is not visible in the optical image; it can be detected accurately only in the Raman spectra. The resulting spectral map is shown in Figure 1b. After the „peak-free" black spectra of the aluminium layer, from 70 μm, the spectra of the approximately 10 μm thick adhesive layer with a strong band of substituted benzene at 1006 cm^{-1} (aromatic ring breathing) and 1616 cm^{-1} (ring stretches) as well as a weak band characteristic of a carbonyl group at 1720 cm^{-1} (C=O stretch) can be observed. This result made it clear that the layer consisted of a resorcinol-formaldehyde resin with a relatively small degree of crosslinking. From 80 μm to 105 μm, the Raman spectrum of the 25 μm thick polyamide layer can be seen. The characteristic bands of the spectra of the layers are highlighted in grey boxes in Figure 1c. The intensity change of characteristic bands of the neighbouring layers can be seen in Figure 1d. The penetration of the adhesive into the polyamide phase is clear from the slower gradual decrease of its characteristic band on the polyamide side. The thickness of this inter-diffusion layer is 5-10 μm.

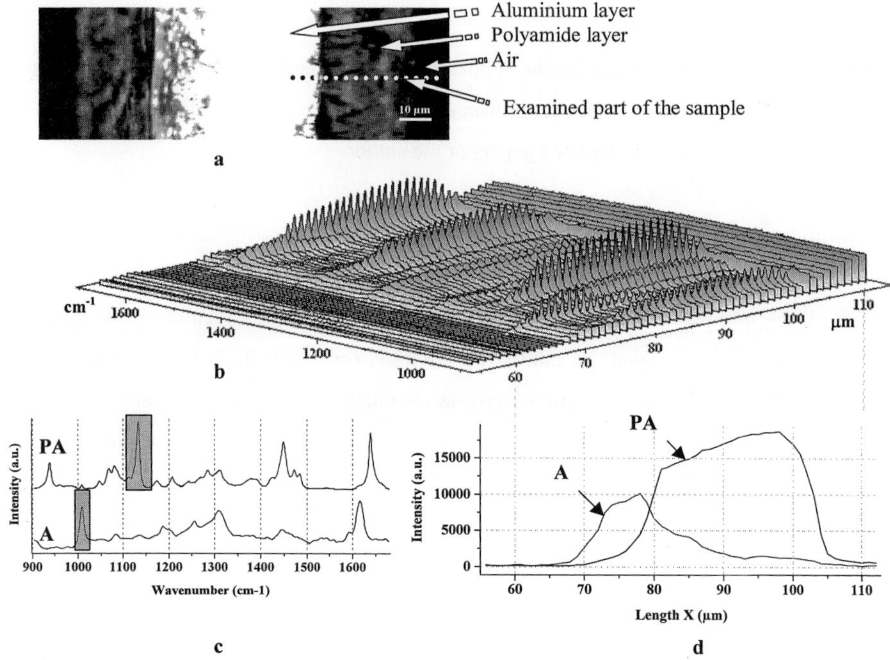

Fig. 1. a, Visual cut-section image of the multilayered packaging; b, Collected Raman spectra map at the examined part of the sample; c, Raman spectra of polyamide (PA) and adhesive (A); characteristic bands used for *d* diagram highlighted in gray boxes; d, Change in the intensity of the characteristic peaks of the adhesive (A) and polyamide (PA) layer

3.2 The Effect of the Exposure to Flame on the Chemical Structure and Properties of a Surface Coating

Surface modification of polymers offers an opportunity for changing the physical and chemical characteristics such as barrier properties, adhesion, mechanical and electrical characteristics. The gas permeability has been decreased by both micrometer and nanometer thick layers inside or outside of the materials. Forming compact SiO_x layer on surface of polypropylene for decreasing the rate of gas permeability can be used not only in packaging technology, but also in other important application fields. However, it is not easy to examine these structures and detect the eventual changes. We have found that not only the identification of coatings but also their chemical reactions can be followed by Raman microscopy. In a recent work we have investigated, how the chemical structure and properties of a coating can be affected by means of flame treatment. Favourable results were obtained in fire retardancy by means of SiO_X layer generated from boroxosiloxane precursor.[16] Flame or plasma treatment on the surface forms a flexible, glassy-like structure decreasing the gas transport, i.e. oxygen from air to the burning material, consequently the combustion of the polymer material can be extinguished.[20]

In Figure 2 the Raman spectra of polypropylene covered by boroxosiloxane (BSil) is shown before and after exposure to flame.

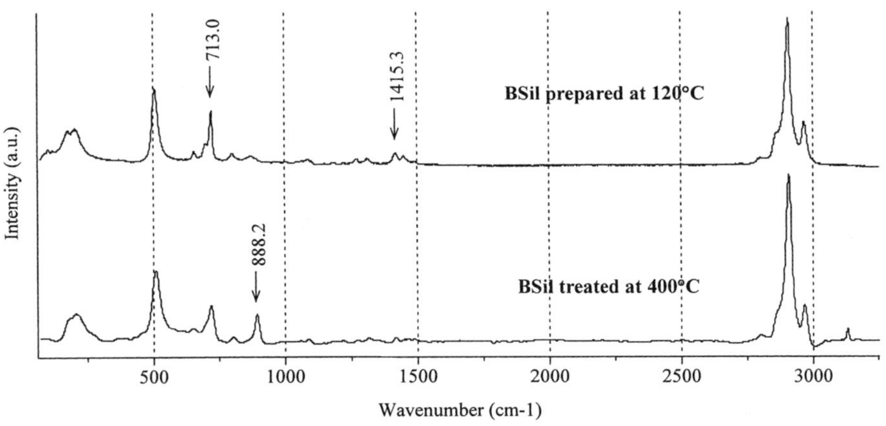

Fig. 2. Raman spectra of boroxosiloxane layer before and after flame treatment.

In the spectra collected after exposure to flame, the intensity of the characteristic bands of the symmetrical stretching of C-Si-C bonds at 713 cm^{-1} and the symmetrical deformation of CH_2 groups at 1415 cm^{-1} decrease. (The intensity in Raman spectra is directly proportional to the occurrence of the structural moieties responsible for the band). A new band appears at 888 cm^{-1} that is characteristic of Si-OH bonds. After flame treatment the carbon content decreases on the surface suggesting the increase of the inorganic character. Furthermore, at the burning temperature of the polymer, the transformation of the precursor is not complete: even after a sustained, intensive flame treatment an organic silicon containing component remains on the surface, with flexible, high performance fire barrier properties.

3.3 Examination of Polymer Nanocomposites

Polymer nanocomposites exhibit novel property combinations in which the filler shape, particle size distribution, dispersion and interfacial adhesion play very important roles. With nanofillers a new set of properties can be achieved, e.g. improved mechanical, thermal, electrical properties, reduced permeation of gases, etc.[21] Consequently, it is very important to investigate how the nanofillers are distributed in nanocomposites.

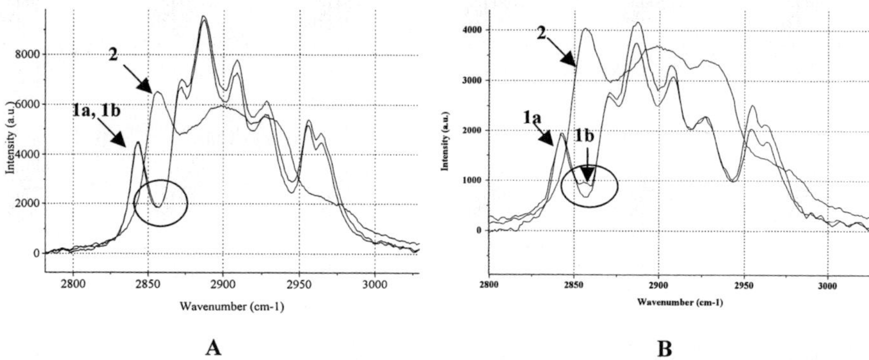

Fig. 3. A) 1. Raman spectrum of polypropylene containing MMT after heat treatment at 250°C (a) and 300°C (b) 2. Raman spectrum of MMT; B) 1. Raman spectrum of polypropylene containing MMT + BSil after heat treatment at 250°C (a) and 300°C (b); 2. Raman spectrum of MMT.

It was proven earlier that not only the gas barrier properties but also the flame retardancy could be improved by nanocomposites. The relationship between these effects and the migration of nanoplatelets in nanocomposite was also suggested.[22] We have investigated, whether the Raman microscope is suitable to follow the migration processes towards the surface induced by heat treatment. Figure 3 shows the Raman spectra of two nanocomposites (1) heat treated at 250°C (a) and 300°C (b) respectively and the organophillic montmorillonite (MMT) (2) as reference. The composition of nano-composites: I), polypropylene containing 1% organophillic montmorillonite (MMT), and II.) polypropylene containing 2% BSil and 1% organophillic

MMT. In Figure 3 the marked part of the spectrum, (measured at an ~0.5 μm thick surface layer) verifies that the Raman microscope is able to detect the slight changes (accumulation of 1% additive on the surface) caused by heat treatment.

The spectra of composites filled with organophillized montmorillonite without boroxo-siloxane show no significant differences when heat treated at 250°C or 300°C (Figure 3 (A)). A characteristic change can be observed, however, at 2860 cm^{-1} in presence of BSil after heat treatment at 300°C (Figure 3 (B)). This change appears at the same wavenumber, at 2858 cm^{-1}, where the strong band of organophillized MMT was observed. This change can be explained by decreasing the compatibility of the polymer and clay platelets. On the other hand, the heat resistant BSil surface treatment promotes migration of layered silicates from the bulk towards the surface. Consequently, their quantity near to the surface will be increased. Due to this enrichment, the characteristic band of organophillized MMT becomes detectable.

3.4 Examination of Semi-crystalline Structure in Polymer Composites

A great benefit of Raman microscopy is that, in principle, it requires no sample preparation. It means that the Raman microscope makes possible the exact spectroscopic monitoring of the thermoanalytical process, for example crystallization. The temperature of the sample stage can be controlled, which allows one to continuously follow the spectral changes occurring with changes in temperature.

In a Raman spectrum, the phase transformation is reflected in the shape and the shift of bands. In Figure 4 the change of the characteristic bands in the spectral range between 800-850 cm^{-1} is presented, where a strong band of CH_2 rocking and CC stretching is changing its shape from a broad diffuse band to a narrow one.[23] The photo in Figure 4 shows the polarizing optical microscopic image of the basalt fibres in polypropylene matrix during the crystallization process. Spectra were accumulated during the crystallization of polypropylene phase in a basalt fibre reinforced composite at the interfacial region and in the bulk. Focusing the exciting laser beam on the surface of the fibre, Raman spectra were detected in every minute whilst the temperature of the sample was decreased continuously with a rate of 1°C/min in the range of the melting point. In case of nucleating effect of basalt fibre, the characteristic band in this region becomes sharp much faster than in the bulk.[24] In this case, however, the examinations performed in other places of the polymer melt did not show any detectable change in the rate of the crystallization, no difference was found in the temperature dependent spectra of interface and bulk. Thus it can be stated that the basalt fibre does not have any nucleating effect, does not act as heterogeneous nucleant and, differently from several other fibres and fillers, does not produce an interfacial, transcrystalline layer in polypropylene.

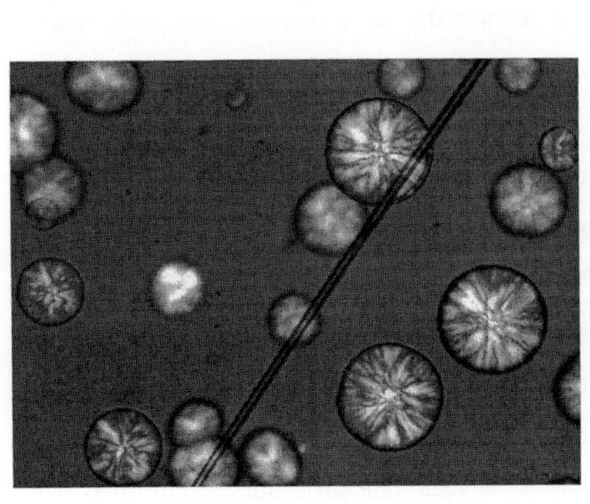

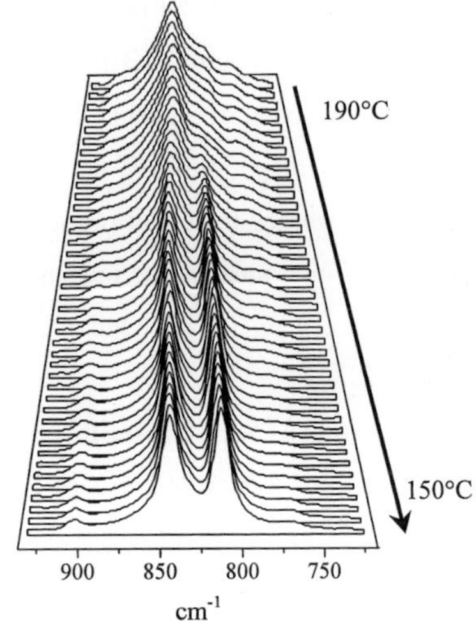

a b

Fig. 4. Crystallization of basalt-reinforced polypropylene examined by Raman microscope.
a) Spherulites observed by a 50x objective at 120°C.
b) Changes in intensity and shape of the characteristic band during crystallization (the arrow shows the temperature decrease in the range of 190-150°C).

4 Conclusions

As it has been discussed above, Raman microscopy offers several alternative possibilities in the field of materials science, especially in the polymer area. The examinations require minimum sample preparation and owing to this, the structure of the sample changes during measurements only slightly or not at all.

This method allows determination of the layer-structure of multilayered packaging materials, the quality and width of the adhesive layer and even the depth of its penetration into the neighbouring polymer layer.

Raman microscopy of boroxosiloxane, a precursor of the surface barrier layer, has verified that the organic character and the flexibility of the surface layer is preserved even after a permanent exposure to intensive flame.

Based on the investigation of the polypropylene samples containing montmorillonite nanoparticles it could be stated that in case of heat or flame treatment the nanoparticles appeared at the surface in abundance if a layer of low compatibility to the polymer matrix covered their surface.

The combination of the heating-cooling sample stage with the Raman microscope made it possible to gain novel information on the crystallization features of the polypropylene composites reinforced by basalt fibre.

Acknowledgment

This work has been financially supported by the Hungarian Research Fund through projects OTKA T32941, NKFP 00169/2001 and NKFP 3A/0036/2002.

[1] J. Corset, P. Dhamelincourt, J. Barbillat, *Chem. Britain,* **1989**, 25(6), 612.
[2] B. Schrader, *Fresenius J. Anal. Chem.* **1990**, 337(7), 824.
[3] W. Schrof, J. Klingler, W. Heckmann, D. Horn, *Colloid Polym. Sci.* **1998**, 276, 577.
[4] R. Tabaksblat, R. J. Meier, B. J. Kip, *Appl. Spectrosc.* **1992**, 46(1), 60.
[5] J. Breitenbach, W. Schrof, J. Neumann, *Pharm. Res.* **1999**, 16(7), 1109.
[6] F. F. M. Demul, A. G. M. Vanwelie, C. Otto, J. Mud, J. Greve, *J. Raman Spectrosc.* **1984**, 15(4), 268.
[7] J. Maruani,"*Molecules in Physics, Chemistry and Biology*" IV. Kluwer, Dordrecht, 1988, p. 87.
[8] P. V. Huong, *Vibrational Spectroscopy* **1996**, 11, 17.
[9] M. Tsuboi, T. Ikeda, T. Ueda, *J. Raman Spectr.* **1991**, 22(11), 619.
[10] C. M. Deeley, R. A. Spragg, T. L. Threlfall, *Spectochimica Acta* **1991**, 47A, 1217.
[11] R. L. McCreery, A. J. Horn, J. Spencer, E. Jefferson, *J. Pharm. Sci.* **1998**, 87(1), 1.
[12] E. D. Williams, L. D. Marks, *Crit Rev. Surf. Chem.* **1995**, 5(4), 275.
[13] P. V. Huong, *J. Pharm. Biom. Anal.* **1986**, 4, 811.
[14] C. Sammon, C. Mura, P. Eaton, J. Yarwood, *Analysis* **2000**, 28(1), 30.
[15] C. Sammon, S. Hajatdoost, P. Eaton, C. Mura, J. Yarwood, *Macromol. Symp.* **1999**, 141, 247.
[16] Gy. Marosi, A. Márton, P. Anna, Gy. Bertalan, B. Marosfői, A. Szép, *Polym. Degrad. Stab.* **2002**, 77, 259.
[17] N. J. Everall, *Appl. Spectrosc.* **2000**, 54(6), 773.
[18] N. J. Everall, *Appl. Spectrosc.* **2000**, 54(10), 1515.
[19] H. Reinecke, S. J. Spells, J. Sacristan, J. Yardwood, C. Mijangos, *Appl. Spectrosc.* **2001**, 55(12), 1660.
[20] I. Ravadits, A. Tóth, Gy. Marosi, A. Márton, A. Szép, *Polym. Degrad. Stab.* **2001**, 74(3), 414.
[21] C. Z ilg, F. Dietsche, B. Hoffman, C. Dietrich, R. Mulhaupt, *Macromol. Symp.* **2001**, 169, 65.
[22] Gy. M arosi, R. Lágner, Gy. Bertalan, P. Anna, A. Tohl, *J. Therm. Anal.* **1996**, 47(4), 1163.
[23] A. S. Nielsen, D. N. Batchelder, R. Pyrz, *Polymer* **2002**, 43, 2671.
[24] C. Shen, A. J. Peacock, R. G. Alamo, T. J. Vickers, L. Mandelkern, C. K. Mann, *Appl. Spectrosc.* **1992**, 46(8), 1226.

Macromol. Symp. **2003**, *202*, 281—290

Study of Nucleation Induced Structure Modification in Isotactic Polypropylene by DMTA and Solid State NMR

Anna Romankiewicz,[1] *Jan Jurga,*[2] *Tomasz Sterzynski**[1]

[1]Institute of Chemical Technology and Engineering,
[2]Institute of Materials Technology, Poznan University of Technology, PL 60-965 Poznan, Poland
Email:tomasz.sterzynski@put.poznan.pl

Summary: Isotactic polypropylene (iPP) modified by heterogeneous nucleation and molten state drawing was investigated using the DMTA and NMR methods. The nucleation was realized by specific α and β nucleating agents, 1,3,2,4-bis(3,4-dimetylobenzylideno) sorbitol (Millad 3988, Miliken) leading to the creation of the α phase, and N,N'-dicyklohexylo-2,6-naftaleno dikarboxy amid (NJ100) as the β phase promoter. The processing induced modification was performed by molten state drawing during an extrusion in the range between the die exit and the calibration unit. An increase of the glass transition temperature of iPP was found to be drawing independent for the β-nucleated samples, and dependent in the case of the α-iPP. Changes in the macromolecular mobility, depending on the α/β iPP structure and molten state drawing, was found by NMR lineshape and second moment measurements.

Keywords: chain mobility; DMTA; glass transition; NMR; poly(propylene) (PP); specific α and β nucleation

Introduction

The heterogeneous nucleation presents a very efficient approach to structure modification for polymers, characterised by a relative low crystallisation rate, and by significant supercooling. This is the case of an isotactic polypropylene.[1-7]

The addition of even of very low amount, like 15 to $30x10^{-4}$ parts, of low molecular substances (*called nucleating agents(NA)*) results in an increase of the nucleation density, and therefore leads to effects like a rise of the crystallisation temperature (*an important processing parameter*), production of fine spherulitic morphology (*an optical effect, but also modification of the properties*), and in specific condition to the modification of the crystal structure, e.g. creation of various crystallographic forms, like α, β or γ crystal phase in polypropylene. For isotactic polypropylene (iPP) and for related copolymers a huge family of nucleating agents was developed and examined.[8-13]

 DOI: 10.1002/masy.200351224

Both crystal phases in iPP (α and β phase) differ by macromolecular chain conformation, crystallographic form, elementary unit density and therefore by molecular mobility, thus consequently by the physical properties. Huge research was published concerning possible applications of iPP crystallised in both crystal forms.[1,2,14-19]

The crystal structure modification, related to the changes of the chain conformation and of the geometry of elementary units, e.g. consequently changes of the interchain distances, may be determined by the measurements of the chain mobility.[20-23]

Significant differences in chain mobility may be awaited if structure of an isotactic polypropylene is modified by both specific nucleating agents and by molten state processing. Especially the different chain arrangement of iPP macromolecules which are in general heterogeneous may be studied by means of broad line NMR spectra giving rise to the multiple transitions occurring in the temperature range around and above glass transition of PP. Three components of the rotating frame spin-lattice relaxation times $T_{1\rho}$ with several relaxation minima in their temperature dependences also confirmed such relaxation behaviour.[24,25] The minimum of T_1 spin-relaxation in laboratory frame being appeared in the temperature range above the glass transition of PP corresponds to the β relaxation which was assigned to the glass transition phenomena.[26,27] The influence of NA on relaxation processes in PP was found out by rotating frame spin-lattice relaxation times $T_{1\rho}$.[25] In this case only the small shifts between β relaxation corresponding to the segmental motion and α relaxation which was ascribed to the motion of the whole amorphous chains were observed. On the other hand broad line NMR spectra provide data which are helpful to interpret also the temperature dependences of the relaxation times. Therefore we used the broad line NMR techniques to estimate the changes due to NA addition to iPP and to processing.

To study the role of both modifying parameters, e.g. of the heterogeneous nucleation and processing the dynamic mechanical thermal analysis (DMTA) and nuclear magnetic resonance (NMR) measurements were realized. Especially the DMTA is commonly applied when the information about the glass transition temperatures (T_g)[28-31] for investigated samples should be delivered, and the NMR for the manifestation of chain mobility in the investigated temperature range.

The aim of our studies was to describe the mechanism of structure modification on the molecular level by the iPP samples, and especially the influence of the nucleating agent on the macromolecular transition.

Materials

An isotactic polypropylene (NOVOLEN 1100H, BASF Germany) with a MFI = 2.4 cm^3/10 min (230°C, 2.16.kg, and density ρ = 910 kg/m^3) was used in our experiments. This polymer was modified by the heterogeneous nucleation using two additives: 1,3,2,4-bis(3,4-dimetylobenzylideno) sorbitol (DMDBS) leading to the creation of the α phase in iPP, and N,N'-dicyklohexylo-2,6-naftaleno dikarboxy amid as a β phase promoter. The DMDBS was Millad 3988 (Miliken), a white powder with a melting temperature T_m = 277°C and crystallization temperature T_{cr} = 214°C. The β-phase nucleate with a commercial name NJ 100 was also a white powder with a T_m = 387°C and T_{cr} = 137°C.

The inclusion of the NJ 100 into the isotactic polypropylene leads to the creation of a high content of the hexagonal β-phase. According to our previous investigations,[14,15,16,31] depending on the concentration of particular nucleating agent and of the processing condition, the β-phase content may reach values over 90%.

Both nucleating agents were incorporated into the iPP matrix in the molten state by extrusion. The high melting temperature of both nucleating agents signifies that during melt mixing both NA where in a solid state, *e.g.* a distribution of the solid nucleating additives, in a form of powder, in the molten iPP matrix has taken place. The studies were carried out for two concentrations of the NA: 0.1 wt. % and 0.5 wt. %, respectively. Pure iPP samples were investigated as well.

Processing

To determine the influence of the molten state processing on the glass transition by iPP some of the samples were extruded and drawn in molten state, in the region between the die exit and the beginning of the cooling bath. The elongation, estimated as the ratio between the average pulling velocity and the average die output velocity, was in the range from λ=1 (reference samples produced by compression moulding without elongation) to λ = 7.11.

Methods

The dynamic mechanical thermal analysis (DMTA) test was performed by torsion with a frequency of 1 Hz, in the temperature region between 240 K and 423 K. The aim of these measurements was to detect the influence of processing and nucleation induced structure modification on the glass transition temperature of iPP, as well on the position of the characteristic relaxation maximum.

The solid-state nuclear magnetic resonance (NMR) broad line investigations were performed with a continuous wave spectrometer[32] operating by frequency of 30 MHz, with frequency sweep and data acquisition performed with PC. The measurement of NMR lineshape was made in the temperature range between 227 K and 403 K, with the task to detect the influence of NA's on the possible changes of temperature dependent NMR second moment.

The Glass Transition Temperature and Relaxation Temperature for Nucleated and Drawn iPP Samples

On Figure 1 the run of the loss factor tan δ, as a function of temperature for both, non-nucleated and nucleated iPP is presented. The NA content (NJ and DMDBS) was 0.1 wt % and 0.5 wt %, respectively. It may be noted that the addition of the NJ (as β phase nucleate in iPP) leads to a shift of the glass transition temperature T_g towards higher values. Depending on the NA content the maximum on the curve tan δ was observed in the range between 275 K and 280 K for the β-nucleated iPP and between 270 K and 275 K for the α-phase iPP.

The crystal phase depending changes of the T_g, may be attributed to the interaction created between the amorphous and crystalline domains in the specific β-nucleated iPP. As it was proposed by Labour et al.[33] and discussed previously[34] the great number of tough tie molecules in the β-structure iPP may be the origin of the chain mobility modification. The macromolecular chains passing partly by the amorphous domains may give a rise of the structure rigidity and therefore of the changes of the T_g observed in this case. Such an assumption may be comprehended as contradictory to the effect of an increased impact resistance of the β-form of iPP.[1,6,7,14,18,19] Although, as it was shown by Aboulfaraj et al,[18] and confirmed,[14,19] the dependability of the impact resistance should be predominantly recognized to the spherulitic morphology, e.g. to the existence of radial and/or tangential lamellae, and to a lower extent to the chain conformation.

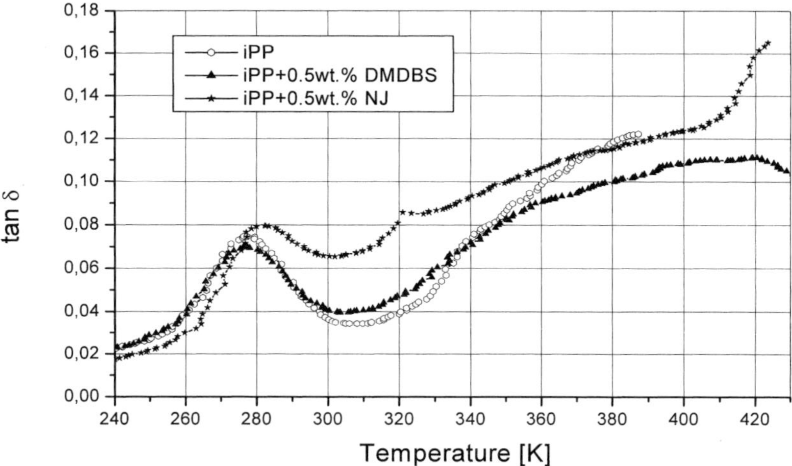

Fig. 1. The loss factor tan δ as a function of temperature for pure iPP, iPP nucleated with DMDBS and NJ.

Similar effects, e.g. an increase of the T_g were observed if in the α-iPP the molten state elongation was applied to modify the crystal order (Figure 2) in both, non-nucleated and α-phase nucleated iPP. As a consequence of drawing a shift of the T_g towards higher temperature of about 283 K, in comparison with the temperature of about 278 K observed for non-stretched samples was noted. In this case a similar explanation may be proposed as before, e.g. the creation of a certain number of tough tie macromolecules in the crystalline domains which provoke the decrease of the mobility in the amorphous phase, e.g. the rise of the T_g.

The increase of the glass transition temperature as well for the β-nucleated iPP samples as for the α-drawn iPP, signify that higher temperature (e.g. higher thermal energy) has to be provided in order to prevail over this mobility transition barrier. On the contrary no changes of the T_g was noted for β-nucleated and subsequently drawn iPP samples (Figure 3). In this case once the higher value of T_g was obtain, due to the β-nucleation, no changes may be observed after drawing. This observation may indicate also that no significant changes of the β-phase content, for the uniaxial molten state drawn iPP, should be awaited.[31,35] Other way, the specific α and/or β nucleation, and the subsequent molten state drawing may be an origin of several changes of the lamellar distribution in the iPP.[36]

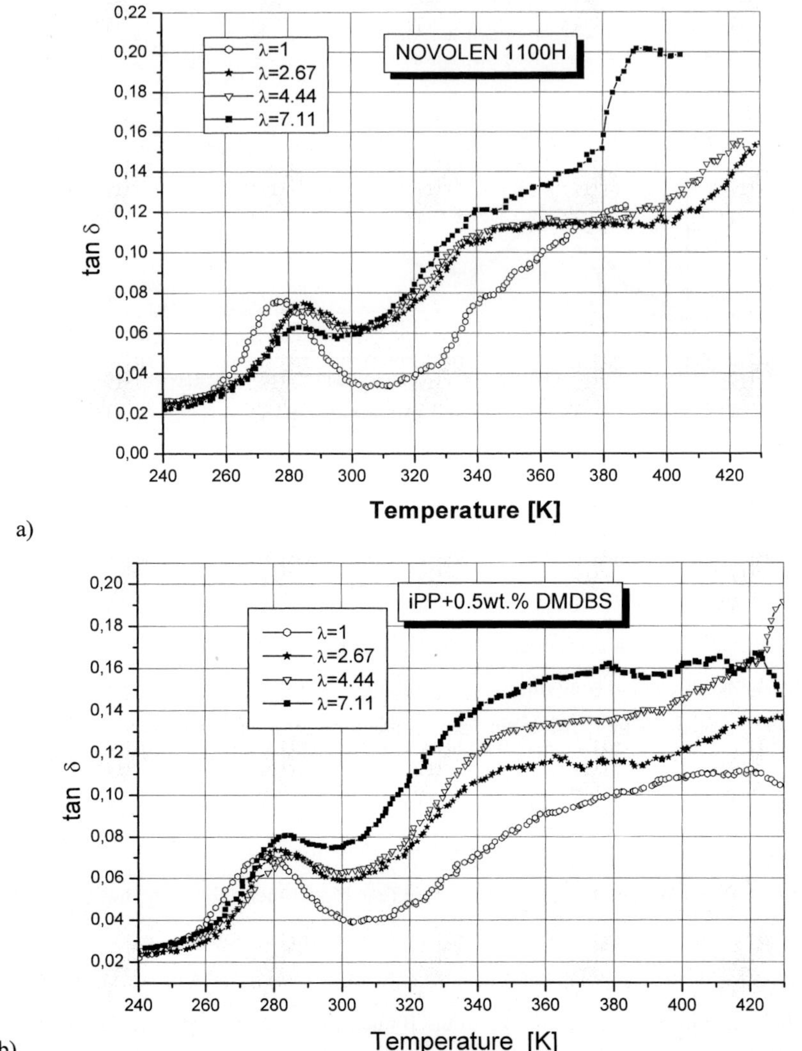

Fig. 2. The loss factor tan δ as a function temperature for molten state elongated samples: a) pure iPP, b) iPP nucleated with 0.5 wt. % of DMDBS.

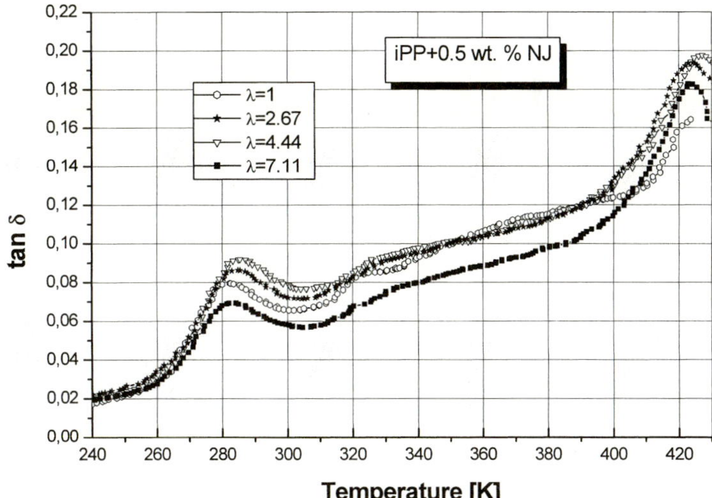

Fig. 3. The loss factor tan δ as a function of temperature for molten state drawn samples of iPP nucleated with 0.5 wt. % of NJ (β-phase iPP).

Another, less expressed maximum on the tan δ-curves may be seen for the temperature range between 320 K and 360 K. This maximum is much wider than those significant for the glass transition, and corresponds probably to the relaxation effect in the iPP. As well the position, as the intensity of the maximum is also nucleation and processing dependent.

NMR Broad Line Investigation

To explain the nature and the character of the appearance of the maximum by 320 K to 360 K, the NMR measurements in the solid state, as a function of temperature were realised.

A significant change of the NMR lineshape (Figure 4), observed for the same temperature range as the maximum on the DMTA curves, allows to relate this structural effect to the changes in the macromolecular mobility. The NMR lineshape, characteristic for the crystalline order in low temperature is transformed into the narrow line in the higher temperature. Between low (233 K) and high temperature one may observe two components of the line which indicate the coexistence of both faces, amorphous and crystalline. Based on the changes of the lineshape one may estimate the second moment which indicate the rigidity of the structure and of the thermally activated changes of mobility.

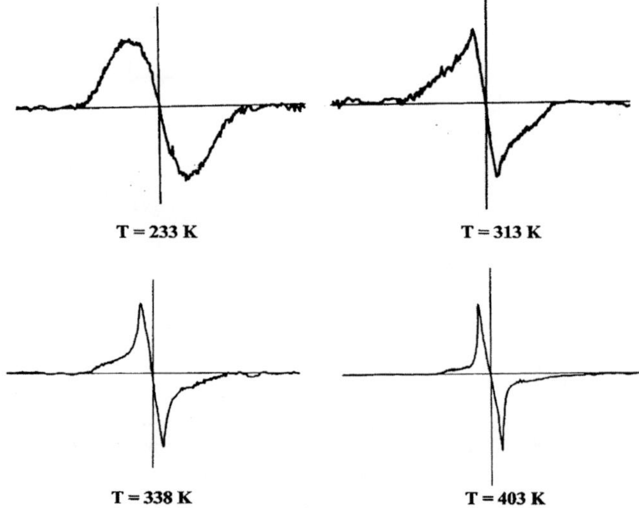

Fig. 4. The NMR line shape for stretched iPP measured by 233K, 313 K, 338 K and 403 K.

Temperature Dependent NMR Second Moment

The temperature dependent run of the NMR second moment for pure iPP samples as well for samples modified with 0.1 wt % of DMDBS and 0.1 wt. % of NJ may be seen on Figure 5.

A clear difference in the macromolecular mobility, represented by the separation of curves of the second NMR moment ΔH^2, for the temperatures below and above the glass transition T_g may be seen. Further, it follows that the temperature dependence of the ΔH^2 is strongly related to the crystalline structure of iPP. Always lower values of ΔH^2 for β - phase iPP indicates a lower chain mobility, what may be understand as a consequence of a different structure packing and organisation of the crystal structure.

An apparent distinction in second moment between α and β-iPP was observed above the T_g, with the exception of the temperature range of about 340 K. In the same temperature range the appearance of a maximum on the tan δ curves was observed, measurements below the T_g.

Such an effect observed in the region between the relaxation and melting temperature, signify a first step from order to disorder transformation in the crystalline phase of the iPP. As effects described above are not observed by WAXS, either by DSC, it may indicate that the changes in the macromolecular mobility, observed in the isotactic polypropylene, independent on the specific crystallographic form, leads to higher disordered without changes of the form and dimensions of the elementary unit.

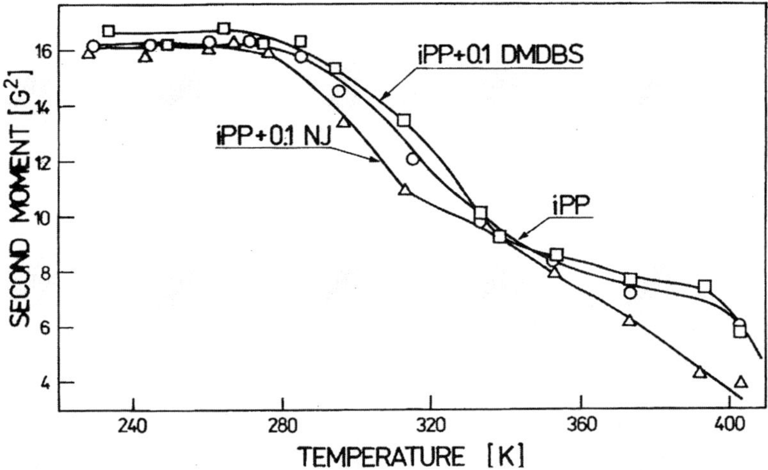

Fig. 5. The second moment as a function of temperature for pure iPP and iPP nucleated with 0.1 wt. % of DMDBS and NJ.

Acknowledgement

Project partly sponsored by Polish National Grant KBN 7T08E04717.

[1] J. Varga, in: *"Polypropylene: Structure, Blends and Composites"*, J. Karger-Kocsis, Ed., Chapman & Hall, London 1995, p.56.
[2] A. Gałęski, in: *"Polypropylene: An A-Z Reference"*, J. Karger – Kocsis, Ed., Kluwer Publishers, Dordrecht 1999, p. 543.
[3] B. Fillon, Lotz, A.Thierry, J. C. Wittmann, *J.Polym. Sci.*, **1993**, *31*, 1395.
[4] P. Jacoby, B. H. Bersted, W. J. Kissel, C. E. Smith, *J. Polym. Sci. Polym. Phys.*, **1986**, *24*, 461.
[5] B. Fillon, J. C. Wittmann, B. Lotz, A. Thierry, *J. Polym. Sci. Polym. Phys.*,**1993**, *31*, 1407.
[6] J. Varga, *J. Macromol. Sci., Phys. B*, **2002**, *41*, 1121.
[7] T. Sterzynski, *Polimery*, **2000**, *45*, 786.
[8] F. L. Binsbergen, B. G. M. De Lange., *Polymer*, **1968**, *9*, 23.
[9] T. Sterzynski, H. Oysaed, *Polym. Eng. Sci.*, **2003**, *43*,
[10] G. Shi, J. Zhang, Z. Qiu, *Macromol. Chem.*,**1992**, *193*, 583.
[11] X. Zhang i G. Shi, *Polymer*, **1994**, *35*, 5067.
[12] B. Lotz, J. C. Wittmann, A. J. Lovinger, *Polymer*, **1996**, *37*, 4979.
[13] J. X. Li, W. L. Cheung, *Polymer*, **1999**, *40*, 2085.
[14] T. Sterzyński, P. Calo, M. Lambla, M. Thomas, *Polym. Eng. Sci.*, **1997**, *37 (12)*, 1917.
[15] J. Karger-Kocsis, P. P. Shang, *J. Thermal. Anal. Cal.*, **1998**, *51*, 237.
[16] T. Sterzyński, M. Lambla, F. Georgi, M. Thomas, *Intern. Polymer Processing*, **1997**, *XII*, *1*, 64.
[17] T. Sterzyński, M. Lambla, H. Crozier, M. Thomas, *Adv. Polymer Tech.*, **1994**, *13*, 25.
[18] M. Aboulfaraj. B. Ulrich, A. Dahoun, C. G'sell, *Polymer*, **1993**, *34*, 4817.
[19] T. Sterzynski, in: *"Performance of Plastics"*, W. Brostow, Ed., Hanser Verlag, Munich 2000, p. 254.
[20] J.Murin, *Czech.J.Phys.* **1981**, *B 31*, 62.
[21] J.Murin and D.Olčák, *Czech. J. Phys.* **1984**, *B34*,247.
[22] D. Olčák, L.Ševčovič, L.Mucha, O.Ďurčová, *Polymer*, **1996**, *28*, 232.
[23] M.A.Gomez, H.Tanaka, A.E.Tonelli, *Polymer*, **1987**, *28*, 2227.
[24] D.Olčák, J.Murin, J.Uhrin, M.Rákoš, W.Schenk, *Polymer*, **1985**, *26*,1455.
[25] D.Olčák, A Stančáková, Špaldonová, O.Katreniaková, *Polymer*, **1995**, *36*, 487.
[26] U.Kienzle, F.Noack, J.von Schütz, *Kolloid Z.-Z. Polym.*, **1970**, *326*, 129.
[27] J.Murin, D.Geschke, P.Holstein, D.Olčák, L.Ševčovič, L.Mucha, *Acta Polymerica*, **1988**, *39*, 389.
[28] J.M. Crissman, *J.Polym.Sci., A2*, **1969**, **7**, 389.
[29] J.Varga, A. Breining, G.W.Ehrenstein, G.Bodor, *Intern. Polymer Processing*, **1999**, *14*, 358.
[30] J. Varga, I.Mudra, G.W.Ehrenstein, *SPE,Inc. Technical Papers,* **1998**, *XLIV*, 3492.
[31] A.Romankiewicz, *PhD Thesis*, Poznan University of Technology, 2002, Poznan, Poland
[32] J.Jurga, *Scientific Instrumentation* **1988**, *4*, 67.
[33] T. Labour, C. Gauthier, R. Seguela, G. Viger, Y. Bomal, G. Orange, *Polymer*, **2001**, *42*, 7127.
[34] A. Romankiewicz, T. Sterzynski, G. Broza, K. Schulte, *„Nucleation and processing induced modification of an isotactic polypropylene (dynamic investigation of specific nucleated iPP)"* in: Polymerwerkstoffe P'2002, Halle (Saale) 2002.
[35] A.Romankiewicz, T.Sterzynski, *in preparation*
[36] A. Romankiewicz, T.Sterzynski, *Macromol. Symp.*, **2002**, *180*, 241.

Aging of Silica-Filled PDMS/PDPS Copolymers in Desiccating Environments: A DSC and NMR Study

Robert S. Maxwell,[1] Bryan Balazs,[1] Rebecca Cohenour,[2] Elizabeth Prevedel[3]*

[1]Lawrence Livermore National Laboratory, Livermore, CA 94551, USA
Email: maxwell7@llnl.gov
[2]Honeywell, FM&T, Kansas City, MO 64141, USA
[3]Department of Chemical Engineering, University of Missouri-Columbia, Columbia, MO 65211, USA

Summary: The involvement of water in the reinforcing mechanism in silica-filled polydimethylsiloxane (PDMS) based composite systems, potentially leaves the polymer composite susceptible to mechanical property changes when exposed to desiccating environments. We have studied the effects of thermal and chemical desiccation on a silica filled PDMS/PDPS copolymer system by DSC and NMR analysis. Our results show that as the polymer was desiccated, segmental dynamics in the polymer network were reduced significantly and changes in melt heats of fusion and stress relaxation were observed. NMR ^{1}H relaxation measurements have measured a change in ^{1}H T_{2e} of 30% after storage in desiccating environments longer than 1 year. The experimental data presented argue that the reduced mobility of the PDMS chains in the interfacial domain after desiccation reduced the overall, bulk, motional properties of the polymer, thus causing an effective "stiffening" of the polymer matrix.

Keywords: desiccation; DSC; filler; NMR; siloxane

Introduction

Silica-filled polydimethylsiloxane (PDMS) composite systems are of broad appeal due to their chemical and environmental resilience and the availability of a wide range of tailorable chemical and mechanical properties.[1-3] This versatility is due, at least in part, to the presence of inorganic filler materials which are well known to significantly alter polymer material mechanical properties, but which may be characterized by complex water speciation and chemistry.[4] This interfacial water, in fact, is thought to be, in part, the reinforcing mechanism of the silica via hydrogen bonding interactions with the siloxane bridging oxygen atoms in the PDMS polymer backbone.

In an effort to gain further insight into the reinforcing mechanisms of inorganic fillers on PDMS based composites, we have initiated experimental and computational studies of the

DOI: 10.1002/masy.200351225

effects of changes in interfacial hydroxyl content on the chain dynamics of adsorbed PDMS chains. Results of computational studies of segmental dynamics at various hydration levels have predicted that removal of water from the silica surface should cause a reduction in the segmental dynamics of the adsorbed polymer chains.[5] These simulations revealed that the polymer-silica contact distance decreased as the level of "water" in the interfacial region decreased. It was also reported that the "water" in the interfacial region seemed to effectively "screen" the long-ranged interactions and mediated the polymer relaxation and reduced polymer "stiffening". A slight diminished contact distance due to the long-ranged electrostatic interactions between hydroxyl surface groups and the oxygen polymer backbone atoms were also revealed upon water removal in these simulations.

In this paper, we report the changes in thermal behavior, as studied by Differential Scanning Calorimetry (DSC), and segmental dynamics, as studied by Nuclear Magnetic Resonance (NMR), as a function of time stored in a desiccating environment. The data obtained in this study show that silica-filled PDMS/PDPS (polydiphenylsiloxane) copolymers undergo time dependent hardening in desiccating environments.

Experimental

The polymer gum used in this study was a random copolymer of dimethyl (DMS), diphenyl (DPS), and methyl vinyl (MVS) siloxanes. The percentages of each monomer unit in the base rubber were 90.7 wt.% PDMS, 9.0 wt.% PDPS, and 0.31 wt.% PVMS (NuSil Corp., Carpenteria, CA). This elastomer was compounded into a reinforced gum by milling with a mixture of 21.6 wt.% fumed silica (Cab-o-Sil M7D, Cabot Corporation, Tuscola, Il), 4.0 wt.% precipitated silica (Hi-Sil 233, PPG Industries Inc., Pittsburgh, Pa), and 6.8 wt.% ethoxy-endblocked siloxane processing aid (Y1587, Union Carbide Corp, Danbury, CT). After bin aging, this reinforced gum was formulated with the addition of a peroxide curing agent. The resulting filled-polymer system was studied in two forms: a fully dense form, and a 50% porous open cell material. The porous samples were formed by milling the reinforced gum with 50 weight percent of 25-40 mesh prilled urea spheres (Sheritt-Gordon Mines Ltd., Canada), which were subsequently rinsed out with water after curing of the polymer.[6,7]

For DSC analysis, samples were subject to a 8 hr, 70 °C bake out and subsequent storage in a nitrogen-purged, desiccated box for 6 months. The samples were then placed in hermetically sealed DSC pans, which were coated with Parylene™ to further protect against moisture ingress and analysis was performed with a TA instruments Model Q1000 DSC under a 50 ml/min He purge and a 3 °C/min ramp rate. DSC analysis of hydrated samples with and without the Parylene™ coating were indistinguishable, suggesting that the contribution to the DSC results was negligible. A container with "Drierite™" was used for storage except when the samples were being analyzed.

For NMR analysis, a small sample of polymer was sealed in a 5 mm NMR tube with a small amount of desiccant [Lithium Hydride (LiH), Phosphorous Pentoxide (P_2O_5), and Molecular Seive (MS)] positioned so that only the polymer was within the reciever coil volume and aged for up to 1 year. Transverse relaxation times were measured periodically with standard spin-echo methods.[6-8] 7μs π/2 pulse lengths, 7-second relaxation delays, and a standard Bruker 5mm TBI probe were used on a Bruker DRX-500 NMR spectrometer [Bruker Biospin, Bilerica, MA]. The echo decay curves were analyzed by measuring the time to reach 1/e of the initial intensity, T_{2e}, after removing the effects of the contribution to the decay of the sol fraction of the polymer (~10% of the echo decay curve).

Dynamic mechanical analysis testing was performed (TA Instruments AR2000 Rheometer, New Castle, Delaware) in parallel plate geometry. Specimens were disks approximately 2 mm in thickness and 8 mm in diameter. The sample was sheared at a frequency of f=6.3 rad/sec. For room temperature shear storage modulus (G') measurements, samples were sheared up to 1% strain level with a static compression force of 1 N.

Results and Discussion

DSC thermograms for the virgin filled polymer and the filled polymer aged for 6 months over Drierite™ are shown in Figure 1. For both samples, T_g was observed at -117 °C, while for desiccated samples, T_g was accompanied by a simultaneous stress relaxation of 0.75 J/g in the non-reversible heat flow measurement. In addition, a double melt peak was observed for both samples. The first melt peak was invariant with desiccation (T_{m1} = -78.5 °C) while the second melt peak shifted slightly to a lower temperature upon desiccation (ΔT_{m2} = - 3.3 °C). The heat of fusion also decreased slightly upon desiccation (ΔH_f = 9.4 J/g hydrated and ΔH_f =8.8 J/g desiccated). Further, the melt as a whole feature shifted from predominantly reversible heat flow for hydrated samples to predominantly non-reversible heat flow for desiccated samples. A small amount of recrystallization heat flow was observed for hydrated samples immediately following the second melt peak. This feature was absent in the desiccated sample. After 24 hours, the DSC pan for the desiccated sample was punctured, exposed to air, and rerun. The DSC thermogram was consistent with a fully hydrated sample, suggesting that the changes observed upon desiccation were reversible, thus due to the action of the desiccant rather than an irreversible chemical reaction.

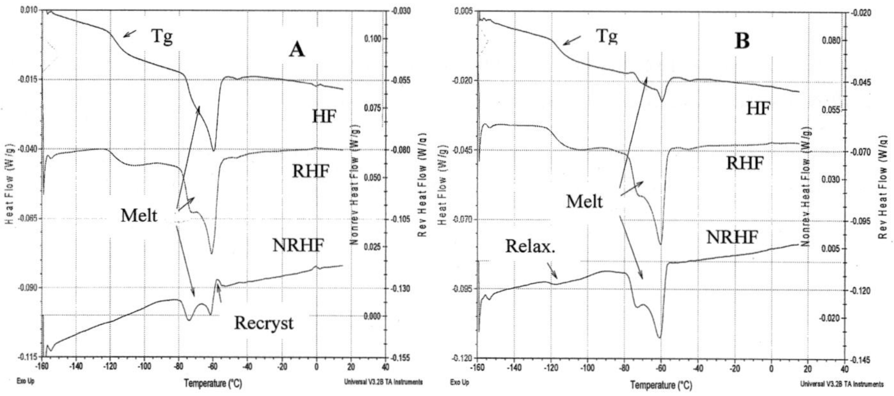

Fig. 1. DSC thermograms of (A) the copolymer at ambient conditions, and (B) the copolymer after 6 months desiccation.

In order to indirectly measure changes in mechanical properties associated with desiccation, NMR relaxation measurements were applied to assess the segmental dynamics changes that might have occurred. ^{1}H NMR relaxation measurements of transverse relaxation times in this polymer system have been directly correlated to the segmental dynamics of the polymer chains and thus the polymer mechanical properties ($1/T_2 \propto$ dynamics $\propto$ crosslink density $\propto$ G').[6-8] This relationship is documented in Figure 2.

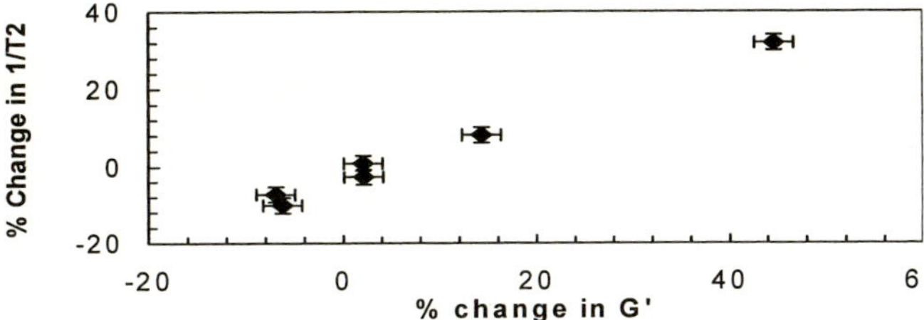

Fig. 2. Relative change in the inverse transverse relaxation time ($1/T_2$) as a function of the relative change in equilibrium storage modulus for various crosslink density copolymers. The crosslinks were modified by exposure to ionizing radiation.

The results of the spin-echo relaxation time experiments on desiccated samples are shown in Figure 3. As can be seen, as the polymer material was aged in the presence of desiccating agents with various relative affinities for water (LiH > P_2O_5 > MS), the polymer segmental dynamics were seen to slow down over the course of one year. Based on the correlation established in Figure 2, this change corresponds to a change in G' after 1 year of ~ 30%. Additionally, the relative stiffening ability of the various desiccating agents was observed to be roughly consistent with their affinities for water.

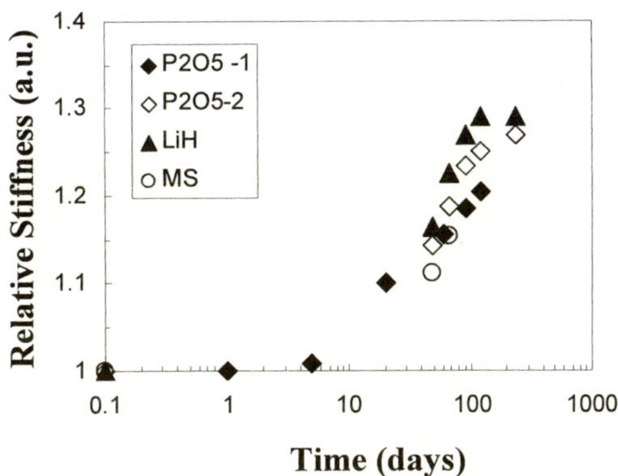

Fig. 3. Relative change in the inverse transverse relaxation time ($1/T_2$) as a function of time the filled PDMS/PDPS copolymer was stored under the desiccating agents indicated.

Conclusions

The DSC and NMR results reported here offer compelling evidence that desiccation of silica-filled PDMS based polymers can lead to dramatic changes in the mechanical properties of the material. Combined with insight from recent molecular modeling results[5] and the sensitivity of NMR to segmental dynamics,[6-8] we believe that this is due to the removal of physisorbed water at the surface. We therefore suggest here the closer polymer-silica contact and the loss of electrostatic "screening" upon dehydration of the silica filler particles as a possible mechanism for polymer "stiffening" seen experimentally in such systems. Once this "screen" has been removed, the polymer directly interacts with the filler and increases the reinforcing mechanism and stiffens the polymer matrix. It is hoped that, with the aid of advance Temperature Programmed Desorption[4] experiments to guide controlled desiccation experiments, improved insight into the aging of these polymers in arid environments might be obtained.

Acknowledgements

This work was performed under the auspices of the U. S. Department of Energy by University of California Lawrence Livermore National Laboratory under contract No. W-7405-Eng-48.

[1] G. Kraus, *Rubber Chem. Tech.* **1965**, 38, 1070.
[2] J. M. Zeigher, F. W. G. Fearon, *Silicon Based Polymer Science: A Comprehensive Resource*, Advances in Chemistry (ACS Press, **1990**, 224.
[3] P. Vondracek, A. Pouchelon, *Rubber Chem. Tech.* **1990**, 63, 202.
[4] L. Dinh, M. Balooch, J. D. LeMay, *J. Coll. Int. Sci.*, **2000**, 230, 432.
[5] R. Gee, R. S. Maxwell, L. Dinh, B. Balazs, Proc. *Mat. Res. Soc.,* **2001**, 710, DD4.2.
[6] A. Chien, R. S. Maxwell, D. Chambers, G. B. Balazs, *Rad. Phys. Chem.* **2000**, 59, 493.
[7] R. S. Maxwell, B. Balazs, *J. Chem. Phys.*, **2002**, 116, 10492.
[8] R. S. Maxwell, B. Balazs, submitted for publication in *Nucl Instr Meth. Phys. Res., B,* **2002**.

Kinetics of Reactions of Coal with Polystyrene: TGA-DSC Approach

P. Straka, J. Náhunková, Z. Brožová*

Institute of Rock Structure and Mechanics, Academy of Sciences of the Czech Republic, V Holešovičkách 41, Prague 8, CZ-182 09, Czech Republic

Summary: Kinetics of thermal reactions of coal with polystyrene by TGA and DSC methods was studied. In the range of about 350-550 °C a thermal degradation of coal proceeds, and gas, tar and coke are evolved. Simultaneously, decomposition of polystyrene occurs (360-470 °C). Unsaturated products of polystyrene decomposition are hydrogenated by coal, because coal is a strong H-donor. Moreover, some aromatic products react with coal tar structures and new aromates are formed. The yield of tar from copyrolysis is then higher in comparison with pyrolysis of coal alone. Kinetic parameters of the process were evaluated and discussed.

Keywords: aromates; coal; copyrolysis; kinetics; polystyrene

Introduction

From the results obtained from copyrolysis of coal with waste polymers in a laboratory and macrolaboratory scale[1-3] it follows that copyrolysis is a good way of recycling. It was found that yields of tar or gas from copyrolysis are significantly higher in comparison with those obtained from pyrolysis of coal. E.g. yield of tar from copyrolysis of waste ABS polymer with coal was very high in comparison with coal alone[4] (Table 1). As tar can be a main product of copyrolysis, attention was paid to thermal decomposition of polymers in the presence of coal and kinetics of reactions was investigated. In this case, kinetics of thermal reactions of coal structures with polystyrene (PS) was studied. Thermogravimetric analysis (TGA) and differential scanning calorimetry (DSC) were applied.

Table 1. Mass balance of copyrolysis (wt.-%).

Feedstock	Coke	Tar	Water	Gas	Losses
coal	71	10	5	10	4
coal + 60 % ABS polymer	34	52	2	9	3

 DOI: 10.1002/masy.200351226

Experimental

Materials

Coal from mine Dukla (Ostrava-Karviná District) was used. Proximate, elemental and petrographic analyses of the coal used are summarized in Table 2. Further, powdered polystyrene (Polymer Laboratories Ltd., U.K.) was used.

Table 2. Proximate (wt-%), elemental (wt.-%) and petrographic (vol.-%) analyses of the coal used. VM – volatile matters, SI – swelling index, V, E, I – content in vitrinite, exinite and inertinite, resp.

Water	Ash (db)[a]	VM (daf)[b]	SI	C	H	N (daf)[b]	S	O	V	E	I
1.6	10.4	33.1	1	81.6	5.5	1.3	0.7	10.9	63	15	22

[a]dry sample; [b]dry ash free basis

Thermogravimetry and Differential Scanning Calorimetry

TGA and DSC measurements with powdered PS and coal with grain size under 0.2 mm were carried out on Perkin-Elmer Pyris TGA 6 and Pyris DSC 7 analyzers. TGA analyses were performed in a nitrogen atmosphere in the temperature range of 25–700 °C with heating rates of 5, 10 and 20 $°C.min^{-1}$. DSC analyses were carried out in the temperature range of 25–550 °C with heating rates of 10, 15 and 20 $°C.min^{-1}$. Kinetic data were evaluated with a Perkin-Elmer software as follows. For reaction order and parameters of the Arrhenius plot calculations a Pyris Series DSC Scanning Kinetics and Pyris Series TGA Decomposition Kinetics were used. For conversion time calculations from both DSC and TGA data a Pyris Series Model Free Kinetics based on generalized descriptions of solid-phase reactions[5] was applied.

NMR Spectrometry

Representative structures of coal used were expressed from ^{13}C CP/MAS NMR parameters. ^{13}C CP/MAS NMR spectra were measured with the spectrometer Bruker DSX 200 in 7 mm ZrO_2 rotor at the frequencies of 50.33 MHz and 200.14 MHz (^{13}C and ^{1}H, resp.). Number of data points was 0.5 K, magic angle spinning frequency 5.0 kHz, "strength" of B_1 field (^{1}H and ^{13}C) was 50.0 kHz. The number of scans for the accumulation of ^{13}C CP/MAS NMR spectra was 3600–7200, repetition delay 3 s and spin lock pulse 1 ms. During the detection a high power dipolar-decoupling was used to eliminate strong heteronuclear dipolar coupling. ^{13}C scale was calibrated by external standard glycine (δ = 176.03 – low field carbonyl signal). For ^{1}H-^{13}C dipolar-dephasing experiments standard pulse sequence was used where cross-polarization period was followed after τ delay by two simultaneous π pulses on both (^{13}C and ^{1}H) channels. Data acquisition starts after second τ delay. 2τ delay was incremented from 2 to 200 µs. 24 increments were performed to obtain dipolar-dephasing dependence. The number of scans amounted to 400. From ^{13}C CP/MAS NMR parameters, coal structures were constructed according to work.[6]

Results and Discussion

First, an influence of the coal presence on PS decomposition was investigated by the DSC method. From the data obtained, the conversion time at conversion degree of 95 %, reaction order, and parameters of the Arrhenius plot were calculated. Results are shown in Table 3.

Table 3. DSC of the decomposition of polystyrene in the presence of coal at the temperature of maximum decomposition (T_{max}). Degree of conversion 95 %, conversion time calculated for heating rates of 10, 15 and 20 °C.min^{-1}. n – reaction order; E_A, ln Z – parameters of the Arrhenius plot calculated for heating rate of 15 °C.min^{-1}.

Mixture (wt.-% PS /wt.-% coal)	T_{max} (°C)	Conversion time (min)	n	E_A (kJ.mole^{-1})	ln Z
100/0	425	121.4	1.25	442.8	72.7
60/40	440	87.6	1.25	459.1	74.0
50/50	440	73.7	1.2	525.5	85.0
40/60	440	32.9	1.2	457.7	73.6
30/70	440	34.7	1.1	314.0	48.8

As the data in Table 3 prove, the conversion time of the PS decomposition was significantly lower in the presence of coal in comparison with that for PS alone. With an increasing share of coal in the mixture the conversion time decreased. It can be deduced that coal improves an endothermic decomposition of PS by an exothermic effect of coal decomposition in the range of 400–500 °C (ΔH -273.8 J.g^{-1}). The DSC maximum of PS shifted to higher temperature in the presence of coal (Fig. 1).

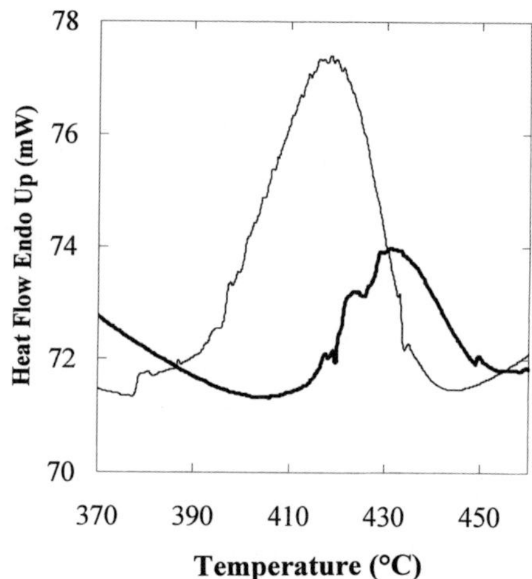

Fig. 1. DSC curves of polystyrene (thin line) and polystyrene with 50 wt.-% of coal (thick line).

It means that conditions of PS decomposition and distribution of products was changed. For this reason, thermal decomposition of PS-coal mixtures were further studied by the TGA method. The temperature of maximum decomposition (T'_{max}), the maximum rate of decomposition (w_{max}), and the percentage of the reacted mixture were measured. Results are summarized in Table 4. Typical courses of decomposition are pictured in Fig. 2.

Table 4. TGA of polystyrene-coal mixtures: thermal decomposition parameters at the heating rate of 10 °C.min^{-1}. T'_{max} – temperature of maximum decomposition, w_{max} – the rate of decomposition at T'_{max}.

Mixture (wt.-% PS/wt.-% coal)	T'_{max} (°C)	w_{max} (mg.min^{-1})	Reacted mixture (wt.-%)
100/0	426.3	0.99	99.9
60/40	441.0	0.45	75.9
50/50	484.8	0.53	65.5
40/60	442.7	0.45	57.5
30/70	447.3	0.21	52.3
0/100	463.5	0.26	35.0

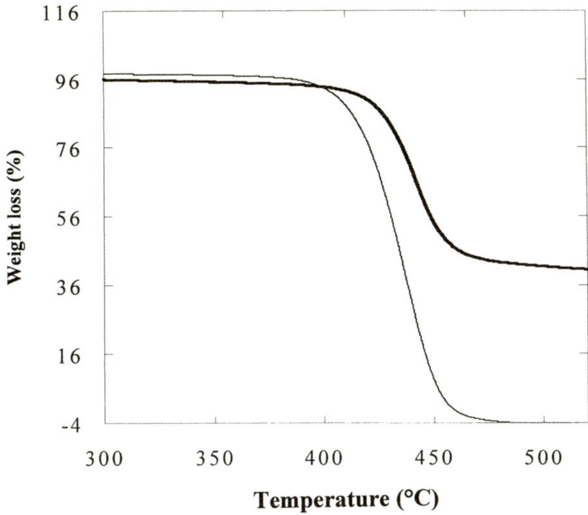

Fig. 2. TGA curves of polystyrene (thin line) and polystyrene with 50 wt.-% of coal (thick line).

From the data in Table 4 it follows that the presence of coal influenced the T'_{max} and w_{max}. It was found that the T'_{max} of PS was significantly lower than those obtained for PS with coal, on the contrary, the w_{max} of PS alone was much higher in comparison with values obtained for PS-coal mixtures. The highest value of T'_{max} at 50 wt.-% of coal in the mixture was observed. The DTG maximum increased by 58–59 °C in this case. Further, the amount of reacted mass decreased linearly with the increasing amount of coal in the mixture. As this decrease was due to a semicoke/coke formation, the increase of the w_{max} with the PS content in the mixture occurred in consequence of tar formation. Practically all PS is decomposed to tar components, but only a part of coal degrades to tar. The yield of tar obtained during copyrolysis and w_{max} then increase with the increasing amount of PS in the mixture. Composition of coal tar changes in the presence of PS, because products of PS decomposition (mainly styrene) are hydrogenated by coal hydrogen and/or, probably, they react with coal degradation structures.

Reactions of PS with coal were further evaluated by the Model Free Kinetics and TGA Decomposition Kinetics. The conversion time (for 95 % of conversion) and the parameters of Arrhenius plot were calculated. Results are showed in Table 5.

Table 5. TGA of decomposition of PS in the presence of coal at temperature of 445 °C. Degree of completion 95 %, coversion time calculated for heating rates of 5, 10, and 20 °C.min^{-1} in the temperature range of 350–500 °C. n – reaction order; E_A, ln Z – parameters of the Arrhenius plot.

Mixture (wt.-% PS/wt.-% coal)	Conversion time (min)	n	E_A (kJ.mole^{-1})	ln Z
100/0	23.9	0.8	199.2	29.7
60/40	29.5	1.1	204.4	30.0
50/50	101.9	1.5	225.0	33.5
40/60	154.9	1.5	272.3	40.9
30/70	85.2	1.9	173.3	24.5

As it is proved by the data in Table 5, the conversion time was much higher with 50–70 wt.-% of coal in the mixture. Reaction order increased from about 1 (PS alone) up to 1.9 (70 wt.-% of coal in the mixture). It means that reactions between PS decomposition products and coal degradation products occur. Two possibilities can be taken into account – hydrogenation and aromates–aromates reactions. As styrene monomer is a main product of polystyrene decomposition (72 %, beside dimer – 11 %, and trimer – 14 %), and also 1,3-diphenyl propene and other unsaturated compounds are formed,[7] hydrogenation of these products takes place, because coal is a strong hydrogen donor (content of hydrogen 5.5 %, Table 2). Moreover, α-methyl styrene is formed[7]. Therefore, styrene, ethyl benzene, toluene, isopropyl benzene, 1,3-diphenyl propane and α-methyl styrene enrich coal tar (beside rests of dimer and trimer). Further, a formation of new aromatics by reactions of aromates from PS with coal tar aromatic structures can be considered, as coal tar structures are derived from coal structure. Because composition of coal (as well as charcoal[8]) is very complicated, its chemical structure was expressed by four representative aromatic/cyclanic clusters (Fig. 3) constructed from ^{13}C CP/MAS NMR parameters.

Cyclanic parts of clusters are cleaved during copyrolysis and aromatic rests (essential constituents of tar[9,10] react with products of PS decomposition by free radical mechanism. New aromates like diphenyl, phenyl naphtalene, benzo fluorenes and benzo fluoranthenes are created,[11] therefore, amount and composition of tar are changed. Probably, that is why the conversion time was much higher in the presence of coal.

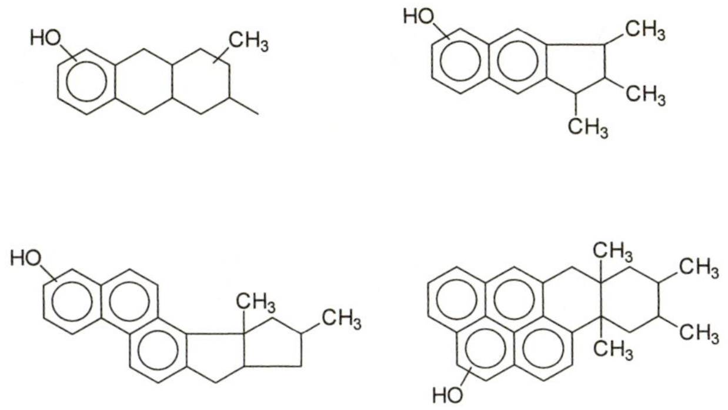

Fig. 3. Representative structures of coal used.

For comparison, decomposition of polypropylene was followed. Results are shown in Table 6.

Table 6. TGA of decomposition of polypropylene in the presence of coal at the temperature of 476 °C. Degree of conversion 95 %, time of conversion was calculated for heating rates of 5, 10, 15 and 20 $°C.min^{-1}$ in the temperature range of 400 – 520 °C. n – reaction order; E_A, ln Z – parameters of the Arrhenius plot.

Mixture (wt.-% PP / wt.-% coal)	Conversion time (min)	n	E_A ($kJ.mole^{-1}$)	ln Z
100/0	36.2	0.6	208.3	29.0
60/40	30.6	0.7	225.5	31.4
50/50	30.4	0.8	245.2	34.6
40/60	29.2	1.1	231.3	32.5
30/70	29.6	1.3	227.2	31.9

From the data in Table 6 it follows that the conversion time of decomposition of the PP was rather shorter in the presence of coal in the temperature range of 400–520 °C. As the reaction order was always about 1, it seems that reactions in the question are more simple in this case. PP decomposes above 400 °C, and propylene trimer is main decomposition product.[8] The presence of coal promotes the formation of monomer and dimer and unsaturated products with low MW (unsaturated hydrocarbons C_2–C_7).[8] These product are hydrogenated by coal. With an increasing content of coal in the mixture the reaction time is then a litle shorter. No such phenomenon was observed in the case of PS. It means that longer conversion times at PS-coal mixtures can be attributed to aromates-aromates reactions.

From above-mentioned findings it follows that coal tars differs from tars obtained from copyrolysis. Due to higher concentration of simple aromates with low substitution a lower aromaticity, substitution degree and density can be expected with copyrolysis tars. We compared these parametrs for tars obtained under high-teperature conditions on a laboratory unit. Coal tars[9] and tars from copyrolysis of coal with polymers of PS type[1,4,12] were taken into account. As Table 7 proves, these parameters were lower in the case of copyrolysis.

Table 7. Parameters of tars.

Tar	Aromaticity	Substitution degree	Density ($g.cm^{-1}$)
dry coal tar	0.69–0.76	0.26–0.61	1.02–1.03
dry copyrolysis tar	0.60–0.71	0.17–0.37	0.96–1.02

Conclusion

Kinetic parameters of thermal reactions of coal with polystyrene by DSC and TGA were determined. On this basis, reactions in question were described. In the range of 350–550 °C a thermal degradation of coal proceeds, simultaneously, decomposition of polystyrene occurs. Because coal is a strong H-donor, unsaturated products of polystyrene decomposition are hydrogenated by coal. Some aromatic products of PS decomposition react with coal tar structures and new aromates are formed. That is why the conversion time of PS decomposition is much higher in the presence of coal. The yield of tar from copyrolysis is then higher in comparison with pyrolysis of coal alone. Also tar composition is changed.

Acknowledgement

Academy of Sciences of the Czech Republic supported this work as the grant project No. S3046004.

[1] Straka P., Buchtele J. and Kovářová J.: Macromolecular Symposia 135, 19 (1998)
[2] Buchtele J. et al.: Research Report, Institute of Rock Structure and Mechanics ASCR, Prague, 1998
[3] Kříž V., Chovancová P. and Buchtele J.: Proc. 2nd Int. Conf. on Metallurgy and Environmental Technologies, 11-12 September, Herlany/Slovakia, 2001, pp.59-66
[4] Bičáková O.: PhD. Thesis, The Institute of Chemical Technology of Prague, Prague, 2002
[5] Vyazovkin S.V. and Lesnikovich A.I.: Journal of Thermal Analysis 36, 599 (1990)
[6] Solumn M.S., Pugmire R.J. and Grant D.M.: Energy and Fuels 3, 187 (1989)
[7] Jakab E., Blazsó M. and Faix O.: Journal of Analytical and Applied Pyrolysis 58-59, 49 (2001)
[8] Jakab E., Várhegyi and Faix O.: Journal of Analytical and Applied Pyrolysis 56, 273 (2000)
[9] Černý J.: Energy and Fuels 5, 781 (1991)
[10] Černý J.: Petroleum and Coal 34, 339 (1992)
[11] Straka P., Náhunková J. and Brožová Z.: Proc. 15th Int. Symp. on Analytical and Applied Pyrolysis, 17-20 September, Leoben/Austria, 2002, p.48
[12] Kozubek E., Buchtele J. and Roubíček V.: Petroleum and Coal 39, 43 (1997)

Macromol. Symp. **2003**, *202*, 307—323

Weak Flocculation of Aqueous Kaolin Suspensions Initiating by NaCMC with Different Molecular Weights

Y. Li,[1] *É.S. Nagy,*[2] *M.N. Esmail,*[1*] *Z. Hórvölgyi*[2*]

[1]Department of Mechanical and Industrial Engineering, Concordia University, de Maisonneuve Blvd. W., H549 Montreal, Quebec H3G 1M8, Canada
E-mail: esmail@encs.concordia.ca
[2]Department of Physical Chemistry, Budapest University of Technology and Economics, H-1521 Budapest, Hungary
E-mail: horvolgyi.fkt@chem.bme.hu

Summary: The present work is an investigation of the effect of NaCMC with different viscosities (molecular weights) on the stability of aqueous kaolin suspensions at pH 5-6. The stabilizing effect of polymers was characterized by measuring the sedimentation volumes (for 2.5% kaolin suspensions) and some important rheological parameters (for 40% and 50% kaolin suspensions). In certain cases the stability of suspensions was also studied in the presence of 0.5-1.0% NaCl. The additives were incorporated into the suspensions separately and simultaneously, as well. In certain cases the effect of mixing order of NaCMCs was also studied. The lower viscosity NaCMC was found to be a better stabilizing agent than its medium viscosity counterpart at the studied polymer concentrations (0.005-1.0%). This was manifested in smaller sedimentation volumes and lower rheological parameters (viscosity, yield stress, degree of thixotropy and elasticity). The lower and the medium viscosity polymer were simultaneously and consecutively added in a mass ratio of 50:1 and 10:1. The resulted observation of low viscosity and yield stress, and more importantly thixotropy and elasticity, can be interpreted in terms of a "site-blocking" type flocculation.

Keywords: anionic polyelectrolyte; kaolin suspension; rheological properties; sedimentation volume; site-blocking flocculation

1 Introduction

The aqueous clay suspensions are of great importance in many fields of industrial applications/e.g. ceramics production, drilling fluids, paper coating.[1,2]

Neutral polymers or polyelectrolytes are frequently used as dispersant or flocculant agents. The combination of different polymers can cause synergetic effects. Due to the practical and theoretical importance, the competitive adsorption of neutral[3-5] and ionic[6-8] macromolecules was intensively studied in the past decades. Macromolecules can cause bridging flocculation if the molecular weight is great enough and the adsorption layer is not saturated. The efficacy of

 DOI: 10.1002/masy.200351227

flocculation can be improved, adding simultaneously a lower and a higher molecular weight polymer[6-8] in the suspension. In most cases cationic[6-8] or oppositely charged[9] polyelectrolytes are used for initiating flocculation resulting in strong floc structure. In the first case, the lower molecular weight polymer enhances the possibility of bridging formation of its higher molecular weight counterpart, blocking the active adsorption sites. Strong floc structure is required e.g. in papermaking[10] due to the high speed paperboard processing. The usage of the mixtures of cationic polymers has been suggested for this purpose.[6-8]

In paper coating liquids, including kaolin, the NaCMC is a commonly used additive.[11-15] There are many conflicting requirements for the coating color. High solid content, e.g., but low shear viscosity and certain degree of thixotropy and elasticity are required. It means that strong flocculation of the pigments should be avoided in the coating colors. That is why the site-blocking based flocculation, concerning the coating colors, is not studied, yet. On the other hand, there is a strong requirement to reach a weak flocculated state of pigments because it can result in advantageous optical properties of the coating layer.[14]

The main purpose of this work is to show that the above mentioned rheological properties are feasible in the aqueous kaolin suspensions using the mixture of a lower and a higher molecular weight NaCMC and certain cases inorganic electrolyte, NaCl. Due to the positively charged edges of kaolinite particles, the anionic polyelectrolytes can cause a weak flocculation based on the site-blocking mechanism. The rheological properties are examined by cone-plate viscometry. In order to get a better understanding of the rheological results, we compare them to the results of sedimentation volume investigations.

2 Experimental

2.1 Materials

Low and medium viscosity (lv and mv) NaCMC (carboxymethylcellulose sodium salt, Sigma), NaCl (sodium chloride, BDH, analytical reagent), kaolin (hydrated aluminium silicate, Sigma K-7375) and distilled water were used in preparation of samples. Every reagent was used as received. The particle size of kaolin was in the range of 0.1-4.0 μm; 90% of particles were smaller than 2 μm. According to the manufacturer, the degree of substitution was 0.7 for the NaCMC. The molecular weight of polymers was determined by using capillary viscometry. Measuring the relative viscosity of aqueous solutions of NaCMCs in the presence of a strong electrolyte (0.1 M NaCl), the intrinsic viscosity, [η], was determined from which the average molecular weight of polymers was calculated. Molecular weights of 50000 Dalton (± 10%) and 151000 Dalton (± 10%) were found for the low and medium viscosity polymers, respectively.

2.2 Methods

2.2.1 Sample Preparation for the Rheological Investigations

The samples were prepared from suitable amounts of the stock solutions (of NaCMC and NaCl), kaolin powder and distilled water. The components were introduced simultaneously, and after completion, stirred using an overhead stirrer at around 800 rpm for five minutes. Samples with 50% kaolin content (and certain cases with 40% kaolin content) were prepared and investigated. In some cases we investigated the effect of mixing order of different molecular weight polymers on the rheological properties of 50% kaolin suspensions. The compositions are shown in Table 1.

Table 1. The compositions of samples prepared for the rheological investigations.
lv: low viscosity; mv: medium viscosity.

Samples	Kaolin (%)	lv NaCMC (%)	mv NaCMC (%)	NaCl (%)
Re0(40)	40	-	-	-
Re1(40)*	40	-	1.0	-
Re2(40)*	40	-	-	1.0
Re3(40)*	40	-	1.0	1.0
Re0(50)	50	-	-	
Re1(50)	50	0.1	-	
Re2(50)	50	0.5	-	-
Re3(50)	50	-	0.5	-
Re4(50)	50	0.1	0.002	-
Re5(50)	50	0.25	0.25	-
Re6(50)**	50	0.5	0.01	-
Re7(50)	50	-	-	1.0
Re8(50)	50	0.5	-	1.0
Re9(50)	50	0.1	0.002	0.5
Re10(50)***	50	0.5	0.01	-
Re11(50)****	50	0.5	0.01	-
Re12(50)***	50	0.5	0.05	-
Re13(50)****	50	0.5	0.05	-

* These samples were prepared from powders. ** The polymers were added at the same time. *** The low viscosity NaCMC was added first. **** The medium viscosity NaCMC was added first.

2.2.2 Rheologica Investigations

The rheological properties were measured at ambient temperature (23 ± 1 °C) using a cone-and-plate Haake-viscometer (RS100). The apparent and plastic viscosities, the Bingham yield stress and thixotropy were examined analysing the steady-state flow curves and hysteresis loops' "area". In the latter case, according to the Green method,[16] the up curves were always determined under steady-state conditions and the down curves were determined as quickly as possible. The elastic behaviour was investigated in certain cases by determining the storage (G') and loss (G") moduli of the samples from the results of oscillation tests.

Table 2. The compositions of samples prepared for the sedimentation volume investigations. lv: low viscosity; mv: medium viscosity.

Samples	Kaolin (%)	lv NaCMC (%)	mv NaCMC (%)	NaCl (%)
Sv0	2.5	-	-	-
Sv1	2.5	0.01	-	-
Sv2	2.5	0.02	-	-
Sv3	2.5	0.04	-	-
Sv4	2.5	-	0.005	-
Sv5	2.5	-	0.01	-
Sv6	2.5	-	0.02	-
Sv7	2.5	-	0.04	-
Sv8	2.5	-	-	0.2
Sv9	2.5	-	-	0.4
Sv10	2.5	-	-	0.6
Sv11	2.5	-	-	0.8
Sv12	2.5	-	-	1.0
Sv13	2.5	0.01	0.01	-
Sv14	2.5	0.02	0.02	-
Sv15	2.5	0.01	-	1.0
Sv16	2.5	0.02	-	1.0
Sv17	2.5	0.04	-	1.0
Sv18	2.5	-	0.01	1.0
Sv19	2.5	-	0.02	1.0
Sv20	2.5	-	0.04	1.0

2.2.3 Sample Preparation for the Sedimentation Volume Investigations

The samples were prepared from suitable amounts of the stock solutions (NaCMC and NaCl) and of kaolin suspension adding the necessary quantity of distilled water in scaled tubes, and homogenized, shaking the tubes thoroughly. For practical reasons, we prepared more diluted (2.5%) kaolin suspensions for the sedimentation volume investigations. The compositions are shown in Table 2.

2.2.4 Sedimentation Volume Measurements

The sedimentation volumes were measured at ambient temperature (23 ± 1 °C). After homogenisation and waiting at least 24 hours, the equilibrated values were determined and used for the characterization of particle-particle interaction (adhesion). As well known, the higher sedimentation volumes correspond to stronger particle-particle adhesion.

3 Results and Discussion

First, in this section, we give the results of the sedimentation volume examinations and provide some explanations. Second, we show the results of the rheological investigations for the 40% and 50% kaolin suspensions, and compare them with one another and with the results of the sedimentation volume studies. The pH of the kaolin samples was found to be between 5 and 6 (assessed by an indicator paper).

3.1 Sedimentation Volumes (V_S)

First, the effect of NaCl on the sedimentation volumes of aqueous kaolin suspensions was studied. By means of this investigation we can get indirect information about the state of surface charges of kaolinite particles. The results are shown in Figure 1. As can be seen the V_S values are decreasing with increasing concentration of the NaCl up to 0.6 M, then remaining at a constant value. It means that the particle-particle adhesion becomes weaker with adding NaCl in kaolin suspension. This can be explained in terms of screening the face-edge electric attractions conforming indirectly our assumption that the particle's edges are positively and the faces are negatively charged at the pH of present investigations.

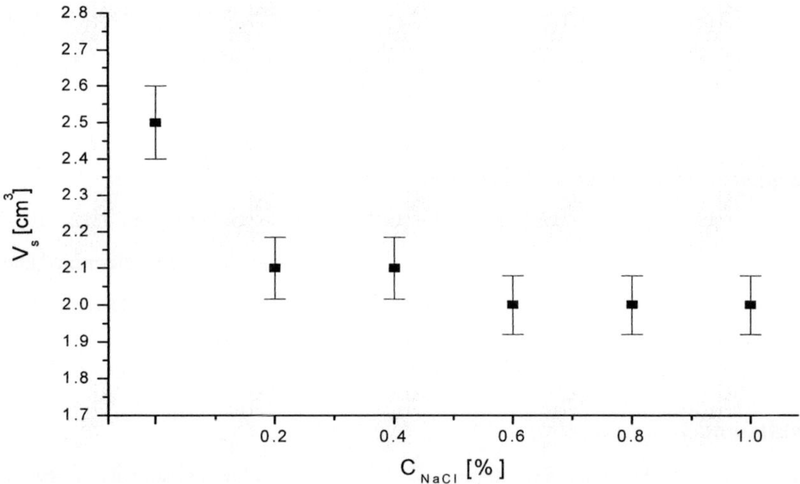

Fig. 1. Sedimentation volumes (V_S) of 2.5% kaolin suspensions as a function of NaCl concentration (C_{NaCl}).

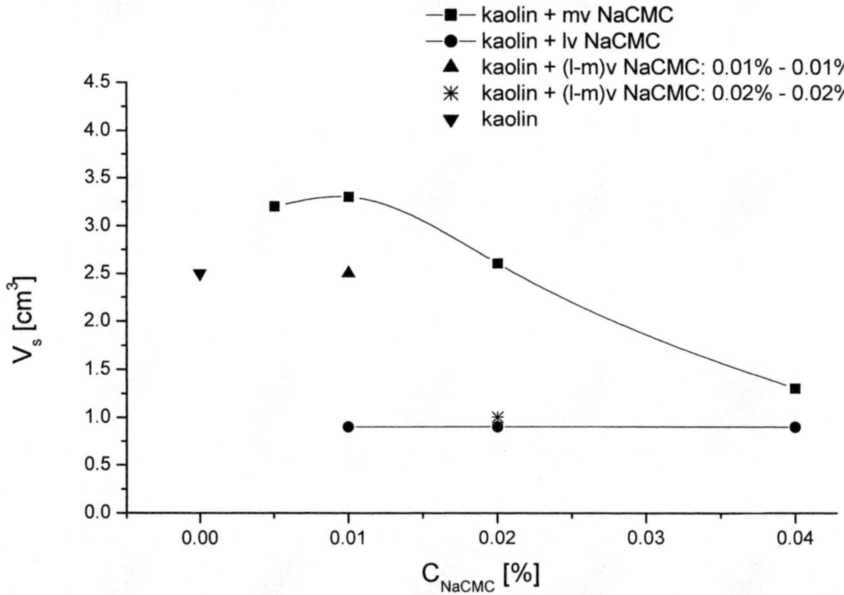

Fig. 2. Sedimentation volumes (V_S) of 2.5% kaolin suspensions as a function of NaCMC concentracion (C_{NaCMC}). lv: low viscosity; mv: medium viscosity.

The sedimentation volumes, obtained for the low and medium viscosity NaCMC, are given in Figure 2. The low viscosity NaCMC shows a stabilizing effect in the whole concentration range, that is the V_S values significantly decrease with increasing concentration of polymer. In the case of the mv NaCMC, however, a maximum of V_S appears at the lowest polymer concentrations due to bridging flocculation. It means that the mv polymer can form bridges between the positively charged edges of different particles. The simultaneous addition of the low and medium viscosity polymers in the suspension leads to intermediate sedimentation volumes, indicating that beside the adsorption of lv NaCMC the "bridge forming adsorption" of mv NaCMC also occurs.

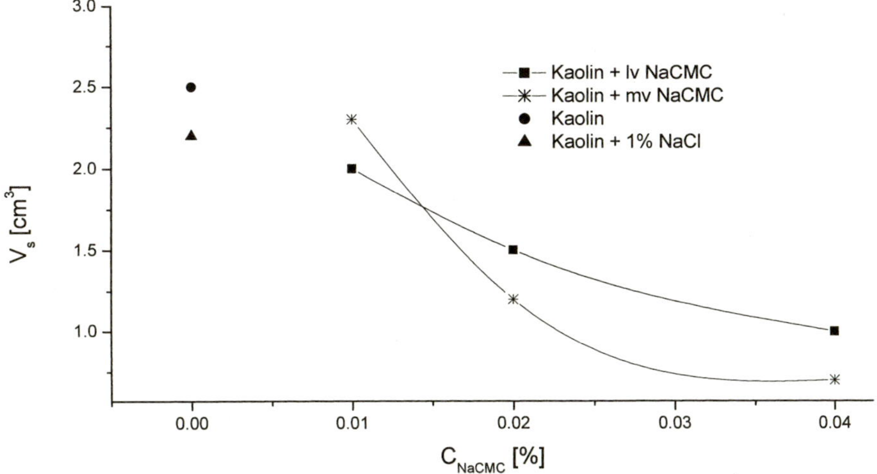

Fig. 3. Sedimentation volumes (V_S) of 2.5% kaolin suspensions as a function of NaCMC concentracion (C_{NaCMC}) in the presence of 1% NaCl. lv: low viscosity; mv: medium viscosity.

The simultaneous addition of NaCl and NaCMC in the kaolin suspensions resulted in interesting observations. As can be seen in Figure 3, the NaCl diminishes the stabilizing capability of lv NaCMC, especially at the lower concentrations of the lv polymer (also see Figure 2). This effect can be attributed to the screening the electrostatic forces by the sodium chloride. On the one hand, the NaCl can decrease the adsorption capability of the negatively charged polymers on the positively charged edges. On the other hand, the sodium-ions also screen the electrostatic repulsion between the polymer-covered (negatively charged) particles. An opposite effect is observed for the mv NaCMC. The NaCl improves the stabilizing capacity

of the polymer, supposedly due to the hindered “bridge forming adsorption”. In the presence of NaCl the mv polymer molecules have a more coiled conformation due to the electrostatic screening between the polymer chains that is not favourable for the bridging formation.[11]
It should be noted that a serious time dependence of the “equilibrated” sedimentation volumes was observed in the presence of polymers. After a repeated homogenisation of the suspensions a significant decrease in the “equilibrated” sedimentation volumes was observed, especially for the higher viscosity polymer, indicating the lack of real equilibrium.

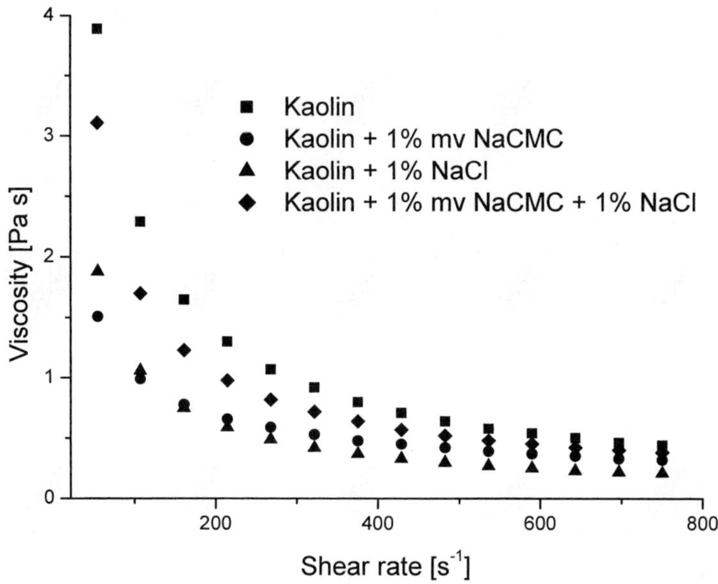

Fig. 4. Apparent viscosities as a function of shear rates obtained for the 40% kaolin composites.

3.2 Rheological Investigations

Every investigated sample showed non-Newtonian, mostly plastic behaviour. The viscosity curves for the 40% kaolin samples can be seen in Figure 4 /samples Re0(40)-Re3(40)/. The determined rheological parameters (the range of apparent viscosities, plastic viscosity, Bingham yield stress, “area” of the hysteresis loop) for both the 40% and 50% kaolin suspensions are given in Table 3 /samples Re0(40)-Re3(40)/ and in Table 4 /samples Re0(50)-Re9(50)/, respectively. Some characteristic results of thixotropy measurements for the 40% and 50% kaolin suspensions appear in Figures 5 and 6. The results concerning te effect of mixing order of different molecular weight polymers can be seen in Table 5 and in Figure 7 /samples Re10(50)-Re13(50)/. The apparent viscosities at some values of shear rate, the Bingham yield values and loop areas concerning the thixotropic behaviour are shown in Table 5. The storage and loss moduli obtained from the oscillation test are depicted in Figure 7.

Table 3. The calculated rheological properties for the 40% kaolin samples.

Samples	Apparent viscosity range (Pa s)	Plastic viscosity (Pa s)	Bingham yield stress (Pa)	Thixotropy ("area" of the loop) (Pa/s)
Re0(40)	3.89-0.44	0.12	240	1810
Re1(40)	1.51-0.32	0.12	150	3.1×10^4
Re2(40)	1.88-0.21	0.05	125	4180
Re3(40)	3.11-0.38	0.06	212	4.7×10^4

The apparent viscosities are appeared at 54 s^{-1} and 750 s^{-1}, respectively. The standard deviation of the averaged values was found to be ± 5% in most cases.

Table 4. The calculated rheological properties for the 50% kaolin samples.

Samples	Apparent viscosity range (Pa s)	Plastic viscosity (Pa s)	Bingham yield stress (Pa)	Thixotropy ("area" of the loop) (Pa/s)
Re0(50)	8.20-0.55	0.10	330	4.20×10^4
Re1(50)	7.26-0.47	0.18	225	-9960
Re2(50)	0.09-0.03	0.03	2.4	750
Re3(50)	3.00-0.30	0.16	116	1.40×10^4
Re4(50)	4.90-0.32	0.13	151	-4630
Re5(50)	0.40-0.08	0.06	16	3040
Re6(50)	0.30-0.08	0.06	15	2590
Re7(50)	10.60-0.60	0.096	390	3.2×10^4
Re8(50)	4.82-0.24	0.07	122	3460
Re9(50)	11.30-0.65	0.11	419	7.44×10^4

The apparent viscosities are appeared at 26 s^{-1} and 750 s^{-1}, respectively. The standard deviation of averaged values of viscosities and yield stresses were found to be ± 10% in most cases. The standard deviation of averaged loop "areas" were ± 15%.

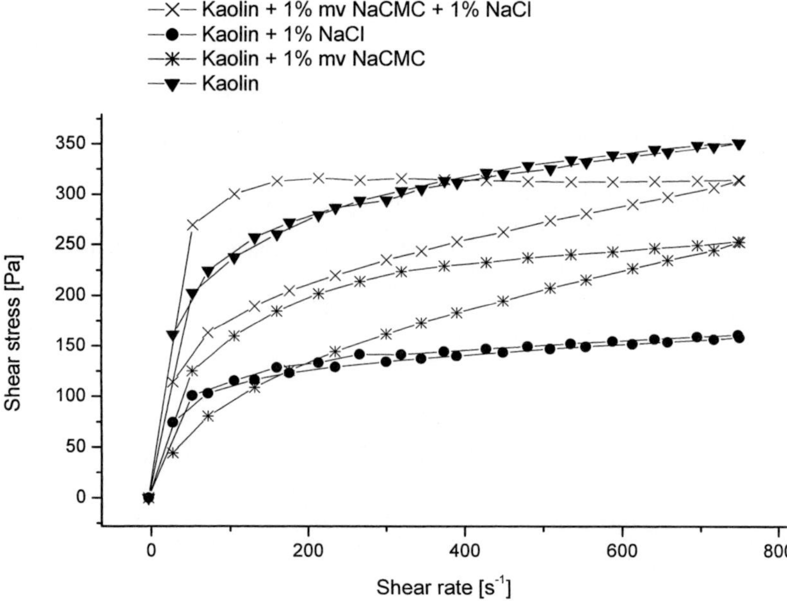

Fig. 5. Thixotropy curves for the 40% kaolin composites. The "down curves" always run under the "up curves".

The section is organized as follows. First, we discuss the results for the kaolin suspensions without additives. Secondly, we deal with the outcome for the kaolin suspensions with individually added NaCl and NaCMCs. Then we interpret the results obtained for the kaolin suspensions with simultaneously and consecutively added lv and mv NaCMCs. And finally, the effects obtained for the kaolin suspensions with simultaneously added (l-m)v NaCMCs and NaCl are interpreted and discussed.

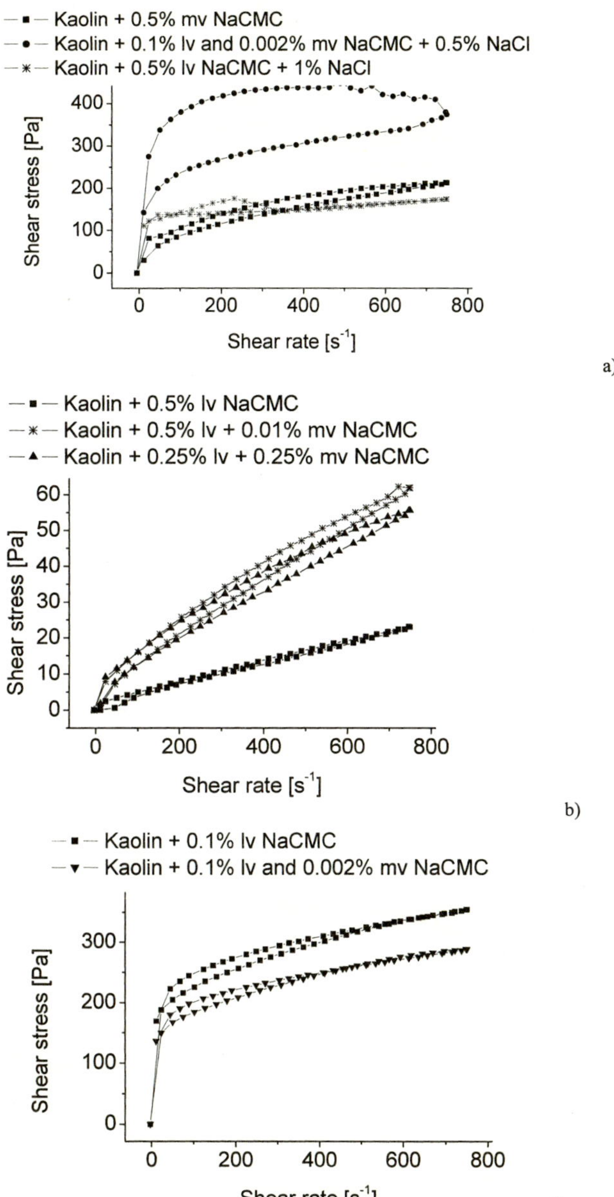

Fig. 6. Some characteristic thixotropy (a. b.) and antithixotropy (c.) curves for the 50% kaolin composites. The "down curves" run above the "up curves" for the case of antithixotropy.

Table 5. The rheological properties for the 50% kaolin suspensions prepared by sequential addition of polymers. For comparison the properties of 50% kaolin suspension is also shown.

	Kaolin	Re10(50)	Re11(50)	Re12(50)	Re13(50)
Shear rate [1/s]	Apparent viscosity [Pas]				
0.27	9.59	0.18	0.22	1.93	0.27
103.80	6.63	0.14	0.17	1.15	0.20
152.00	5.04	0.12	0.15	0.86	0.17
200.20	4.09	0.11	0.13	0.71	0.16
	Bingham yield value [Pa]				
	850	18	25	125	25
	Loop area [Pa/s]				
	-	2614	3732	12550	5301

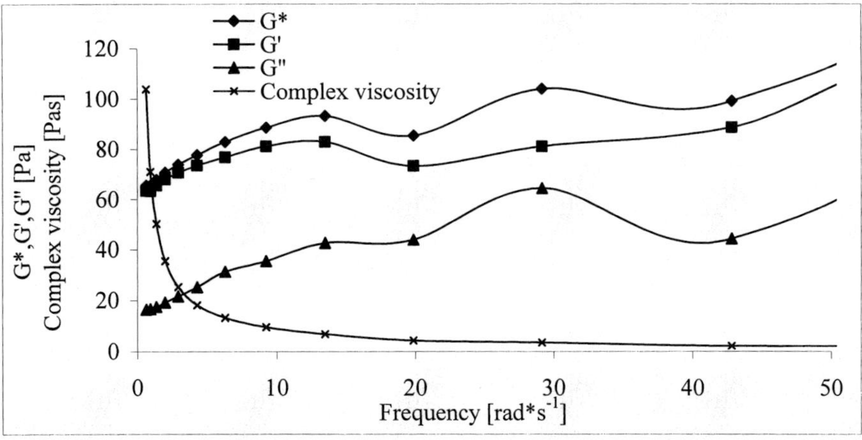

Fig. 7. Frequency sweep oscillation test of the 50% kaolin suspension containing the medium and low viscosity NaCMCs in a 1:10 mass ratio /sample Re12(50)/. The low viscosity NaCMC was added first. (G*:complex modulus of torsional shear, G': storage modulus, G": loss modulus).

3.2.1 Kaolin Suspensions without Additives/Samples: Re0(40) and Re0(50)/

The kaolin suspensions show relatively high viscosity and yield stress values, as expected (see Tables 3 and 4). A modest thixotropy was observed for the sample with the lower kaolin concentration (Table 3 and Fig 5), indicating the presence of a weak aggregated structure due to the face-edge attraction of particles (card-house structure). The "area" of hysteresis loop was significantly higher for the case of the more concentrated kaolin suspension (Table 4).

3.2.2 Kaolin Suspensions with Individually added NaCl and NaCMCs /samples: Re1(40), Re1(50), Re2(50), Re3(50), Re2(40) and Re7(50)/

The NaCl decreases the viscosity and the yield stress of the 40% kaolin suspension (Table 3/Re2(40)/ and Figure 4) showing an increase in the stability of suspension. This is in good agreement with the results of the sedimentation volume investigation (Figure 1). Significant effect, however, is not observed for the case of the denser suspension (Table 4/Re7(50)/). In this case the majority amount of water might be bounded in the liosphere around the particles, and the incorporated inorganic electrolyte can act as a "salting-out" agent causing destabilization of the suspension. Therefore, the observed behaviour can be attributed to two opposite effects: screening of the face-edge attraction and salting-out. The more significant thixotropy obtained for the 40% kaolin suspensions in the presence of NaCl, are presumably also due to the salting-out effect (Table 3). The salting-out effect can be related to the diminished hydration repulsion[17] between the particles.

The separately added polymers, independently of their molecular weights, always acted as stabilizing agents causing lower viscosity and yield stress parameters in comparison with the results of the kaolin suspensions without additives (Figure 4, Tables 3 and 4). There was an important difference, however, between the effect of lv and mv polymers (Table 4). The low viscosity NaCMC was found to be a better thinning (and stabilizing) agent than the mv NaCMC. This observation correlates well to the results of the sedimentation volume investigation and indicates "bridge-forming adsorption" of the higher viscosity polymer. In line with the above finding, it was established that the mv NaCMC in the 40% kaolin dispersion significantly increased the thixotropy (see the loop "area" column in Table 3). This can be interpreted in terms of a flocculation initiating by the mv NaCMC. It means that some amount of mv polymer takes part in bridge formation. A slight "viscosity shock" was also found earlier after adding NaCMC in kaolin based coating colors.[14] But the reason for this effect (and for the presumable bridge formation) is not understood well. Maybe, the solid-liquid adsorption layer is unsaturated in the initial stage of polymer incorporation into the kaolin suspension and as a consequence of the irreversible nature of polymer adsorption[18] the polymer bridges cannot be eliminated easily during the mixing. *Vide supra* the time dependent "equilibrated" sedimentation volumes. When the samples were prepared from NaCMC solutions (50% kaolin suspensions), time dependence of viscosities was not found. However, when NaCMC powder was directly added in kaolin (40% kaolin suspensions), a continuous decrease in viscosity was found. It means that the polymer bridges were gradually breaking during the storage. (In this case, the measurements were carried out after a two-week storage.).

An increase in thixotropy was not observed in the case of the 50% kaolin suspension in the presence of the mv polymer (Table 4 /sample Re3(50/). The significantly greater loop "area" value (1.4 x 10^4 Pa/s), however, in relation to that of the suspension prepared from the lv polymer, also indicates a weak flocculated state. The creep recovery curve (not shown here) also confirms the above statement.[19] It shows real elasticity for the 50% kaolin suspension in the presence of 0.5% mv NaCMC but it reveals liquid-like behaviour even at lower stress in the presence of 0.5% lv polymer.[19] Interestingly, the lv NaCMC caused anti-thixotropy (Table 4 and Figure 6c) at a concentration of 0.1%.

3.2.3 Kaolin Suspensions with Simultaneously Added lv and mv NaCMCs/Samples: Re4(50), Re5(50) and Re6(50)/

The mixture of the different viscosity polymers also increased the stability of kaolin suspensions leading to lower viscosity and yield stress values (Table 4). A small amount (0.01%) of mv NaCMC in the mixture (Re6(50)), however, increased the value of every rheological parameter in comparison with suspension of 0.5% lv NaCMC, presumably due to the bridge flocculation. Although, the apparent viscosities for this sample *ca.* were three times greater than the viscosities of the sample with 0.5% lv NaCMC (sample Re2(50)), these values are low enough, especially at higher shear stress, and do not affect significantly the runnability. On the other hand, the improved thixotropy can be advantageous after the coating. This effect can be attributed to the site-blocking bridging mechanism. Interestingly, the rheological behaviour of the suspension does not change significantly if we alter the mass ratio of the lower and medium viscosity polymers drastically (from 50:1 to 1:1; sample Re5(50)).

3.2.4 The Effect of Mixing Order of Different Molecular Weight Polymers (Samples: Re10(50)-Re13(50))

We studied the effect of mixing order of different molecular weight polymers on the rheological behaviour of 50% kaolin dispersions in order to get more information about the site-blocking flocculation in the present systems. The most important results concerning the rotational viscometry are collected in Table 5. Complete oscillation test was accomplished for sample Re12(50). The linear viscoelasticity range could only be found for this case (0-50 Pa), hence, the elastic and visous properties could be investigated at 25 Pa (Figure 7). As can be seen, the values of storage modulus (G') significantly exceed the loss modulus values in the investigated frequency range (0-50 rad.s^{-1}). The visual observations also revealed distinct differences among the samples. Sample Re12(50) showed a gelatinous (gel-like), colloidally homogeneous structure for more than one month but the other samples separated into two phases within 1-2 days after the preparation. Summarizing the results in Table 5 and Figure 7 from practical viewpoint: the increased amount of mv NaCMC in the sample, provided the lv NaCMC was added first into the suspension, resulted in advantageous rheological properties, i.e. relatively low viscosity at higher shear rate, significant thixotropy and elasticity, as well. The improved rheological properties can be attributed to the site-blocking aggregation mechanism that takes place more effectively if we add the lv polymer first into the suspension and increase the mass ratio of mv NaCMC (sample Re12(50)).

3.2.5 Kaolin Suspensions with Simultaneously Added (l-m)v NaCMCs and NaCl/Samples: Re3(40), Re8(50), Re9(50)/

When NaCl is added in the samples in the presence of polymer molecules, a drastic increase is observed in the rheological parameter values (Tables 3 and 4) even for the 40% kaolin suspension (sample Re3(40)) in the presence of 1% mv NaCMC. The observations are in good agreement with the results of the sedimentation volume investigations except for the latter case. It seems that the NaCl cannot suppress the "bridge forming adsorption" of the mv polymer in the dense suspensions presumably due to the increased number of particle-particle and particle-macromolecule collisions. The incorporation of NaCl into the dense kaolin suspensions in the presence of polymers always increases the particle-particle adhesion resulting in higher viscosity and yield values (Tables 3-4) and in an increased degree of thixotropy (Tables 3-4 and Figures 5-6). Interestingly, the anti-thixotropy found for the kaolin suspension in the presence of 0.1% lv and 0.002% mv NaCMCs (sample Re4(50), Fig. 6c), was entirely eliminated by adding NaCl in the composite (Figure 6a).

It seems, however, that the NaCl at these relatively high concentrations (0.5% and 1%) cannot control gently the weak, site-blocking flocculation in the present systems. It always increases the viscosity drastically because of the strong electrostatic screening of particle-particle repulsion. However, additional study concerning the effect of inorganic electrolytes on this phenomenon is required.

4 Conclusion

Weak flocculation of the aqueous kaolin suspension (pH 5-6) was initiated by addition of NaCMCs with different molecular weights. The rheological and sedimentation volume investigations revealed that the lower molecular weight polymer alone increases the stability of the kaolin suspensions and decreases the viscosity but neither thixotropy nor elasticity was observed. However, the stabilizing ability of the higher molecular weight polymer was found to be weaker and consequently the suspension viscosity was higher, though thixotropy and elasticity was observed due to the bridging flocculation. However, when the lower and higher molecular weight polymers were simultaneously and consecutively added in the kaolin suspension (in a weight ratio of 50:1 and 10:1), the viscosity was lowered significantly, especially at higher shear rates, and the sample showed thixotropy and in certain cases elasticity, as well. The reason for this behaviour can be the bridge forming adsorption of the higher molecular weight NaCMC initiating by the lower molecular weight NaCMC by a site-blocking mechanism. Moreover, the adsorbed lower molecular weight polymers assure a significant repulsion between the particles resulting in a weak and very specific flocculated state of such composite, i.e. there is a not too deep secondary energy minimum in the total pair-potential energy curve.

The effect of an inorganic electrolyte, NaCl, was also studied on the stability of the aqueous kaolin suspensions (pH 5-6). We came to the conclusion that the bridge-forming adsorption of the higher molecular weight polymer was hindered in the presence of NaCl in the case of dilute suspensions (2.5% kaolin content). This effect was not observed, however, for denser kaolin suspensions (40 and 50% kaolin content) presumably due to the more frequent particle-particle collisions. The NaCl was not found to be a suitable agent for the gentle control of aggregation at the applied concentrations (0.5% and 1%), since it caused strong destabilization of the denser kaolin suspensions.

Acknowledgment

Support for this work in the form of grants from NSERC (to M.N.E.) and from OTKA (T030457 to Z.H.) is gratefully acknowledged.

[1] K. M. Beazly, in: "*Rheometry: Industrial Applications (Materials Science Research Studies Series, C.R. Tottle, Ed.)*", K. Walter, Ed., Research Studies Press (Division of John Wiley & Sons LTD), Chichester,**1980**, 339.
[2] J. C. Husband, in: "*Proc. Tappi Coating Conf., 1996, Nashville, TN, United States*", Tappi Press, Atlanta, **1996**, 99.
[3] F. Csempesz, S. Rohrsetzer, *Colloids Surfaces*, **1984**, 11, 173.
[4] F. Csempesz, S. Rohrsetzer, P. Kovács, *Colloids Surfaces*, **1987**, 24, 101.
[5] F. Csempesz, K. F-Csáki, Langmuir, **2000**, 16, 5917.
[6] A. Swerin, G. Glad-Nordmark, L. Ödberg, *J. Pulp Paper Sci.*, **1997**, 23(8), J389.
[7] A. Swerin, L. Ödberg, L. Wågberg, *Colloids Surfaces A: Physicochem. Eng. Asp,.* **1996**, 113, 25.
[8] H. Tanaka, A. Swerin, L. Ödberg, S. B. Park, *J. Pulp Paper Sci,.* **1997**, 823, J359.
[9] R. Aksberg, L. Ödberg, *Nordic Pulp Paper Res. J,.* **1990**, 4, 168.
[10] S. Main, P. Simonson, *Tappi J.*, **1999**, 82(4), 78.
[11] L. Jarnström, G. Ström, P. Stenius, *Tappi J,.* **1987**, 70(9), 101.
[12] H. El-Saied, A. H. Basta, S. Y. El-Sayed, F. Morsy, *Pigment and Resin Technology*, **1996**, 25(4), 15.
[13] M. Mäkinen, D. Eklund, in: "*Proc. Tappi Coating Conf. 1996, Nashville, TN, United States*", Tappi Press, Atlanta, **1996**, 61.
[14] T. Persson, L. Jarnström, M. Rigdahl, *Tappi J.*, **1997**, 80(2), 117.
[15] X. Q. Wang, J. Grön, D. Eklund, in: "*Proc. Tappi Coating Conf. 1996, Nashville, TN, United States*", Tappi Press, Atlanta, **1996**, 79.
[16] H. Green, "*Industrial Rheology and Rheological Structures*", John Wiley and Sons Inc, New York, 1949.
[17] B. V. Derjaguin, N. V. Churaev, *Colloids Surfaces*, **1989**, 41, 223.
[18] E. Dickinson, L. Eriksson, *Adv. Colloid Interface Sci.*, **1991**, 34, 1.
[19] Y. Li, MsC Thesis, Department of Mechanical and Industrial Engineering, Concordia University, Montreal, Quebec, Canada 2001.

Application of Modified Wastes from Phenol-Formaldehyde Resin and Expanded Polystyrene in Sewage Treatment Processes

Wioletta M. Bajdur[1], *Wiesław W. Sułkowski**[2]

[1]Institute of Environmental Engineering, Technical University of Częstochowa, ul. Armii Krajowej 36B, 42-200 Częstochowa, Poland
[2]Department of Environmental Chemistry and Technology, Institute of Chemistry, University of Silesia, Szkolna 9, 40-006 Katowice, Poland
Email: wsulkows@uranos.cto.us.edu.pl

Summary: A solution to environmental pollution by polymer plastic wastes can be their chemical modification in to useful products. Such a new solution is the obtaining of effective flocculants for sewage treatment from chemically modified phenol-formaldehyde resin production wastes (SE and NS novolak) and expanded polystyrene wastes. Comparative analysis of flocculation properties was performed for amino derivatives of novolak wastes, synthesized sulphonated derivatives of novolak and expanded polystyrene wastes, of standard polyacryloamide and for Praestol commercial polyelectrolyte. Amino derivatives and sulphonated derivatives of polymer plastic wastes, having properties of anionic type polyelectrolytes, exhibit good flocculation properties in purification processes of sewages with a chemical composition close to that of found in the water circulating system power plant, coal-mine, and steel plant. Application of synthesised flocculants caused a decrease of turbidity, concentration of solved impurities and improved quality parameter of purified water. It was found that synthesised polyelectrolytes could be used in industrial water treatment processes.

Keywords: functionalization of polymers; flocculation; phenol-formaldehyde resin wastes; polyelectrolytes; polystyrene wastes; sewage treatment; water soluble-polymers

Introduction

It is a well known fact, that storage of polymer plastic wastes is a serious global hazard to the environment. Perfect properties of plastics obtained from polymers as substrates are responsible for the fact that we are in permanent contact with polymers. Dumps and dumping grounds are full of polymer plastics, and their indestructibility previously regarded as an advantage has become their essential fault. Quantitative composition of polymer plastic wastes on dumping grounds approximately corresponds to the amount of produced polymer plastics considering their time of use. Dumping grounds consist mainly of wastes from

 DOI: 10.1002/masy.200351228

packaging. In particular they are products made from expanded polystyrene of low density, and therefore of high capacity. Recently, the increase of the amount of polymer plastic waste processed into energy in combustion processes or processed into other useful wares is observed.[1-4] This is caused by intensive progress of studies on polymer plastic waste management and recycling of used polymer plastic ware, and particulary by law regulations in respective countries. Burdensome to environment are also faulty manufactured units, although the plants try to use them. Another problem is connected with useful products as, e.g., phenol-formaldehyde resins, which are produced to bind waste phenol and diminish its hazard to environment. This product, potentially useful, is stored on dumping grounds until it gets utilized. It may be said, that elaboration of polymer and polymer plastic wastes management technology is of great economic significance. Chemical modification of polymer wastes to polyelectrolytes can be a way of polymer and polymer plastic waste management. Nowadays the increase of environment pollution and enormous growth of power consumption in high-developed countries is observed. This led to the introduction of low-waste or no-waste technologies. Besides the elaboration of new technologies, such as used ware recycling in the mid eighties, it should be mentioned that from the beginning of 1970 intensive studies on waste usage as source of chemical and energetic raw material have been conducted in highly-developed countries. Utilization of polymer wastes instead of new synthesized polymers for ware production will limit material and energy consumption and pollutant emission to the environment. It is known, that polymer sorbents replace active carbons and unlike them can be regenerated with the use of an electrolyte solution or organic solvent.[5] The best known polymer sorbent on market is Amberlite produced by Rohm and Haas Co. However, this is a hydrophobic sorbent made from non-waste polymers and copolymers. Nowadays, the studies are conducted of the synthesis of sorbents (used mainly in sewage and water treatment processes) with the use of phenol-formaldehyde resin production wastes[6-8] and polystyrene wastes.[1-4] Modification of phenol-formaldehyde resins is essential due to the various kinds of pollution they form during the process of production waste storage. Influence of humidity, solar energy, acids, bases and other external agents, will cause inevitable transfer of phenol and formaldehyde to the environment. The formation of these wastes is independent of the production technology. Phenol-formaldehyde resin wastes are sometimes large manufactured units of novolak synthesis. They are the result of deflection from a process production parameter and as a result the obtained product properties are sub-standard. Management of these products is a serious problem for a production plant and hazard to the environment.. An advantageous solution is the modification of phenol-formaldehyde resins by chemical methods to products which can be used as flocculants. This solution saves raw material and energy and reduces environmental pollution. Highly diluted polyelectrolytes (mostly 0,1% to 0,01% solutions) used to aid flocculation in the sewage treatment process are transferred to the sludge and then are deposited in sedimentation lagoons or are used with sludge in the agriculture and power industry. To obtain effective polyelectrolytes, the synthesis of amino derivatives of phenol-formaldehyde resins, sulphonated derivatives of SE and NS novolak wastes, and expanded polystyrene wastes was performed. The synthesized compounds were utilized in the industrial sewage treatment processes. Comparative analysis of flocculation properties of synthesized polyelectrolytes from novolak and expanded polystyrene wastes, of synthesized standard polyacryloamide and of commercial Praestol-polyelectrolyte was performed.

Materials and Methodology

The novolak waste from "ERG" Plastic Factory in Pustków, nitric acid, sulphuric acid, hydrochloric acid, calcium carbonate, sodium carbonate, ammonia, stannic chloride and silver sulphate from POCh, Gliwice were used for chemical modification. Aluminium sulphate (POCh, Gliwice), as a coagulant (1% solution), commercial Praestol 2515 polyelectrolyte (anionic commercial polyelectrolyte syntesized from polyacryloamide[9]), Jaroszów clay for preparation of high turbidity water according to PN-71/C-04583 standard, coal mine water and blast-furnace circulation water from metallurgical plant were used in these studies.

Nitro derivatives of novolak were obtained by nitration of linear structure novolak with nitrating acid HNO_3/H_2SO_4. The amino derivatives were obtained by reduction of novolak nitro derivatives with mixture of $SnCl_2/H_2O$ and concentrated HCl. The obtained products were separated and purified.[6,7]

Sulphonated derivatives of novolaks were obtained by sulphonation of novolak wastes with concentrated sulphuric acid, according to the well-known sulphonation method of aromatic compound. The excess of sulphuric acid was next removed in a process of liming by means of calcium carbonate. Water-soluble product was obtained. Products were precipitated in form of sodium salts in reaction with Na_2CO_3.[8]

Sulphonated derivatives of expanded polystyrene waste were obtained by sulphonation of polystyrene waste with concentrated sulphuric acid, according to the well-known sulphonation method of aromatic compound.[3,4] The excess of sulphuric acid was removed in a process of liming with $CaCO_3$. Products were precipitated as sodium salts in reaction with Na_2CO_3, or a product of sulphonation reaction was passed through the column filled with Dowex-1 anion exchanger.

Intrinsic viscosity was determined by means of Ubbelohde viscometer with Pollena K capillary no. I ((K=0,02510 and A_p.103) at 298± 0,01 K. Measurements were conducted in a Julabo Labortechnik GMBH thermostat. Bulk density was measured according to PN-64/C-98054 standard. Identification of polymer characteristic groups and nitro- and amino- groups was done on a Spectrum One IR spectrometer. Samples were prepared in form of tablets using 1 mg of investigated compound and 100 mg of KBr. Spectra were recorded in 4000-400 cm^{-1} range. Studies of the flocculation process were conducted according to PN-71/C-04583 standard. Study of the flocculation process started after selection of an optimum dose of coagulant and then of flocculent. Before this, at the beginning of each measurement the turbidity of the investigated water was determined. Then an optimal dose of coagulant was chosen. For a known dose of coagulant an optimum dose of flocculent was then found by measuring the turbidity of the tested water. The turbidity was measured by means of a Turb 550 IR according to ISO 797/DIN 27027 standard and US EPA recommendation.

Results and Discussion

Chemical modification of the environmentally hazardous novolak and expanded polystyrene wastes gave flocculation materials useful for purification of sewages from power plant, coal mines and steel plant water circulating systems. Properties of synthesized polyelectrolytes are much better than those of polyacryloamide commercial flocculants. Comparative analysis of flocculation properties was conducted for synthesized polyelectrolytes containing maximum functional groups in a polymer. It was found, that the amount of functional groups per one constitutional unit has a great effect on the flocculation and coagulation processes.[3,4,6-9] In the process of chemical modification the amino and sulphonated derivatives of SE and NS novolak and of expanded polystyrene wastes were obtained. Nitration of SE novolak wastes (with increased by 2% content of phenol in relation to commercial product) or SE (with increased by 2% content of phenol) gave nitro derivatives containing 5.21% and 5.30% of nitrogen, respectively. IR confirmed presence of nitrogen groups. Spectra of nitro derivatives exhibit a characteristic asymmetrical stretch vibration bands of –NO at 1541 cm^{-1} and 1537 cm^{-1} and symmetrical stretch vibration band of –NO at 1344 cm^{-1} and 1340 cm^{-1}. Nitro derivatives of SE and NS novolaks were reduced with a reducing mixture (tin salt and hydrochloric acid). The obtained novolak amino derivatives contain 3.02% of nitrogen (SE novolak amino derivative) and 2.95% of nitrogen (NS novolak amino derivative), which corresponds to the presence one NH_2 amino group in two constitutional units. Reduction of SE and NS novolaks nitro derivatives is carried out an identical manner, which is confirmed by the presence of amino groups.

Amino derivatives of NS and SE novolaks are well soluble in 3% water solutions of KOH and NaOH.[6,7]

Sulphonated derivatives of SE and NS novolak wastes were obtained by sulphonation with sulphuric acid. Obtained products contain 10.62% of sulphur (SE novolak) and 9.15% of sulphur (NS novolak). It coresponds to one sulfo group per 3 constitutional units. IR spectra of sulphonated derivatives of SE and NS novolaks exhibit a characteristic asymmetrical stretch vibration bands of S=O sulfo group at 1300 cm^{-1}. It was found, that sodium salts of sulphonated novolaks derivatives are very well soluble in water and can be used as flocculents.[8]

Sulphonation of expanded polystyrene wastes in excess of concentrated sulphuric acid, gave products with a maximum sulphur concentration equal to 13.80% . This corresponds to one sulfo group per one constitutional unit.

IR analysis confirmed the presence of sulpho- groups. IR spectra of these products exhibit a characteristic asymmetrical and symmetrical stretch vibration bands of S=O sulpho group at 1370-1070 cm^{-1} range. It was found, that sodium salts of sulphonated expanded polystyrene wastes are very well soluble in water and can be used as flocculants.[3,4]

Polyacryloamide (M_η= $3{,}5 \bullet 10^5$) obtained in radical homogenise polymerisation or precipitation polymerization of acrylamide was used as a standard polyelectrolyte.[9] Commercial Praestol polyelectrolyte was also used as comparative material.

The obtained modified phenol-formaldehyde resin production wastes and expanded polystyrene wastes, and synthesized standard polyacryloamide were used for study of the flocculation process. Coal mine sewages, water from the steel plant circulating system, and high turbidity model water prepared from tap water and Jaroszów clay were used in studies of the flocculation process. Chemical composition of Jaroszów clay close to the chemical composition of sewages from the power plant was the following: SiO_2 (54-68%); Al_2O_3 (29-40%); Fe_2O_3 (1,5-2,9%); K_2O (1,2-2,4%). Water of high turbidity was prepared according to the Polish Standard – 71/C-04538.

Such choice of water for analysis resulted from the fact that both amino derivatives of phenol-formaldehyde resins and sulphonated derivatives of these resins and polystyrene wastes are polyelectrolytes of the anionic type. Such kind of polyelectrolytes are best for aiding the coagulation process of coal mine water and industrial sewages. Commercial Praestol polyelectrolyte and synthesized polyacrylamide were used as comparative material. Studies were conducted according to standard PN-79/C-04619/03. At first the optimal dose of the coagulant was selected and then the flocculent one.

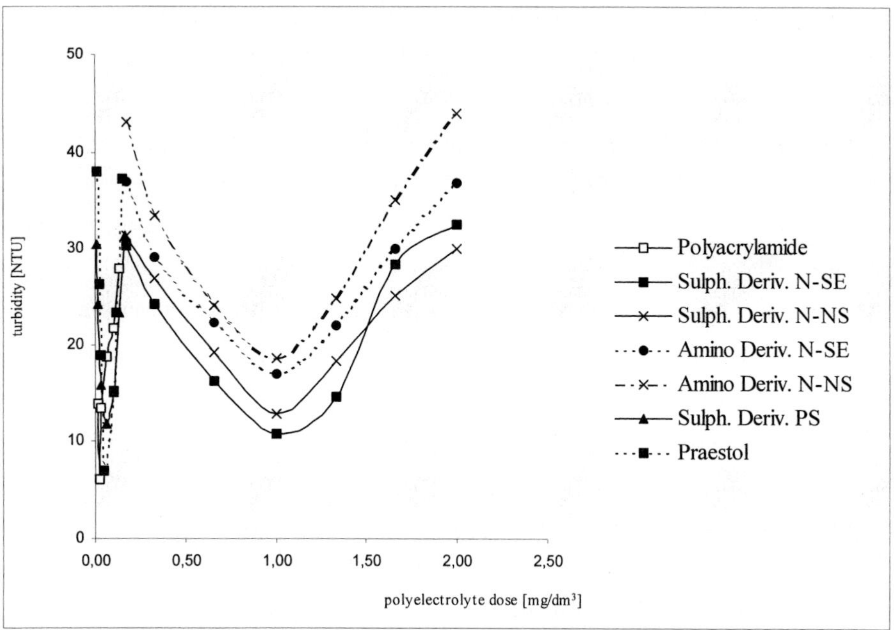

Fig. 1. Dependence of turbidity on optimal doses of synthesized standard polyacrylamide, sulphonated derivatives of SE novolak (sulphonated derivatives N-SE) and NS novolak (sulphonated derivatives N-NS) - phenol-formaldehyde resins wastes, amino derivatives of SE novolak (amino derivatives N-SE) and NS novolak (amino derivatives N-NS) - phenol-formaldehyde resins wastes, sulphonated derivatives of expanded polystyrene wastes (sulphonated derivatives PS), and commercial Praestol polyelectrolyte during purification of model water with initial turbidity of 160 NTU, at coagulant dose of 50.0 mg/dm^3 and pH=6,95.

It was found from flocculation studies with the use of sulphonated derivatives of SE and NS novolak wastes, and amino derivatives of SE and NS novolak wastes that the aiding effects of the coagulation process are comparative with those of commercial Praestol polyelectrolyte and synthesized polyacrylamide. These effects were obtained for Praestol polyelectrolyte and synthesized polyacrylamide having similar values of intrinsic viscosity but with ten times greater concentration of polyelectrolytes, due to the lower molecular weight of the obtained polymers (Fig. 1,2,3). It can be said, that amino and sulphonated derivatives of SE and NS novolak wastes appeared to be the most efficient polyelectrolytes for model water, chemically corresponding to sewages from the power plant (turbidity decrease was 72.2% for sulphonated derivative of SE novolak, 67.1% for sulphonated derivative of NS novolak, 56.3% for amino derivative of SE novolak, 52.2% for amino derivative of NS novolak) (Table 1). Similarly amino and sulphonated derivatives of novolak production wastes are efficient in a broad range of concentrations in comparison to standard polyacryloamide and commercial Praestol polyelectrolyte, and also to sulphonated derivatives of expanded polystyrene wastes.

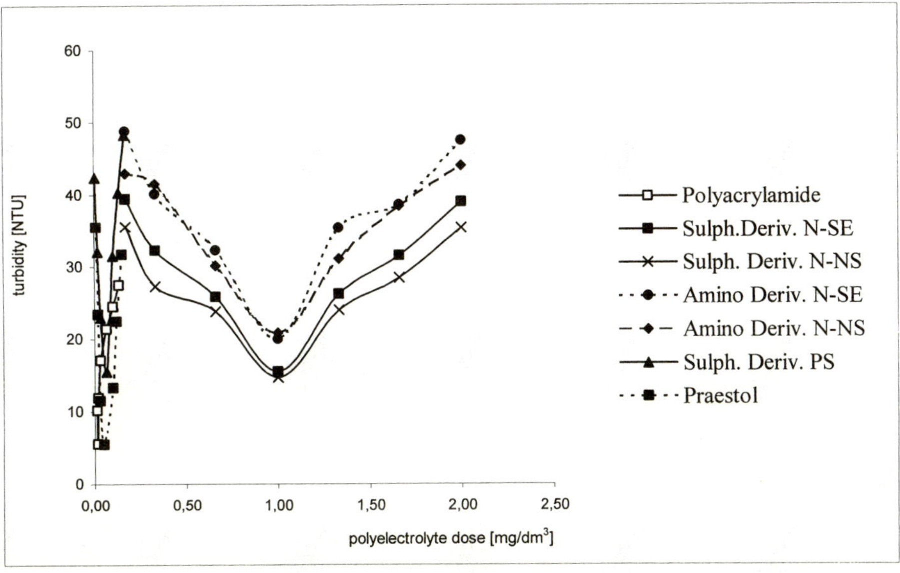

Fig. 2. Dependence of turbidity on optimal doses of synthesized standard polyacrylamide, sulphonated derivatives of SE novolak (sulphonated derivatives N-SE) and NS novolak (sulphonated derivatives N-NS) - phenol-formaldehyde resins wastes, amino derivatives of SE novolak (amino derivatives N-SE) and NS novolak (amino derivatives N-NS) - phenol-formaldehyde resins wastes, sulphonated derivatives of expanded polystyrene wastes (sulphonated derivatives PS), and commercial Praestol polyelectrolyte during purification of coal mine water with initial turbidity 177 NTU, at coagulant dose of 66.7 mg/dm^3 and pH=6.96.

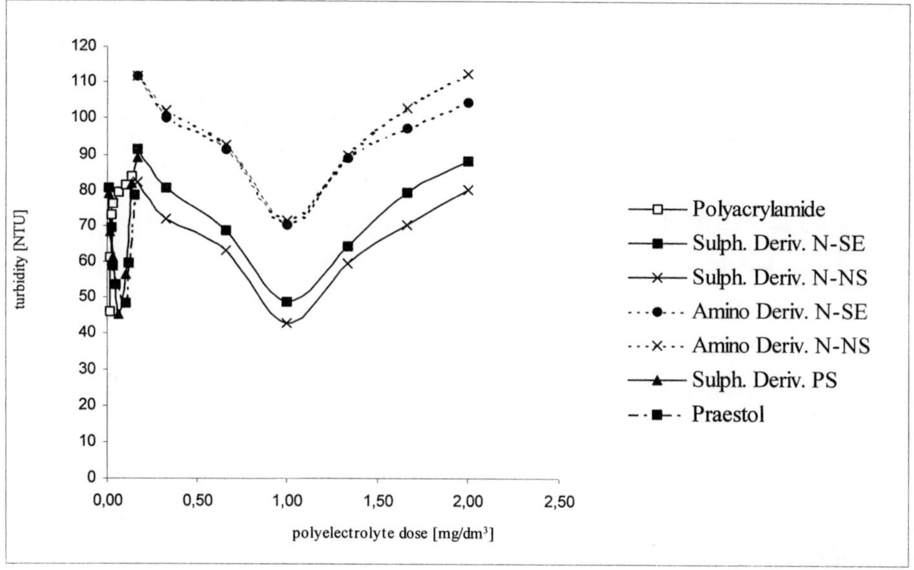

Fig. 3. Dependence of turbidity on optimal doses of synthesized standard polyacrylamide, sulphonated derivatives of SE novolak (sulphonated derivatives N-SE) and NS novolak (sulphonated derivatives N-NS) - phenol-formaldehyde resins wastes, amino derivatives of SE novolak (amino derivatives N-SE) and NS novolak (amino derivatives N-NS) - phenol-formaldehyde resins wastes, sulphonated derivatives of expanded polystyrene wastes (sulphonated derivatives PS), and commercial Praestol polyelectrolyte during purification of water from steel plant circulation system with initial turbidity 247 NTU, at coagulant dose of 100.0 mg/dm^3 and pH=6.94.

It was found from flocculation studies with the use of sulphonated derivatives of expanded polystyrene wastes, that aiding effects of the coagulation process are comparative with that of commercial Praestol polyelectrolyte and synthesized polyacrylamide obtained for slightly lower concentrations. This is particularly visible in the case of water from the steel plant circulating system (Fig. 3, Table 1). Decrease of coal mine water turbidity was 20% higher when Praestol was used. Sulphonated derivatives of expanded polystyrene wastes like commercial Praestol polyelectrolyte and standard polyacrylamide are efficient in a narrow range of polyelectrolyte concentrations.

Table 1. The highest decrease of turbidity for analyzed waters in flocculation processes with synthesized polymers and copolymers obtained from modified plastic wastes and commercial polyelectrolyte; in % of initial turbidity value.

Number	Polyelectrolyte type	The highest decrease of turbidity of analyzed waters [%]* for polyelelectrolytes concentration giving the highest decrease of turbidity [mg/dm^3]		
		model water	water from metallurgical plant circulation system	water from coal mine
1	polyacrylamide	80,2 0,020	61,0 0,017	85,9 0,066
2	commercial polyelectrolyte Praestol 2515	82,0 0,050	59,1 0,100	86,11 0,050
3	sulphonated derivatives of SE novolak	72,2 1,000	58,9 1,000	60,7 1,000
4	sulphonated derivatives of NS novolak	67,1 1,000	63,7 1,000	62,9 1,000
5	amino derivatives of SE novolak	56,3 1,000	40,6 1,000	49,5 1,000
6	amino derivatives of SE novolak	52,2 1,000	39,8 1,000	47,5 1,000
7	sulphonated derivatives of expanded polystyrene wastes	69,7 0,066	61,9 0,066	60,9 0,066

*numerator- % turbidity decrease, denominator – polyelectrolyte concentration [mg/dm^3]

Coal-mine water and water from the steel plant circulating system was analysed in the laboratories of each plant. The following analyses were conducted: reaction, solvated oxygen, biochemical oxygen demand (BZT), chemical oxygen demand (ChZT), oxygen consumption, ether extract, detergents, ammonia nitrogen, sulphates, chlorides, total iron, manganese, phenols, cyanides, solvated parts and suspension, total hardness of water for studies of synthesised flocculants efficiencies, and total hardness of water after flocculation studies (these studies were conducted 24 hours after the flocculation process). Studies were performed according to standards.[10] Results show (Table 2, 3), that the use of chemically modified polymer wastes, similarly to sulphonated derivatives of expanded polystyrene wastes and phenol-formaldehyde resin production wastes and amino derivatives of these resins give a significance decrease of all the analysed factors. However, the observed decreases are not equal for all flocculants. Particularly important and visible are the content decreases of sulphates, chlorides, and solvated parts for the coal mine water and of cyanides, ammonia nitrogen, chlorides, solvated parts, and ChZT for water from steel plant circulating system. This is due to the fact that all contamination factors higher than those permitted for water drained to water race before the flocculation process are significantly below the permitted values or below the detection threshold after the flocculation process. The obtained values of contamination factors of this water after the coagulation process show that purified sludges can be disposed to water races. It was found from the changes of the values of contamination factors of analysed water before and after the flocculation process with the use of chemically modified polymer plastic wastes that the best results were obtained for sodium salts of sulphonated derivatives of expanded polystyrene waste.

Table 2. Permissible values of pollution indicators in sewage, values for water from metallurgical plant circulation system, values for the water coagulationand flocculation processes with modified plastic wastes and synthesized model polymers.

Indicator	Permissible value of indicator for sewage	Value before treatment	Value after treatment with sulfone derivative of expanded polystyrene	Value after treatment with sulfone derivative of novolak NS	Value after treatment with amine derivative of novolak NS	Value after treatment with polyacryl-amide
Reaction (pH)	6,5-9,0	6,9	7,38	7,62	7,45	6,60
Solvated oxygen (mgO_2/dm^3)	-	49,0	8,2	7,5	7,9	7,0
Phenols (mg/dm^3)	0,5	5,1	< 0,005	< 0,005	< 0,005	< 0,005
Cyanides (mg/dm^3)	0,1	1,8	< 0,005	< 0,005	< 0,005	< 0,005
Oxygen consumption-dichromate method (mgO_2/dm^3)	150,0	185,4	< 10,0	25,5	157,8	59,9
Oxygen consumption-permanganate method (mgO_2/dm^3)	-	49,0	4,6	3,1	59,6	25,0
Ether extract (mg/dm^3)	50,0	12,9	11,5	10,5	8,0	15,5
Ammonia nitrogen (mg/dm^3)	6,0	284,0	0,18	0,30	0,41	31,3
Sulphates ($mgSO_4/dm^3$)	500,0	141,8	60,9	71,2	95,5	553,1
Chlorides ($mgCl/dm^3$)	1000,0	1386,0	33,6	12,5	42,3	1423,0
Total hardness ($mval/dm^3$)	70,0	13,67	2,3	2,2	2,3	15,4
Suspension- total amount (mg/dm^3)	50,0	13,2	20,8	32,0	10,8	19,6

Table 3. Permissible values of pollution indicators in sewage, values for water from coal mine, values for the water coagulation and flocculation processes with modified plastic wastes and synthesised model polymers.

Indicator	Permissible value of indicator for sewage	Value before treatment	Value after treatment with sulfone derivative of expanded polystyrene	Value after treatment with sulfone derivative of novolak NS	Value after treatment with amine derivative of novolak NS	Value after treatment with polyacryl-amide
Reaction (pH)	6,5-9,0	7,0	7,16	7,34	7,21	6,85
Solvated oxygen (mgO_2/dm^3)	-	7,7	8,3	7,7	8,1	7,5
Five-day biochemical oxygen demand (mgO_2/dm^3)	30,0	4,9	1,1	0,9	1,9	1,3
Oxygen consumption- dichromate method (mgO_2/dm^3)	150,0	15,8	< 10,0	<10,0	170,8	<10,0
Oxygen consumption- permanganate method (mgO_2/dm^3)	-	6,5	3,8	3,0	63,6	2,3
Ether extract (mg/dm^3)	50,0	5,0	2,5	8,0	2,0	1,0
Detergents (mg/dm^3)	5,0	<0,1	< 0,1	<0,1	< 0,1	0,57
Ammonia nitrogen (mg/dm^3)	6,0	0,18	0,23	0,13	0,50	0,10
Sulphates ($mgSO_4/dm^3$)	500,0	1510,0	28,0	59,7	51,4	92,2
Chlorides ($mgCl/dm^3$)	1000,0	1710,0	12,5	19,2	13,5	47,1
Total iron ($mgFe/dm^3$)	10,0	0,45	< 0,50	<0,50	< 0,50	< 0,50
Manganese ($mgMn/dm^3$)	-	0,85	< 0,15	< 0,15	< 0,15	< 0,15
Total hardness ($mval/dm^3$)	70,0	50,7	2,1	2,3	2,4	2,9
Solvated parts- total amount (mg/dm^3)	2000,0	5120,0	168,0	124,0	280,0	336,0
Suspension- total amount (mg/dm^3)	50,0	78,0	11,4	20,8	12,8	22,8

It is also observed that a small increase of investigated water basicity, is probably due to the use of sodium salts of sulphonated derivatives or with the use of diluted KOH solutions of amine derivatives. However, the pH values do not exceed the values permitted for sewages disposed to water race.

Polyacrylamide, used as a polyelectrolyte turned out to be very effective for the turbidity decrease of the analysed water (Fig. 1,2,3 Table 1) at a very low concentration of this polyelectrolyte. Decrease of the values of the investigated contamination factors of water from the coagulation process with the use of this polyelectrolyte is not so marked as in the case of the used chemically modified polymer wastes. It is particularly visible for total amounts of solvated parts and suspension.

It can be said, that sulphonated and amino derivatives of SE and NS novolak wastes, and sulphonated derivatives of expanded polystyrene, (but not always as efficient as the Praestol and standard polyacryloamide) aid the flocculation process. In the case of phenol-formaldehyde resins derivatives the concentrations should be significantly higher. In despite of this they often exhibit better efficiency than the standard polyelectrolytes.

Chemical modification was conducted for different expanded polystyrene wastes and for different NS-type phenol-formaldehyde resins from unsuccessful manufactured unit. The obtained results were reproducible.

Conclusions

It was found that the phenol-formaldehyde resins wastes and expanded polystyrene wastes can be used as substrates to obtain effective polyelectrolytes. This offers a new solution for their management. It was ascertained, that amino and sulphonated derivatives of phenol-formaldehyde resins production wastes and sulphonated derivatives of expanded polystyrene exhibit good flocculation properties comparable to those of standard polyacryloamide and commercial Praestol polyelectrolyte. They can be used as polyelectrolytes aiding flocculation and improving sedimentation conditions of water having properties analogous to those of sewage from the power plant, the coal mine, and the water from the steel plant circulating system.

This result is an additional argument for further studies concerning processes of chemical modification of polymer plastic wastes and for search of new application areas for these modification products.

[1] Y. Inagaki, M. Kuromiya, T. Noguchi, H. Watanabe, Langmuir, **1999**, *15*, 4171-4175.
[2] J. Simitzis, D. Fountas, J. Appl. Polym. Sci., **1995**, *55*, 879-887.
[3] W. Bajdur, J. Pajączkowska, B. Makarucha, A. Sułkowska, W. Sułkowski, Eur. Polym. J., **2002**, *38*, 299-304.
[4] W. Bajdur, W. Sułkowski, Chemia i Inżynieria Ekologiczna, **2000**, *7*, 119-127 [in Polish].
[5] R. S. Harland, R. K. Prud'homme, Polyelectrolytes Gels – Propertis, Preparation and Application, American Chemical Society, Washington DC, 1992.
[6] W. Bajdur, W. Sułkowski, Inżynieria i Ochrona Środowiska, **1998**, *1*, 191-200 [in Polish].
[7] W. Bajdur, W. Sułkowski, Chemia i Inżynieria Ekologiczna, **2000**, *7*, 129-134 [in Polish].
[8] W. Bajdur, W. Sułkowski, Polymer Recycling, **2001**, *6*, 71-76.
[9] W. Bajdur, W. Sułkowski, Zeszyty Naukowe Politechniki Śląskiej, Chemia, **2001**, *146*, 243-246 [in Polish].
[10] Dziennik Ustaw Nr 116 (Polish Acts), Supplement to the Decree of the Ministry of Environmental Protection, Natural Resources and Forestry, from 05.11.1991, Data of Environmental Pollutant Concentration in Water, Warszawa, Poland [in Polish].